Accretion is ... ntal importance in astrophysics Acc ... on as a source of energy in binary star s ... ctive galactic nuclei. The authors assume a basic knowledge of physics in order to describe the physical processes at work in accretion discs. The first three chapters explain why accretion is a source of energy, and then present the gas dynamics and plasma concepts necessary for astrophysical applications. The next three chapters then develop accretion in stellar systems, including accretion on to compact objects. Three further chapters give extensive treatment of accretion in active galactic nuclei, and the concluding chapter describes thick accretion discs. The second edition is a complete revision of the earlier account. In particular it gives much greater attention to active galaxies and quasars, where the accretion model is now accepted as the central energy source. The treatment is at a level appropriate for graduate students.

In this series

1. Active Galactic Nuclei
   edited by C. Hazard and S. Mitton
2. Globular Clusters
   edited by D. A. Hanes and B. F. Madore
3. Low Light-level Detectors in Astronomy
   by M. J. Eccles, M. E. Sim and K. P. Tritton
4. Accretion-driven Stellar X-ray Sources
   edited by W. H. G. Lewin and E. P. J. van den Heuvel
5. The Solar Granulation
   by R. J. Bray, R. E. Loughhead and C. J. Durrant
6. Interacting Binary Stars
   edited by J. E. Pringle and R. A. Wade
7. Spectroscopy of Astrophysical Plasmas
   by A. Dalgarno and D. Layzer
8. Accretion Power in Astrophysics, First Edition
   by J. Frank, A. R. King and D. J. Raine
9. Gamma-ray Astronomy
   by P. V. Ramama Murthy and A. W. Wolfendale
10. Quasar Astronomy
    by D. W. Weedman
11. X-ray Emission from Clusters of Galaxies
    by C. L. Sarazin
12. The Symbiotic Stars
    by S. J. Kenyon
13. High Speed Astronomical Photometry
    by B. Warner
14. The Physics of Solar Flares
    by E. Tandberg-Hanssen and A. G. Emslie
15. X-ray Detectors in Astronomy
    by G. W. Fraser
16. Pulsar Astronomy
    by A. Lyne and F. Graham-Smith
17. Molecular Collisions in the Interstellar Medium
    by D. Flower
18. Plasma Loops in the Solar Corona
    by R. J. Bray, L. E. Cram, C. J. Durrant and R. E. Loughhead
19. Beams and Jets in Astrophysics
    edited by P. A. Hughes
20. The Observation and Analysis of Stellar Photospheres
    by D. Gray
21. Accretion Power in Astrophysics, Second Edition
    by J. Frank, A. R. King and D. J. Raine

Cambridge astrophysics series
Editors: R. F. Carswell, D. N. C. Lin *and* J. E. Pringle

Accretion power in astrophysics
Second edition

The authors (from the left JF, ARK and DJR) at IAU Colloquium 129, Paris, 1990. (*Photograph by Jean Mouette.*)

ACCRETION POWER IN ASTROPHYSICS

SECOND EDITION

JUHAN FRANK
Department of Physics and Astronomy,
Louisiana State University, Baton Rouge

ANDREW KING
Astronomy Group, University of Leicester

and **DEREK RAINE**
Astronomy Group, University of Leicester

Published by the Press Syndicate of the University of Cambridge
The Pitt Building, Trumpington Street, Cambridge CB2 1RP
40 West 20th Street, New York, NY 10011-4211, USA
10 Stamford Road, Oakleigh, Victoria 3166, Australia

© Cambridge University Press 1985, 1992

First published 1985
Second edition 1992

Printed in Great Britain at the University Press, Cambridge

A catalogue record of this book is available from the British Library

Library of Congress cataloguing in publication data

Frank, J.
Accretion power in astrophysics/J. Frank, A. R. King, and D. J. Raine.
– 2nd ed.
  p. cm. – (Cambridge astrophysics series: 20)
Includes bibliographical references and index.
ISBN 0 521 40306 5 (hardback). – ISBN 0 521 40863 6 (pbk.)
1. Accretion (Astrophysics) I. King, A. R. II. Raine, Derek, J., 1946– . III. Title. IV. Series.
QB466.A25F73 1992
523.8′41–dc20   91-27460   CIP

ISBN 0 521 40306 5 hardback
ISBN 0 521 40863 6 paperback

UP

Contents

Preface to the first edition	*page* xiii
Preface to the second edition	xv

1 Accretion as a source of energy	**1**
1.1 Introduction	1
1.2 The Eddington limit	2
1.3 The emitted spectrum	5
1.4 Accretion theory and observation	6

2 Gas dynamics	**8**
2.1 Introduction	8
2.2 The equations of gas dynamics	8
2.3 Steady adiabatic flows; isothermal flows	11
2.4 Sound waves	12
2.5 Steady, spherically symmetric accretion	14

3 Plasma concepts	**22**
3.1 Introduction	22
3.2 Charge neutrality, plasma oscillations and the Debye length	22
3.3 Collisions	25
3.4 Thermal plasmas: relaxation times and mean free path	29
3.5 The stopping of fast particles by a plasma	31
3.6 Transport phenomena: viscosity	33
3.7 The effect of strong magnetic fields	36
3.8 Shock waves in plasmas	40

4 Accretion in binary systems	**46**
4.1 Introduction	46
4.2 Interacting binary systems	46
4.3 Roche lobe overflow	47
4.4 Roche geometry and binary evolution	50
4.5 Disc formation	54
4.6 Viscous torques	58
4.7 The magnitude of viscosity	61
4.8 Accretion in close binaries: other possibilities	62

5 Accretion discs — 67
5.1 Introduction — 67
5.2 Radial disc structure — 67
5.3 Steady thin discs — 71
5.4 The local structure of thin discs — 75
5.5 The emitted spectrum — 77
5.6 The structure of steady α-discs ('the standard model') — 80
5.7 Steady discs: confrontation with observation — 85
5.8 Time dependence and stability — 96
5.9 Tides, resonances and superhumps — 106
5.10 Discs around young stars — 114
5.11 Spiral shocks — 115

6 Accretion on to a compact object — 117
6.1 Introduction — 117
6.2 Boundary layers — 117
6.3 Accretion on to magnetized neutron stars and white dwarfs — 121
6.4 Accretion columns: the white dwarf case — 133
6.5 Accretion column structure for neutron stars — 151
6.6 X-ray bursters — 161
6.7 Black holes — 166
6.8 Accreting binary systems with compact components — 170

7 Active galactic nuclei — 171
7.1 Observations — 171
7.2 The distances of active galaxies — 175
7.3 The sizes of active nuclei — 179
7.4 The mass of the central source — 180
7.5 Models of active nuclei — 183
7.6 The gas supply — 185
7.7 Black holes — 189
7.8 Accretion efficiency — 192

8 Accretion discs in active nuclei — 199
8.1 The nature of the problem — 199
8.2 Radio, millimetre and infrared emission — 201
8.3 X-rays and discs — 202
8.4 The broad and narrow, permitted and forbidden — 204
8.5 The narrow line region — 205
8.6 The broad line region — 209

9 Accretion power in active nuclei — 218
9.1 Introduction — 218
9.2 Extended radio sources — 218
9.3 Compact radio sources — 223
9.4 The nuclear continuum — 229
9.5 Applications to discs — 232

Contents

9.6	Magnetic fields	234
9.7	Newtonian electrodynamic discs	236
9.8	The Blandford–Znajek model	239
9.9	Circuit analysis of black hole power	242
10	**Thick discs**	**245**
10.1	Introduction	245
10.2	Equilibrium figures	246
10.3	The limiting luminosity	252
10.4	Newtonian vorticity-free torus	254
10.5	Thick accretion discs	257
10.6	Dynamical stability	261
10.7	Astrophysical implications	263
	Appendix	266
	Problems	270
	Bibliography	282
	List of symbols	290
	Object index	291
	Index	

To the memory of
SBJF
1952–1990

Preface to the first edition

The subject of this book is astrophysical accretion, especially in those circumstances where accretion is believed to make an important contribution to the total light of an astrophysical system. Our discussion therefore centres mainly on close binary systems containing compact objects and on active nuclei. The reader is assumed to possess a basic knowledge of physics at first degree level, but only a rudimentary experience of astronomy is required. We have tried to concentrate on those features, particularly the basic physics, that are probably more firmly established; but the treatment is necessarily somewhat heterogeneous. For example, there is by now a tolerably coherent line of argument showing that the formation of an accretion disc is very likely in many close binaries, and giving a plausible picture of what such a disc is like, at least in some simple cases. In other areas, such as accretion on to the surface of a compact object, or in active nuclei, we are not so fortunate, and we must work back and forth between theory and observation. Our aim is that the book should provide a systematic introduction to the subject for graduate students. We hope it may also serve as a reference for interested astronomers in other fields, and that selected material will be suitable for undergraduate options in astronomy.

In Chapters 2 and 3 we present introductory material on fluid dynamics and plasma physics. Many excellent texts exist in these areas, but they tend to be too detailed for our needs; we have tried to extract just those basic ideas necessary for the subsequent discussion, and to set them in an astrophysical context. We also need basic concepts of radiation mechanisms and radiative transfer theory. These we have not attempted to expound systematically since there are many books written for astrophysics students which are suitable. For convenience we have collected some of the necessary results in an appendix. The astrophysics of stellar accretion is dealt with in Chapters 4 to 6. Chapters 7 and 8 set the observational scene for two models of active nuclei (or possibly for two aspects of a single model) considered in Chapters 9 and 10. In the main we have not given references to sources, since this would yield an enormous list out of keeping with the spirit of the book. Detailed references can be found in the reviews cited in the bibliography.

A final note on units and notation. For the most part astrophysicists use units based on the cm, g, s (cgs system) when they are not indulging themselves in archaic astronomical conventions. The system one rarely sees in astrophysics is the otherwise standard mks (SI) system. For ease of comparison with the astrophysical literature we have quoted numerical values in cgs units. A special problem arises in electromagnetism; here the formulae are different in the two systems. We have adopted the compromise of giving the formulae in cgs units with a multiplicative conversion factor to mks units in square brackets. These factors always involve the quantities ε_0, μ_0, or c, and no confusion should arise with the use of square brackets in algebraic formulae. Also, we have followed the normal astrophysical usage of the symbols '$\sim$' and '$\cong$' in algebraic formulae; the former standing for 'is of the order of' of the latter for 'is approximately equal to'.

The idea for this book grew out of discussions with Dr Simon Mitton, whom we should like to thank for his encouragement and advice. We have benefited from the comments of our students, on whom some of the early versions of much of this material were tested. We also thank our scientific colleagues for much useful advice. We are grateful to Diane Fabian for her help in seeing our final efforts through the Press.

Preface to the second edition

In the years since the first edition of this book appeared the study of astrophysical accretion has developed rapidly. Perhaps the most fundamental change has been the shift in attitude over active galaxies and quasars: the view that accretion is the energy source is now effectively standard, and the emphasis is much more on close comparison of observation and theory. This change, and the less spectacular but still profound one which has occurred in the study of close binary accretion, have been largely brought about by the wealth of new data accumulated in the interval. In X-rays, the ability of EXOSAT to observe continuously for as much as 3 to 4 days was a dramatic advance. In the optical, new instrumentation has produced far tighter observational constraints on theory. Despite these challenges, the basic outlines of the theory are still recognizably the same.

Of course our understanding is very incomplete. As the most glaring example, we still have essentially no idea what drives disc accretion; and there are new problems such as the dynamical stability of thick discs, or the nature of fieldline-threading in magnetic binaries. But it is now difficult to deny that some close binaries possess discs approximately conforming to theoretical ideas; or that some kind of anisotropic accretion occurs in active galactic nuclei. Encouragingly, accretion theory is increasingly integrated into wider pictures of the relevant systems. The process is well advanced for close binaries, particularly for the secular evolution of cataclysmic variables, and is in its early stages for active galaxies.

We were therefore very glad to have the chance to revise our book, and extremely grateful to the many colleagues who made suggestions for improvements. Inevitably the vast expansion of the subject has obliged us to be selective, and we have had to omit or curtail discussion of some topics. This is particularly true of fairly specialized areas such as quasi-periodic oscillations in low-mass X-ray binaries, or the jets in SS 433 (incidentally the subjects of two of EXOSAT's more spectacular discoveries). We have completely rewritten Section 4.4, adding a discussion of secular binary evolution. In Chapter 5 on accretion discs we have rewritten Section 5.7 on the confrontation with observations, particularly of low-mass X-ray binaries, and added three new sections. Section 5.9 deals with tides and resonances and the phenomenon of superhumps. Much of accretion disc theory is relevant to star formation, and Section 5.10 gives a brief introduction. In Section 5.11 we discuss accretion via spiral shocks. In Chapter 6, Sections 6.3 (accretion on to magnetized neutron stars and white dwarfs) and 6.4 (white dwarf column accretion) have been extensively revised.

The material on active galaxies has been subject to substantial rearrangement, reflecting the changing emphasis in the subject. In Chapter 7 we have added a new section on the gas supply to the central engine. Chapter 8 includes an extended discussion of the broad line region. New material on X-ray emission has been added in Chapters 8 and 9, which now include a brief discussion of two-temperature discs (or ion tori) and slim discs.

Finally, Chapter 10 now includes a discussion of the instability of thick discs to global non-axisymmetric modes. This is treated in a new Section 10.6, and Section 10.7 on astrophysical applications has been rewritten.

We have also added a selection of problems, of varying degrees of difficulty, which we hope will make the book more useful for teaching purposes.

We would particularly like to thank Mitch Begelman, Jean-Pierre Lasota, Takuya Matsuda, Robert Connon Smith and Henk Spruit for pointing out errors in the first edition and suggesting new material.

1
Accretion as a source of energy

1.1 Introduction

For the nineteenth century physicists, gravity was the only conceivable source of energy in celestial bodies, but gravity was inadequate to power the Sun for its known lifetime. In contrast, in the latter half of the twentieth century it is to gravity that we look to power the most luminous objects in the Universe, for which the nuclear sources of the stars are wholly inadequate. The extraction of gravitational potential energy from material which accretes on to a gravitating body is now known to be the principal source of power in several types of close binary systems, and is widely believed to provide the power supply in active galactic nuclei and quasars. This increasing recognition of the importance of accretion has accompanied the dramatic expansion of observational techniques in astronomy, in particular the exploitation of the full range of the electromagnetic spectrum from the radio to X-rays and γ-rays. At the same time, the existence of compact objects has been placed beyond doubt by the discovery of the pulsars, and black holes have been given a sound theoretical status. Thus, the new role for gravity arises because accretion on to compact objects is a natural and powerful mechanism for producing high-energy radiation.

Some simple order-of-magnitude estimates will show how this works. For a body of mass M and radius R_* the gravitational potential energy released by the accretion of a mass m on to its surface is

$$\Delta E_{acc} = GMm/R_* \tag{1.1}$$

where G is the gravitation constant. If the accreting body is a neutron star with radius $R_* \sim 10$ km, mass $M \sim M_\odot$, the solar mass, then the yield ΔE_{acc} is about 10^{20} erg per accreted gram. We would expect this energy to be released eventually mainly in the form of electromagnetic radiation. For comparison, consider the energy that could be extracted from the mass m by nuclear fusion reactions. The maximum is obtained if, as is usually the case in astrophysics, the material is initially hydrogen, and the major contribution comes from the conversion, (or 'burning'), of hydrogen to helium. This yields an energy release

$$\Delta E_{nuc} = 0.007\, mc^2 \tag{1.2}$$

where c is the speed of light, so we obtain about 6×10^{18} erg g^{-1} or about one twentieth of the accretion yield in this case.

It is clear from the form of equation (1.1) that the efficiency of accretion as an energy release mechanism is strongly dependent on the compactness of the accreting object: the larger the ratio M/R_*, the greater the efficiency. Thus, in treating accretion on to objects of stellar mass we shall certainly want to consider neutron stars ($R_* \sim 10$ km) and black holes with radii $R_* \sim 2GM/c^2 \sim 3\,(m/M_\odot)$ km (see Section 7.7). For white dwarfs with $M \sim M_\odot$, $R_* \sim 10^9$ cm, nuclear burning is more efficient than accretion by factors 25–50. However, it would be wrong to conclude that accretion on to white dwarfs is of no great importance for observations, since the argument takes no account of the timescale over which the nuclear and accretion processes act. In fact, when nuclear burning does occur on the surface of a white dwarf, it is likely that the reaction tends to 'run away' to produce an event of great brightness but short duration, a *nova* outburst, in which the available nuclear fuel is very rapidly exhausted. For almost all of its lifetime no nuclear burning occurs, and the white dwarf (may) derive its entire luminosity from accretion. Binary systems in which a white dwarf accretes from a close companion star are known as *cataclysmic variables* and are quite common in the Galaxy. Their importance derives partly from the fact that they provide probably the best opportunity to study the accretion process in isolation, since other sources of luminosity, in particular the companion star, are relatively unimportant.

For accretion on to a 'normal', less compact, star, such as the Sun, the accretion yield is smaller than the potential nuclear yield by a factor of several thousand. Even so, accretion on to such stars may be of observational importance. For example, a binary system containing an accreting main-sequence star has been proposed as a model for the so-called symbiotic stars.

For a fixed value of the compactness, M/R_*, the luminosity of an accreting system depends on the rate $\dot{M}$ at which matter is accreted. At high luminosities, the accretion rate may itself be controlled by the outward momentum transferred from the radiation to the accreting material by scattering and absorption. Under certain circumstances, this can lead to the existence of a maximum luminosity for a given mass, usually referred to as the Eddington luminosity, which we discuss next.

1.2 The Eddington limit

Consider a steady spherically symmetrical accretion; the limit so derived will be generally applicable as an order-of-magnitude estimate. We assume the accreting material to be mainly hydrogen and to be fully ionized. Under these circumstances, the radiation exerts a force mainly on the free electrons through Thomson scattering, since the scattering cross-section for protons is a factor $(m_e/m_p)^2$ smaller, where $m_e/m_p \approx 5 \times 10^{-4}$ is the ratio of the electron and proton masses. If S is the radiant energy flux (erg s^{-1} cm^{-2}) and $\sigma_T = 6.7 \times 10^{-25}$ cm^2 is the Thomson cross-section, then the outward radial force on each electron equals the rate at which it absorbs momentum, $\sigma_T S/c$. If there is a substantial population of elements other than hydrogen, which have retained some bound electrons, the effective cross-section, resulting from the absorption of photons in spectral lines, can exceed σ_T considerably. The attractive electrostatic Coulomb force between the electrons and protons means that as they move out the electrons drag the protons with them. In effect, the radiation pushes out electron–proton pairs against the total gravitational force $GM(m_p + m_e)/r^2 \approx GMm_p/r^2$ acting on each pair at a radial distance r from the centre. If the luminosity of the accreting source is L (erg s^{-1}), we have $S = L/4\pi r^2$ by spherical symmetry, so the net inward force on an electron–proton pair is

$$\left(GMm_p - \frac{L\sigma_T}{4\pi c}\right)\frac{1}{r^2}.$$

There is a limiting luminosity for which this expression vanishes, the Eddington limit,

$$L_{Edd} = 4\pi GMm_p c/\sigma_T \qquad (1.2a)$$
$$\cong 1.3 \times 10^{38}(M/M_\odot)\,\text{erg s}^{-1}. \qquad (1.2b)$$

At greater luminosities the outward pressure of radiation would exceed the inward gravitational attraction and accretion would be halted. If all the luminosity of the source were derived from accretion this would switch off the source; if some, or all, of it were produced by other means, for example nuclear burning, then the outer layers of material would begin to be blown off and the source would not be steady. For stars with a given mass-luminosity relation this argument yields a maximum stable mass.

Since L_{Edd} will figure prominently later, it is worth recalling the assumptions made in deriving expressions (1.2a, b). We assumed that the accretion flow was *steady* and *spherically symmetric*. A slight extension can be made here without difficulty: if the accretion occurs only over a fraction f of the surface of a star, but is otherwise dependent only on radial distance, r, the corresponding limit on the accretion luminosity is fL_{Edd}. For a more complicated geometry, however, we cannot expect (1.2a, b) to provide more than a crude estimate. Even more crucial was the restriction to *steady* flow. A dramatic illustration of this is provided by a supernova, in which L_{Edd} is exceeded by many orders of magnitude. Our other main assumptions were that the accreting material was largely hydrogen and that it was fully ionized. The former is almost always a good approximation, but even a small admixture of heavy elements can invalidate the latter. Almost complete ionization is likely to be justified however in the very common case where the accreting object produces much of its luminosity in the form of X-rays, because the abundant ions can usually be kept fully stripped of electrons by a very small fraction of the X-ray luminosity. Despite these caveats, the Eddington limit is of great practical importance, in particular because certain types of system shows a tendency to behave as 'standard candles' in the sense that their typical luminosities are close to their Eddington limits.

For accretion powered objects the Eddington limit implies a limit on the steady accretion rate, $\dot{M}$(g s^{-1}). If all the kinetic energy of infalling matter is given up to radiation at the stellar surface, R_*, then from (1.1) the *accretion luminosity* is

$$L_{acc} = GM\dot{M}/R_*. \qquad (1.3)$$

4 Accretion as a source of energy

It is useful to re-express (1.3) in terms of typical orders of magnitude: writing the accretion rate as $\dot{M} = 10^{16} \dot{M}_{16} \text{ g s}^{-1}$ we have

$$L_{\text{acc}} = 1.3 \times 10^{33} \dot{M}_{16} (M/M_\odot)(10^9 \text{ cm}/R_*) \text{ erg s}^{-1} \quad (1.4a)$$
$$= 1.3 \times 10^{36} \dot{M}_{16} (M/M_\odot)(10 \text{ km}/R_*) \text{ erg s}^{-1}. \quad (1.4b)$$

The reason for rewriting (1.3) in this way is that the quantities $(M/M_\odot)$, $(10^9 \text{ cm}/R_*)$ and $(M/M_\odot)$, $(10 \text{ km}/R_*)$ are of order unity for white dwarfs and neutron stars respectively. Since $10^{16} \text{ g s}^{-1} (\sim 1.5 \times 10^{-10} M_\odot \text{ yr}^{-1})$ is a typical order of magnitude for accretion rates in close binary systems involving these types of star, we have $\dot{M}_{16} \sim 1$ (in (1.4a, b), and the luminosities $10^{33} \text{ erg s}^{-1}$, $10^{36} \text{ erg s}^{-1}$ represent values commonly found in such systems. Further, by comparison with (1.2b) it is immediately seen that for steady accretion $\dot{M}_{16}$ is limited by the values $\sim 10^5$ and 10^2 respectively. Thus, accretion rates must be less than about 10^{21} g s^{-1} and 10^{18} g s^{-1} in the two types of system if the assumptions involved in deriving the Eddington limit are valid.

For the case of accretion on to a black hole it is far from clear that (1.3) holds. Since the radius does not refer to a hard surface but only to a region into which matter can fall and from which it cannot escape, much of the accretion energy could disappear into the hole and simply add to its mass, rather than be radiated. The uncertainty in this case can be parametrized by the introduction of a dimensionless quantity η, the *efficiency*, on the right hand side of (1.3):

$$L_{\text{acc}} = 2\eta GM\dot{M}/R_* \quad (1.5a)$$
$$= \eta \dot{M} c^2 \quad (1.5b)$$

where we have used $R_* = 2GM/c^2$ for the black hole radius. Equation (1.5b) shows that η measures how efficiently the rest mass energy, c^2 per unit mass, of the accreted material is converted into radiation. Comparing (1.5b) with (1.2) we see that $\eta = 0.007$ for the burning of hydrogen to helium. If the material accreting on to a black hole could be lowered into the hole infinitesimally slowly – scarcely a practical proposition – all of the rest mass energy could, in principle, be extracted and we should have $\eta = 1$. As we shall see in Chapter 7 the estimation of realistic values for η is an important problem. A reasonable guess would appear to be $\eta \sim 0.1$, comparable to the value $\eta \sim 0.15$ obtained from (1.5a) for a solar mass neutron star. Thus, despite its extra compactness, a stellar mass black hole may be no more efficient in the conversion of gravitational potential energy to radiation than a neutron star of similar mass.

As a final illustration here of the use of the Eddington limit we consider the nuclei of active galaxies and the closely related quasars. These are probably the least understood class of object for which accretion is thought to be the ultimate source of energy. The main reason for this belief comes from the large luminosities involved: these systems may reach $10^{47} \text{ erg s}^{-1}$, or more, varying by factors of order 2 on timescales of weeks, or less. With the nuclear burning efficiency of only $\eta = 0.007$, the rate at which mass is processed in the source could exceed $250 M_\odot \text{ yr}^{-1}$. This is a rather severe requirement and it is clearly greatly reduced if accretion with an efficiency $\eta \sim 0.1$ is postulated instead. The accretion rate

1.3 The emitted spectrum

required is of order 20 $M_\odot$ yr^{-1}, or less, and rates approaching this might plausibly be provided by a number of the mechanisms considered in Chapter 7. If these systems are assumed to radiate at less than the Eddington limit, then accreting masses exceeding about $10^9 M_\odot$ are required. White dwarfs are subject to upper limits on their masses of 1.4 $M_\odot$ and neutron stars cannot exceed about 3 $M_\odot$; thus, only massive black holes are plausible candidates for accreting objects in active nuclei.

1.3 The emitted spectrum

We can now make some order-of-magnitude estimates of the spectral range of the emission from compact accreting objects, and, conversely, suggest what type of compact object may be responsible for various observed behaviour. We can characterize the continuum spectrum of the emitted radiation by a temperature $T_{\rm rad}$ defined such that the energy of a typical photon, $h\bar{\nu}$, is of order $kT_{\rm rad}$, $T_{\rm rad} = h\bar{\nu}/k$, where we do not need to make the choice of $\bar{\nu}$ precise. For an accretion luminosity $L_{\rm acc}$ from a source of radius R, we define a blackbody temperature, $T_{\rm b}$, as the temperature the source would have if it were to radiate the given power as a blackbody spectrum:

$$T_{\rm b} = (L_{\rm acc}/4\pi R_*^2 \sigma)^{1/4}. \tag{1.6}$$

Finally, we define a temperature $T_{\rm th}$ that the accreted material would reach if its gravitational potential energy were turned entirely into thermal energy. For each proton–electron pair accreted, the potential energy released is $GM(m_{\rm p}+m_{\rm e})/R_* \approx GMm_{\rm p}/R_*$, and the thermal energy is $2 \times \frac{3}{2}kT$; therefore

$$T_{\rm th} = GMm_{\rm p}/3kR_*. \tag{1.7}$$

Note that some authors use the related concept of the viral temperature, $T_{\rm vir} = T_{\rm th}/2$, for a system in mechanical and thermal equilibrium. If the accretion flow is optically thick, the radiation reaches thermal equilibrium with the accreted material before leaking out to the observer and $T_{\rm rad} \sim T_{\rm b}$. On the other hand, if the accretion energy is converted directly into radiation which escapes without further interaction (i.e. the intervening material is optically thin), we have $T_{\rm rad} \sim T_{\rm th}$. This occurs in certain types of shock wave that may be produced in some accretion flows and we shall see in Chapter 3 that (1.7) provides an estimate of the shock temperature for such flows. In general, the radiation temperature may be expected to lie between the thermal and blackbody temperatures, and, since the system cannot radiate a given flux at less than the blackbody temperature, we have

$$T_{\rm b} \lesssim T_{\rm rad} \lesssim T_{\rm th}.$$

Of course, these estimates assume that the radiating material can be characterized by a single temperature. They need not apply, for example, to a non-Maxwellian distribution of electrons radiating in a fixed magnetic field, such as we shall meet in Chapter 9.

Let us apply the limits (1.6), (1.7) to the case of a solar mass neutron star. The upper limit (1.7) gives $T_{\rm th} \sim 5.5 \times 10^{11}$ K, or, in terms of energies, $kT_{\rm th} \sim 50$ MeV.

6 Accretion as a source of energy

To evaluate the lower limit, T_b, from (1.6), we need an idea of the accretion luminosity, L_{acc}; but T_b is, in fact, very insensitive to the assumed value of L_{acc}, since it is proportional to the fourth root. Thus we can take $L_{acc} \sim L_{Edd} \sim 10^{38}$ erg s^{-1} for a rough estimate; if, instead, we were to take a typical value $\sim 10^{36}$ erg s^{-1} (equation (1.6)) this would change T_b only by a factor of ~ 3. We obtain $T_b \sim 10^7$ K or $kT_b \sim 1$ keV, and so we expect photon energies in the range

$$1 \text{ keV} \lesssim h\bar{\nu} \lesssim 50 \text{ MeV}$$

as a result of accretion on to neutron stars. Similar results would hold for stellar mass black holes. Thus we can expect the most luminous accreting neutron star and black hole binary systems to appear as medium to hard X-ray emitters and possibly as γ-ray sources. There is no difficulty in identifying this class of object with the luminous galactic X-ray sources discovered by the first satellite X-ray experiments, and added to by subsequent investigations.

For accreting white dwarfs it is probably more realistic to take $L_{acc} \sim 10^{33}$ erg s^{-1} in estimating T_b (cf. (1.4a)). With $M = M_\odot$, $R = 5 \times 10^8$ cm, we obtain

$$6 \text{ eV} \lesssim h\bar{\nu} \lesssim 100 \text{ keV}.$$

Consequently, accreting white dwarfs should be optical, ultraviolet and possibly X-ray sources. This fits in neatly with our knowledge of cataclysmic variable stars, which have been found to have strong ultraviolet continua by the Copernicus and IUE satellite experiments. In addition, some of them are now known to emit a small fraction of their luminosity as thermal X-ray sources We shall see that in many ways cataclysmic variables are particularly useful in providing observational tests of theories of accretion.

1.4 Accretion theory and observation

So far we have discussed the amount of energy that might be expected by the accretion process, but we have made no attempt to describe in detail the flow of accreting matter. A hint that the dynamics of this flow may not be straightforward is provided by the existence of the Eddington limit, which shows that, at least for high accretion rates, forces other than gravity can be important. In addition, it will emerge later that, certainly in many cases and probably in most, the accreting matter possesses considerable angular momentum per unit mass which, in realistic models, it has to lose in order to be accreted at all. Furthermore, we need a detailed description of the accretion flow if we are to explain the observed spectral distribution of the radiation produced: crudely speaking, in the language of Section 1.3, we want to know whether T_{rad} is closer to T_b or T_{th}.

The two main tools we shall use in this study are the equations of gas dynamics and the physics of plasmas. We shall give a brief introduction to gas dynamics in Section 2.1–2.4 of the next chapter, and treat some aspects of plasma physics in Chapter 3. In addition, the elements of the theory of radiative transfer are summarized in the Appendix. The reader who is already familiar with these subjects can omit these parts of the text. The rest of the book divides into three somewhat distinct parts. First, in Chapters 4 to 6 we consider accretion by stellar mass objects in binary systems. In these cases, we often find that observations provide fairly direct evidence for the nature of the systems. For example, there is sometimes direct evidence for the importance of angular momentum and the existence of accretion *discs*. This contrasts greatly with the subsequent discussion of active nuclei in Chapters 7 to 10. Here, the accretion theory arises at the end of a sequence of plausible, but not unproblematic, inductions. Furthermore, there appears to be no absolutely compelling evidence for, or against, the existence of accretion discs in these systems. Thus, whereas we normally use the observations of stellar systems to test the theory, for active nuclei we use the theory, to some extent, to illustrate the observations. This is particularly apparent in the final part of the book, where, in the last two chapters, we discuss two quite different models for powering an active nucleus by an accretion disc around a supermassive black hole.

2

Gas dynamics

2.1 Introduction

All accreting matter, like most of the material in the Universe, is in a gaseous form. This means that the constituent particles, usually free electrons and various species of ions, interact directly only by *collisions*, rather than by more complicated short-range forces. In fact, these collisions involve the electrostatic interaction of the particles and will be considered in more detail in Chapter 3. On average, a gas particle will travel a certain distance, the *mean free path*, λ, before changing its state of motion by colliding with another particle. If the gas is approximately uniform over length scales exceeding a few mean free paths, the effect of all these collisions is to randomize the particle velocities about some mean velocity, the velocity of the gas, **v**. Viewed in a reference frame moving with velocity **v**, the particles have a Maxwell–Boltzmann distribution of velocities, and can be described by a temperature T. Provided we are interested only in length scales $L \gg \lambda$ we can regard the gas as a continuous *fluid*, having velocity **v**, temperature T and density ρ defined at each point. We then study the behaviour of these and other fluid variables as functions of position and time by imposing the laws of conservation of mass, momentum and energy. This is the subject of gas dynamics. If we wish to look more closely at the gas, we have to consider the particle interactions in more detail; this is the domain of plasma physics, or, more strictly, plasma kinetic theory, about which we shall have something to say in Chapter 3. Note that the equations of gas dynamics may not always be applicable. For example, these equations may themselves predict large changes in gas properties over lengthscales comparable with λ; under these circumstances the fluid approximation is invalid and we must use the deeper but more complicated approach of the plasma kinetic theory.

2.2 The equations of gas dynamics

Here we shall write down the three conservation laws of gas dynamics, which, together with an equation of state and appropriate boundary conditions, describe any gas dynamical flow. We shall not give the derivations, which can be found in many books, for example Landau & Lifshitz, 1959, but merely point out the significance of the various terms.

Given a gas with, as before, a velocity field **v**, density ρ and temperature T, all defined as functions of position, **r**, and time, t, conservation of mass is ensured by the *continuity equation*:

$$\frac{\partial \rho}{\partial t} + \nabla \cdot (\rho \mathbf{v}) = 0. \tag{2.1}$$

Because of the thermal motion of its particles the gas has a pressure, P, at each point. An equation of state relates this pressure to the density and temperature. Astrophysical gases, other than the degenerate gases in white dwarfs and neutron stars and the cores of 'normal' stars, have as equation of state the *perfect gas law*:

$$P = \rho k T / \mu m_{\mathrm{H}}. \tag{2.2}$$

Here $m_{\mathrm{H}} \sim m_{\mathrm{p}}$ is the mass of the hydrogen atom and μ is the mean molecular weight, which is the mean mass per particle of gas measured in units of m_{H}, or, equivalently, the inverse of the number of particles in a mass m_{H} of the gas. Hence, $\mu = 1$ for neutral hydrogen, $\frac{1}{2}$ for fully ionized hydrogen, and something in between for a mixture of gases with cosmic abundances, depending on the ionization state.

Gradients in the pressure in the gas imply forces since momentum is thereby transferred. Other, as yet unspecified, forces acting on the gases are represented by the force density, the force per unit volume, $\mathbf{f}$. Conservation of momentum for each gas element then gives the *Euler equation*:

$$\rho \frac{\partial \mathbf{v}}{\partial t} + \rho \mathbf{v} \cdot \nabla \mathbf{v} = -\nabla P + \mathbf{f}. \tag{2.3}$$

This has the form (mass density) × (acceleration) = (force density) and is, in fact, simply an expression of Newton's second law for a continuous fluid. The term $\rho \mathbf{v} \cdot \nabla \mathbf{v}$ on the left hand side of (2.3) represents the convection of momentum through the fluid by velocity gradients. The presence of this term means that *steady* motions are possible in which the time derivatives of the fluid variables vanish, but $\mathbf{v}$ is non-zero. An example of an external force is gravity: in this case $\mathbf{f} = -\rho \mathbf{g}$, where $\mathbf{g}$ is the local acceleration due to gravity. Another example would be the force due to an external magnetic field. Further important contributions to $\mathbf{f}$ can come from *viscosity*, which is the transfer of momentum along velocity gradients by random motions of the gas, especially turbulence and thermal motions. The inclusion of viscosity usually considerably complicates the momentum balance equation, so it is fortunate that in many cases it may be neglected. We anticipate some later results by stating that viscous effects are chiefly important in flows which show either large shearing motions or steep velocity gradients.

The third, and most complicated, conservation law is that of *energy*. An element of gas has two forms of energy: an amount $\frac{1}{2}\rho v^2$ of kinetic energy per unit volume, and internal or thermal energy $\rho \varepsilon$ per unit volume, where ε, the internal energy per unit mass, depends on the temperature, T, of the gas. According to the equipartition theorem of elementary kinetic theory, each degree of freedom of each gas particle is assigned a mean energy $\frac{1}{2}kT$. For a monatomic gas the only degrees of freedom are the three orthogonal directions of translational motion and

$$\varepsilon = \tfrac{3}{2} k T / \mu m_{\mathrm{H}}. \tag{2.4}$$

Molecular gases have additional internal degrees of freedom of vibration or rotation. In reality, cosmic gases are not quite monatomic and the effective number of degrees of freedom is not quite three; but in practice (2.4) is usually a good approximation.

The energy equation for the gas is

$$\frac{\partial}{\partial t}(\tfrac{1}{2}\rho v^2 + \rho\varepsilon) + \nabla \cdot [(\tfrac{1}{2}\rho v^2 + \rho\varepsilon + P)\mathbf{v}] = \mathbf{f}\cdot\mathbf{v} - \nabla\cdot\mathbf{F}_{\text{rad}} - \nabla\cdot\mathbf{q}. \tag{2.5}$$

The left hand shows a family resemblance to the continuity equation (2.1), with the expected difference that the conserved quantity ρ is replaced by $(\tfrac{1}{2}\rho v^2 + \rho\varepsilon)$. The last term in the square brackets represents the so-called pressure work. Two new quantities appear on the right hand side: first, the radiative flux vector $\mathbf{F}_{\text{rad}} = \int d\nu \int d\Omega \mathbf{n} I_\nu(\mathbf{n},\mathbf{r})$ where I_ν is the specific intensity of radiation at the point $\mathbf{r}$ in the direction $\mathbf{n}$ and the integrals are over frequency, ν, and solid angle, Ω (see the Appendix). The term $-\nabla\cdot\mathbf{F}_{\text{rad}}$ gives the rate at which radiant energy is being lost by emission, or gained by absorption, by unit volume of the gas. In general, the specific intensity, I_ν, is itself governed by a further equation, the conservation of energy equation for the radiation field. Fortunately, we can often approximate the radiative losses quite simply. For example, let j_ν (erg s^{-1} cm^{-3} sr^{-1}) be the rate of emission of radiation per unit volume per unit solid angle; j_ν is the emissivity of the gas and is usually given as a function of ρ, T (and ν), but might also depend on external magnetic fields or the radiation field itself (examples are given in the Appendix). If the gas is optically thin, so that radiation escapes freely once produced and the gas itself reabsorbs very little, the volume loss is just $-\nabla\cdot\mathbf{F}_{\text{rad}} = -4\pi\int j_\nu d\nu$. For a hot gas radiating thermal bremsstrahlung (or 'free-free radiation'), this has the approximate form constant $\times \rho^2 T^{1/2}$. At the opposite extreme, if the gas is very optically thick, as in the interior of a star, then $\mathbf{F}_{\text{rad}}$ approximates the blackbody flux and $-\nabla\cdot\mathbf{F}_{\text{rad}}$ is given by the Rosseland approximation $\mathbf{F}_{\text{rad}} = -(16\sigma/3\kappa_R\rho)T^3\nabla T$ where κ_R is a weighted average over frequency of the opacity. This Rosseland approximation is discussed in any book on stellar structure (see the Appendix).

The second new quantity in the energy equation (2.5) is the conductive flux of heat, $\mathbf{q}$. This measures the rate at which random motions, chiefly those of electrons, transport thermal energy in the gas and thus act to smooth out temperature differences. Standard kinetic theory (cf. equation (3.42)), shows that for an ionized gas obeying the requirement $\lambda \ll T/|\nabla T|$

$$\mathbf{q} \approx -10^{-6} T^{5/2} \nabla T \text{ erg s}^{-1} \text{ cm}^{-2}. \tag{2.6}$$

(See Section 3.6 for a discussion of transport processes.) Obviously the term $-\nabla\cdot\mathbf{q}$ raises the order of differentiation of T in the energy equation, so it is again fortunate that, in many cases, temperature gradients are small enough that this term can be omitted from (2.5).

The system of equations (2.1–2.6), supplemented, if necessary, by the radiative transfer equation and the specification of $\mathbf{f}$, give, in principle, a complete description of the behaviour of a gas under appropriate boundary conditions. Of course, in practice one cannot hope to solve the equations in the fearsome generality in which

2.3 Steady adiabatic flows; isothermal flows

Let us consider first *steady* flows, for which time derivatives are put equal to zero, and let us specialize to the case in which there are no losses through radiation and no thermal conduction.

Our three conservation laws of mass, momentum and energy then become

$$\nabla \cdot (\rho \mathbf{v}) = 0, \tag{2.7a}$$

$$\rho(\mathbf{v} \cdot \nabla)\mathbf{v} = -\nabla P + \mathbf{f}, \tag{2.7b}$$

$$\nabla \cdot [(\tfrac{1}{2}\rho v^2 + \rho \varepsilon + P)\mathbf{v}] = \mathbf{f} \cdot \mathbf{v}. \tag{2.7c}$$

Substituting the first of these equations in the third implies

$$\rho \mathbf{v} \cdot \nabla(\tfrac{1}{2}v^2 + \varepsilon + P/\rho) = \mathbf{f} \cdot \mathbf{v}, \tag{2.8}$$

while (2.7b), the Euler equation, shows that $\mathbf{f} \cdot \mathbf{v} = \rho \mathbf{v}(\mathbf{v} \cdot \nabla)\mathbf{v} + \mathbf{v} \cdot \nabla P = \rho \mathbf{v} \cdot (\tfrac{1}{2}v^2) + \mathbf{v} \cdot \nabla P$; hence eliminating $\mathbf{f} \cdot \mathbf{v}$ from (2.8) we get

$$\rho \mathbf{v} \cdot \nabla(\varepsilon + P/\rho) = \mathbf{v} \cdot \nabla P,$$

or, expanding $\nabla(P/\rho)$ and rearranging,

$$\mathbf{v} \cdot [\nabla \varepsilon + P \nabla(1/\rho)] = 0.$$

By the definition of the gradient operator, this means that, if we travel a small distance along a streamline of the gas, i.e. if we follow the velocity $\mathbf{v}$, the increments $d\varepsilon$ and $d(1/\rho)$ in ε and $1/\rho$ must be related by

$$d\varepsilon + P\,d(1/\rho) = 0.$$

But from the expression for the internal energy (2.4) and the perfect gas law (2.2) this requires that

$$\tfrac{3}{2}dT + \rho T\,d(1/\rho) = 0,$$

which is equivalent to

$$\rho^{-1} T^{3/2} = \text{constant}$$

or

$$P\rho^{-5/3} = \text{constant} \tag{2.9}$$

using (2.2).

Equation (2.9) describes the so-called *adiabatic* flows. Although we have demonstrated only that the combination $P\rho^{-5/3}$ is constant along a given streamline, in many cases it is assumed that this constant is the same for each streamline, i.e. it is the same throughout the gas. This condition is equivalent to setting the entropy of the gas constant. The resulting flows are called *isentropic*. Note that adiabatic and isentropic are often used synonymously in the literature.

12 Gas dynamics

In a sense, our derivation of the adiabatic law (2.9) is 'back-to-front', since thermodynamic laws go into the construction of the energy equation (2.5). It is presented here to demonstrate the consistency of (2.5) with expectations from thermodynamics. If our gas were not monatomic, so that the numerical coefficient in (2.4) differed from $\frac{3}{2}$, we would obtain a result like (2.9), but with a different exponent for ρ:

$$P\rho^{-\gamma} = \text{constant.} \tag{2.10}$$

In this form γ is known as the *adiabatic index*, or the *ratio of specific heats*.

A further important special type of flow results from the assumption that the gas temperature T is constant throughout the region of interest. This is called *isothermal flow*, and is obviously equivalent to postulating some unspecified physical process to keep T constant. This, in turn, means that the energy equation (2.5) is replaced in our system describing the gas by the relation $T = $ constant. Formally, this latter requirement can be written, using the perfect gas law (2.2), as

$$P\rho^{-1} = \text{constant},$$

which has the form of (2.10) with $\gamma = 1$.

2.4 Sound waves

An obvious class of solution to our gas equations is that corresponding to *hydrostatic equilibrium*. In this case, in addition to the restriction to steady flow, and the absence of losses assumed in Section 2.3 above, we take $\mathbf{v} = 0$. Then the only equations remaining to be satisfied are (2.7b), which reduces to

$$\nabla P = \mathbf{f},$$

together with an explicit expression for $\mathbf{f}$, and the perfect gas law (2.2). Solutions of this type are, for example, appropriate to stellar, or planetary, atmospheres in radiative equilibrium.

Let us assume that we have such a solution, in which P and ρ are certain functions of position, P_0 and ρ_0, and consider small perturbations about it. We set

$$P = P_0 + P', \quad \rho = \rho_0 + \rho', \quad \mathbf{v} = \mathbf{v}'$$

where all the primed quantities are assumed small, so that we can neglect second and higher order products of them. In place of the energy equation (2.5), we assume that the perturbations are adiabatic, or isothermal: in reality, either of these cases can occur. Thus

$$P + P' = K(\rho + \rho')^\gamma, \quad K = \text{constant} \tag{2.11}$$

with $\gamma = \frac{5}{3}$ (adiabatic) or $\gamma = 1$ (isothermal). Linearizing the continuity equation (2.1) and the Euler equation (2.3), and using the fact that $\nabla P_0 = \mathbf{f}$, we get

$$\frac{\partial \rho'}{\partial t} + \rho_0 \nabla \cdot \mathbf{v}' = 0, \tag{2.12a}$$

$$\frac{\partial \mathbf{v}'}{\partial t} + \frac{1}{\rho_0} \nabla P' = 0. \tag{2.12b}$$

Sound waves

From (2.11) P is purely a function of ρ, so $\nabla P' = (dP/d\rho)_0 \nabla \rho'$ to first order, where the subscript zero implies that the derivative is to be evaluated for the equilibrium solution, i.e. $(dP/d\rho)_0 = dP_0/d\rho_0$. Thus, (2.12b) becomes

$$\frac{\partial \mathbf{v}'}{\partial t} + \frac{1}{\rho_0}\left(\frac{dP}{d\rho}\right)_0 \nabla \rho' = 0. \tag{2.13}$$

Eliminating $\mathbf{v}'$ from (2.13) and (2.12a) by operating with $\nabla \cdot$ and $\partial/\partial t$ respectively and then subtracting, gives

$$\frac{\partial^2 \rho'}{\partial t^2} = c_s^2 \nabla^2 \rho', \tag{2.14}$$

where we have defined

$$c_s = \left(\frac{dP}{d\rho}\right)_0^{1/2}. \tag{2.15}$$

Equation (2.14) will be recognized as the *wave equation*, with the wave speed c_s. It is easy, now, to show that the other variables P', $\mathbf{v}'$ obey similar equations; this implies that small perturbations about hydrostatic equilibrium propagate through the gas as sound waves with speed c_s. From (2.11), (2.15) we see that the *sound speed* c_s can have two values:

$$\text{adiabatic:} \quad c_s^{\text{ad}} = \left(\frac{5P}{3\rho}\right)^{1/2} = \left(\frac{5kT}{3\mu m_H}\right)^{1/2} \propto \rho^{1/3}, \tag{2.16a}$$

$$\text{isothermal:} \quad c_s^{\text{iso}} = \left(\frac{P}{\rho}\right)^{1/2} = \left(\frac{kT}{\mu m_H}\right)^{1/2}. \tag{2.16b}$$

The sound speeds c_s^{ad}, c_s^{iso}, are basic quantities which can be defined locally at any point of a gas. Note first that both c_s^{ad} and c_s^{iso} are of the order of the mean thermal speed of the ions of the gas, cf. equation (2.4). Numerically,

$$c_s \cong 10 \, (T/10^4 \, \text{K})^{1/2} \, \text{km s}^{-1} \tag{2.17}$$

where c_s stands for either sound speed.

Since c_s is the speed at which pressure disturbances travel through the gas, it limits the rapidity with which the gas can respond to pressure changes. For example, if the pressure in one part of a region of the gas of characteristic size L is suddenly changed, the other parts of the region cannot respond to this change until a time of order L/c_s, the sound crossing time, has elapsed. Conversely, if the pressure in one part of the region is changed on a timescale much longer than L/c_s the gas has ample time to respond by sending sound signals throughout the region, so the pressure gradient will remain small. Thus, if we consider *supersonic flow*, where the gas moves with $|\mathbf{v}| > c_s$, then the gas cannot respond on the flow time $L/|\mathbf{v}| < L/c_s$, so pressure gradients have little effect on the flow. At the other extreme, for *subsonic flow* with $|\mathbf{v}| < c_s$, the gas can adjust in less than the flow time, so to a first approximation the gas behaves as if in hydrostatic equilibrium.

14 Gas dynamics

These properties can be inferred directly from an order-of-magnitude analysis of the terms in the Euler equation (2.3). For example, for supersonic flow we have

$$\frac{|\rho(\mathbf{v}\cdot\nabla)\mathbf{v}|}{|\nabla P|} \sim \frac{v^2/L}{P/\rho L} \sim \frac{v^2}{c_s^2} > 1$$

and pressure gradients can be neglected in a first approximation.

A very important property of the sound speed is its dependence on the gas density (2.16a). This means that regions of higher than average density have higher than average sound speeds, a fact which gives rise to the possibility of *shock waves*. In a shock the fluid quantities change on lengthscales of the order of the mean free path λ and this is represented as a *discontinuity* in the fluid. Shock waves are important in physics and astrophysics and we shall return to them in Section 3.8.

2.5 Steady, spherically symmetric accretion

Let us now attack a real accretion problem and show how all of the apparatus we have developed in Sections 2.1–2.4 can be put to use. We consider a star of mass M accreting spherically symmetrically from a large gas cloud. This would be a reasonable approximation to the real situation of an isolated star accreting from the interstellar medium, provided that the angular momentum, magnetic field strength and bulk motion of the interstellar gas with respect to the star could be neglected. For other types of accretion flows, such as those in close binary systems and models of active galactic nuclei, spherical symmetry is rarely a good approximation, as we shall see. Nonetheless, the spherical accretion problem is of very great significance for the theory, as it introduces some important concepts which have much wider validity. Furthermore, it is possible to give a fairly exact treatment, allowing us to gain insight into more complicated problems.

Let us ask what we might hope to discover by analysing this problem. First, we should expect to be able to predict the steady accretion rate $\dot{M}$ (g s^{-1}) on to our star, given the ambient conditions (the density $\rho(\infty)$ and the temperature $T(\infty)$) in the parts of the gas cloud far from the star and some boundary conditions at its surface. Second, we might hope to learn how big a region of the gas cloud is influenced by the presence of the star. These questions can be answered in a natural and physically appealing way. In addition, we shall obtain an understanding of the relation between the gas velocity and the local sound speed which can be carried over quite generally to more complicated accretion flows.

To treat the problem mathematically we take spherical polar coordinates (r, θ, ϕ) with origin at the centre of the star. The fluid variables are independent of θ and ϕ by spherical symmetry, and the gas velocity has only a radial component $v_r = v$. We take this to be negative, since we want to consider infall of material; $v > 0$ would correspond to a stellar wind. For steady flow, the continuity equation (2.1) reduces to

$$\frac{1}{r^2}\frac{d}{dr}(r^2 \rho v) = 0 \qquad (2.18)$$

using the standard expression for the divergence of a vector in spherical polar

Steady, spherically symmetric accretion

coordinates. This integrates to $r^2 \rho v =$ constant. Since $\rho(-v)$ is the inward flux of material, the constant here must be related to the (constant) accretion rate $\dot{M}$; the relation is

$$4\pi r^2 \rho(-v) = \dot{M}. \tag{2.19}$$

In the Euler equation the only contribution to the external force, **f**, is from gravity, and this has only a radial component

$$f_r = -GM\rho/r^2$$

so that (2.3) becomes

$$v\frac{dv}{dr} + \frac{1}{\rho}\frac{dP}{dr} + \frac{GM}{r^2} = 0. \tag{2.20}$$

We replace the energy equation (2.5) by the polytropic relation (2.10):

$$P = K\rho^\gamma, \quad K = \text{constant}. \tag{2.21}$$

This allows us to treat both approximately adiabatic ($\gamma \cong \frac{5}{3}$) and isothermal ($\gamma \cong 1$) accretion simultaneously. After the solution has been found, the adiabatic or isothermal assumption should be justified by consideration of the particular radiative cooling and heating of the gas. For example, the adiabatic approximation will be valid if the timescales for significant heating and cooling of the gas are long compared with the time taken for an element of the gas to fall in. In reality, neither extreme is quite satisfied, so we expect $1 < \gamma < \frac{5}{3}$. In fact, the treatment we shall give is valid for $1 < \gamma < \frac{5}{3}$: the extreme values require special consideration (see e.g. (2.28), (2.29)). The interested reader may consult the article by Holzer & Axford (1970).

Finally, we can use the perfect gas law (2.2) to give the temperature

$$T = \mu m_\text{H} P/\rho k \tag{2.22}$$

where $P(r)$, $\rho(r)$ have been found.

The problem therefore reduces to that of integrating (2.20) with the help of (2.21) and (2.19) and then identifying the unique solution corresponding to our accretion problem. We shall integrate (2.20) shortly, but it is instructive to see how much information can be extracted without explicit integration, since this technique is very useful in other cases when analytic integration is not possible, or not straightforward. We write first

$$\frac{dP}{dr} = \frac{dP}{d\rho}\frac{d\rho}{dr} = c_s^2 \frac{d\rho}{dr}.$$

Hence, the term $(1/\rho)(dP/dr)$ in the Euler equation (2.20) is $(c_s^2/\rho)(d\rho/dr)$. But, from the continuity equation (2.18),

$$\frac{1}{\rho}\frac{d\rho}{dr} = -\frac{1}{vr^2}\frac{d}{dr}(vr^2).$$

16   Gas dynamics

Therefore, (2.20) becomes

$$v\frac{dv}{dr} - \frac{c_s^2}{vr^2}\frac{d}{dr}(vr^2) + \frac{GM}{r^2} = 0,$$

which, after a little rearrangement, gives

$$\tfrac{1}{2}\left(1 - \frac{c_s^2}{v^2}\right)\frac{d}{dr}(v^2) = -\frac{GM}{r^2}\left[1 - \left(\frac{2c_s^2 r}{GM}\right)\right]. \tag{2.23}$$

At first sight, we appear to have made things worse by these manipulations, since c_s^2 is, in general, a function of r. However, the physical interpretation of c_s as the sound speed, plus the structure of equation (2.23), in which factors on either side can in principle vanish, allow us to sort the possible solutions of (2.23) into distinct classes and to pick out the unique one corresponding to our problem. First, we note that at large distances from the star the factor $[1 - (2c_s^2 r/GM)]$ on the right hand side must be negative, since c_s^2 approaches some finite asymptotic value $c_s^2(\infty)$ related to the gas temperature far from the star, while r increases without limit. This means that for large r the right hand side of (2.23) is positive. On the left hand side, the factor $d(v^2)/dr$ must be negative, since we want the gas far from the star to be at rest, accelerating as it approaches the star with r decreasing. These two requirements $(d(v^2)/dr < 0$, the r.h.s. of (2.23) > 0) are compatible only if at large r the gas flow is *subsonic*, i.e.

$$v^2 < c_s^2 \quad \text{for large } r. \tag{2.24}$$

This is, of course, a very reasonable result, as the gas will have a non-zero temperature and hence a non-zero sound speed far from the star. As the gas approaches the star, r decreases and the factor $[1-(2c_s^2 r/GM)]$ must tend to increase. It must eventually reach zero, unless some way can be found of increasing c_s^2 sufficiently by heating the gas. This is very unlikely, since the factor reaches zero at a radius given by

$$r_s = \frac{GM}{2c_s^2(r_s)} \cong 7.5 \times 10^{13}\left(\frac{T(r_s)}{10^4 \text{ K}}\right)^{-1}\left(\frac{M}{M_\odot}\right) \text{cm} \tag{2.25}$$

where we have used (2.17) to introduce the temperature. The order of magnitude $r_s \cong 7.5 \times 10^{13}$ cm in (2.25) is so much larger than the radius, R_*, of any compact object ($R_* \lesssim 10^9$ cm) that very high temperatures would be required to make r_s smaller than R_*. In fact, it is clear from (2.16) and (1.7) that a gas temperature of order T_{th} is required: this can be achieved, for example, in a standing shock wave close to the stellar surface. We shall have much to say about this possibility later on, but it does not enter our analysis here as it requires discontinuous jumps in ρ, T, P, etc. A similar analysis of the signs in (2.23) for $r < r_s$ shows that the flow must be supersonic near the star:

$$v^2 > c_s^2 \quad \text{for small } r. \tag{2.26}$$

The discussion above shows that the problem we are considering is not

Steady, spherically symmetric accretion

mathematically well posed if we only give the ambient conditions at infinity. To specify the problem correctly we need a condition at or near the stellar surface also. Here we have imposed (2.26), which will have the effect of picking out just one solution (Type 1 below). Without (2.26) there would be another possible solution (Type 3).

The existence of a point r_s satisfying the implicit equation (2.25) is of great importance in characterizing the accretion flow. The direct mathematical consequence is that at $r = r_s$ the left hand side of (2.23) must also vanish: this requires

$$\text{either} \quad v^2 = c_s^2 \quad \text{at } r = r_s, \tag{2.27a}$$

$$\text{or} \quad \frac{d}{dr}(v^2) = 0 \quad \text{at } r = r_s. \tag{2.27b}$$

All solutions of (2.23) can now be classified by their behaviour at r_s, given by either (2.27a) or (2.27b), together with their behaviour at large r; for example, (2.24). This is very easy to see if we plot $v^2(r)/c_s^2(r)$ against r (Fig. 1).

From the figure it is clear that there are just six distinct families of solutions:

Type 1: $v^2(r_s) = c_s^2(r_s)$, $v^2 \to 0$ as $r \to \infty$
$(v^2 < c_s^2, r > r_s; v^2 > c_s^2, r < r_s)$;

Type 2: $v^2(r_s) = c_s^2(r_s)$, $v^2 \to 0$ as $r \to 0$
$(v^2 > c_s^2, r > r_s; v^2 < c_s^2, r > r_s)$;

Type 3: $v^2 < c_s^2$ everywhere, $\frac{d}{dr}(v^2) = 0$ at r_s;

Type 4: $v^2 > c_s^2$ everywhere, $\frac{d}{dr}(v^2) = 0$ at r_s;

Type 5: $\frac{d}{dr}(v^2) = \infty$ at $v^2 = c_s^2(r_s)$; $r > r_s$ always;

Type 6: $\frac{d}{dr}(v^2) = \infty$ at $v^2 = c_s^2(r_s)$; $r < r_s$ always.

There is just one solution for each of Types 1 and 2: these are called *trans-sonic* as they make a transition between sub- and supersonic flow at r_s; r_s itself is known as the *sonic point* for these solutions. The occurrence of sonic points is a quite general feature of gas dynamical problems. Types 3 and 4 represent flow which is everywhere sub- or supersonic. Types 5 and 6 do not cover all of the range of r and are double-valued in the sense that there are two possible values of v^2 at a given r. We exclude these last two for these reasons, although they can represent *parts* of a correct solution if shocks are present. Types 2 and 4 must be excluded since they are supersonic at large r, violating (2.24), while Type 3 is subsonic at small r, violating (2.26). A solution of Type 2 with $v > 0$ describes a stellar wind: note that (2.23) is unchanged for $v \to -v$. Solutions of Type 3 with $v > 0$ give the so-called stellar 'breeze' solutions which are everywhere subsonic; if $v < 0$ this is a slowly sinking 'atmosphere'.

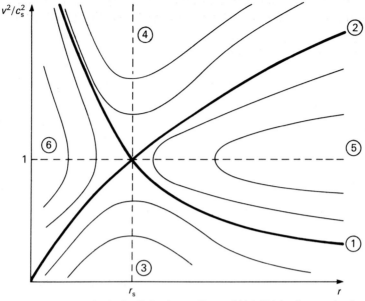

Figure 1. Spherical adiabatic gas flows $v^2(r)/c_s^2(r)$ in the gravitational field of a star. For $v < 0$ these are accretion flows, while for $v > 0$ they are winds or 'breezes'. The two trans-sonic solutions 1, 2 divide the remaining solutions into the families 3–6 described in the text.

We are left finally with just the Type 1 solution: this has all the properties we want and is the unique solution to our problem. The sonic point condition (2.27a) will lead us to the goal of relating the accretion rate $\dot{M}$ to the conditions at infinity.

With the question of uniqueness settled, we now integrate (2.20) directly, using the fact that (2.21) makes ρ a function of P:

$$\frac{v^2}{2} + \int \frac{dP}{\rho} - \frac{GM}{r} = \text{constant}.$$

From (2.21) we have $dP = K\gamma\rho^{\gamma-1}\,d\rho$, and performing the integration, we obtain (for $\gamma \neq 1$)

$$\frac{v^2}{2} + \frac{K\gamma}{\gamma-1}\rho^{\gamma-1} - \frac{GM}{r} = \text{constant}.$$

But $K\gamma\rho^{\gamma-1} = \gamma P/\rho = c_s^2$, and we obtain the *Bernoulli integral*:

$$\frac{v^2}{2} + \frac{c_s^2}{\gamma-1} - \frac{GM}{r} = \text{constant}. \tag{2.28}$$

(The strictly isothermal ($\gamma = 1$) case gives a logarithmic integral.) From the known property of our physical solution (Type 1) we have $v^2 \to 0$ as $r \to \infty$, so the constant in (2.28) must be $c_s^2(\infty)/(\gamma-1)$, where $c_s(\infty)$ is the sound speed in the gas far away from the star. The sonic point condition now relates $c_s(\infty)$ to $c_s(r_s)$, since (2.27a), (2.25) imply $v^2(r_s) = c_s^2(r_s)$, $GM/r_s = 2c_s^2(r_s)$, and the Bernoulli integral gives

$$c_s^2(r_s)\left[\tfrac{1}{2}+\frac{1}{\gamma-1}-2\right] = \frac{c_s^2(\infty)}{\gamma-1}$$

or

$$c_s(r_s) = c_s(\infty)\left(\frac{2}{5-3\gamma}\right)^{1/2}. \tag{2.29}$$

We now obtain $\dot{M}$ from (2.19)

$$\dot{M} = 4\pi r^2 \rho(-v) = 4\pi r_s^2 \rho(r_s) c_s(r_s) \tag{2.30}$$

since $\dot{M}$ is independent of r. Using $c_s^2 \propto \rho^{\gamma-1}$ we find

$$\rho(r_s) = \rho(\infty)\left[\frac{c_s(r_s)}{c_s(\infty)}\right]^{2/(\gamma-1)}.$$

Putting this and (2.30) into (2.29) gives, after a little algebra, the relation we are looking for between $\dot{M}$ and conditions at infinity:

$$\dot{M} = \pi G^2 M^2 \frac{\rho(\infty)}{c_s^3(\infty)}\left[\frac{2}{5-3\gamma}\right]^{(5-3\gamma)/2(\gamma-1)}. \tag{2.31}$$

Note that the dependence on γ here is rather weak: the factor $[2/(5-3\gamma)]^{(5-3\gamma)/2(\gamma-1)}$ varies from unity in the limit $\gamma = \tfrac{5}{3}$ to $e^{3/2} \cong 4.5$ in the limit $\gamma = 1$. For a value $\gamma = 1.4$, which would be typical for the adiabatic index of a part of the interstellar medium, the factor is 2.5.

Equation (2.31) shows that accretion from the interstellar medium is unlikely to be an observable phenomenon; reasonable values would be $c_s(\infty) = 10$ km s^{-1}, $\rho(\infty) = 10^{-24}$ g cm^{-3}, corresponding to a temperature of about 10^4 K and number density near 1 particle cm^{-3}. Then (2.31) gives (with $\gamma = 1.4$)

$$\dot{M} \cong 1.4 \times 10^{11} \left(\frac{M}{M_\odot}\right)^2 \left(\frac{\rho(\infty)}{10^{-24}}\right)\left(\frac{c_s(\infty)}{10 \text{ km s}^{-1}}\right)^{-3} \text{ g s}^{-1}. \tag{2.32}$$

From (1.46) even accreting this on to a neutron star yields L_{acc} only of the order 2×10^{31} erg s^{-1}; at a typical distance of 1 kpc this gives far too low a flux to be detected.

To complete the solution of the problem to find the run of all quantities with r we could now get $v(r)$ in terms of $c_s(r)$ from (2.30), using $c_s^2 = \gamma P/\rho \propto \rho^{\gamma-1}$:

$$(-v) = \frac{\dot{M}}{4\pi r^2 \rho(r)} = \frac{\dot{M}}{4\pi r^2 \rho(\infty)}\left[\frac{c_s(\infty)}{c_s(r)}\right]^{2/(\gamma-1)}.$$

Substituting this into the Bernoulli integral (2.28) gives an algebraic relation for $c_s(r)$; the solution of this then gives $\rho(r)$ and $v(r)$. In practice, the algebraic equation for $c_s(r)$ has fractional exponents and must be solved numerically. However, the main features of the r-dependence can be inferred by looking at the Bernoulli integral (2.28). At large r the gravitational pull of the star is weak and all quantities have their 'ambient' values ($\rho(\infty)$, $c_s(\infty)$, $v \cong 0$). As one moves to smaller r, the inflow velocity increases until $(-v)$ reaches $c_s(\infty)$, the sound speed at infinity. The only term in (2.28) capable of balancing this increase is the gravity term GM/r; since $c_s(r)$ does not greatly exceed $c_s(\infty)$ this must occur at a radius

$$r \cong r_{\text{acc}} = \frac{2GM}{c_s^2(\infty)} \cong 3 \times 10^{14} \left(\frac{M}{M_\odot}\right)\left(\frac{10^4 \text{ K}}{T(\infty)}\right) \text{ cm.} \tag{2.33}$$

At this point $\rho(r)$ and $c_s(r)$ begin to increase above their ambient values. At the sonic point $r = r_s$ (see 2.27a) the inflow becomes supersonic and the gas is effectively in free fall: from (2.28) $v^2 \gg c_s^2$ implies

$$v^2 \cong 2GM/r = v_{\text{ff}}^2$$

with v_{ff}^2 the free-fall velocity. The continuity equation (2.19) now gives

$$\rho \cong \rho(r_s)\left(\frac{r_s}{r}\right)^{3/2} \quad \text{for } r \lesssim r_s.$$

Finally, we can, in principle, get the gas temperature, using the perfect gas law and the polytropic relation

$$T \cong T(r_s)\left(\frac{r_s}{r}\right)^{[3/2](\gamma-1)} \quad \text{for } r \lesssim r_s.$$

However, the steady increase in T for decreasing r predicted by this equation is probably unrealistic: radiative losses must begin to cool the gas, so a better energy equation than (2.21) is needed at this point.

The radius r_{acc} defined by (2.33) has a simple interpretation: at a radius r the ratio of internal (thermal) energy to gravitational binding energy of a gas element of mass m is

$$\frac{\text{thermal energy}}{\text{binding energy}} \sim \frac{mc_s^2(r)}{2}\frac{r}{GMm} \sim \frac{r}{r_{\text{acc}}} \quad \text{for } r \gtrsim r_{\text{acc}}$$

since $c_s(r) \sim c_s(\infty)$ for $r > r_{\text{acc}}$. Hence, for $r \gg r_{\text{acc}}$ the gravitational pull of the star has little effect on the gas. We call r_{acc} the accretion radius: it gives the range of influence of the star on the gas cloud which we sought at the outset. Note that in terms of r_{acc} the relation (2.31) giving the steady accretion rate can be rewritten as

$$\dot{M} \sim \pi r_{\text{acc}}^2 c_s(\infty) \rho(\infty). \tag{2.34}$$

Dimensionally r_{acc} must have a form like (2.33); however, since the proper specification of the accretion flow involves a 'surface' condition like (2.26) the numerical factor in the formula for r_{acc} is in general undetermined, and the concept of an 'accretion radius' is not well defined. A Type 3 solution for the same $c_s(\infty)$, $\rho(\infty)$ would give a smaller accretion rate $\dot{M}$ than (2.31). If an $\dot{M}$ greater than the

Steady, spherically symmetric accretion

value (2.31) is externally imposed (e.g. by mass exchange in a binary system) the flow must become supersonic near the star and must involve discontinuities (i.e. shocks).

We have treated the problem of steady spherical accretion at some length. The main conclusions we can draw from this study and apply generally are:

1. The steady accretion rate $\dot{M}$ is determined by ambient conditions at infinity (equation (2.31)) and a 'surface' condition (e.g. equation (2.26)). For accretion by isolated stars from the interstellar medium, the resulting value of $\dot{M}$ is too low to be of much observational importance. Clearly, we must look to close binaries to find more powerful accreting systems.
2. The star's gravitational pull seriously influences the gas's behaviour only inside the accretion radius r_{acc}.
3. A steady accretion flow with $\dot{M}$ greater or equal to the value (2.31) must possess a sonic point; i.e. the inflow velocity must become supersonic near the stellar surface.

The immediate consequence of point 3 is that, since for a star (although not for a black hole) the accreting material must eventually join the star with a very small velocity, some way of stopping the highly supersonic accretion flow must be found. Consideration of how this stopping process can work leads us naturally into the area of plasma physics, which we touched on briefly at the beginning of this chapter. In the next chapter we shall develop in more detail the plasma concepts we shall need.

3
Plasma concepts

3.1 Introduction

Whenever we need to consider the behaviour of a gas on lengthscales comparable to the mean free path between collisions, we must use the ideas of plasma physics. In this chapter we shall briefly introduce some of the concepts that will be important to our study of accretion.

A *plasma* differs from an atomic or molecular gas in that it consists of a mixture of two gases of electrically charged particles: an electron gas and an ion gas, with very different particle masses m_e and m_i.

The electrons and ions interact with each other through their electrostatic Coulomb attractions and repulsions. These Coulomb forces decrease only slowly ($\propto r^{-2}$) with distance and do not have a characteristic length scale. Thus, a plasma particle interacts with many others at any one instant, and this makes the description of collisions more complicated than in atomic or molecular gases, where the interparticle forces are very short-range. A further complication arises from the great difference in particle masses m_e and m_i. Since collisions between particles of very different masses can transfer only a small fraction of the kinetic energy of order $m_e/m_i \ll 1$, it is possible for electrons and ions to have significantly different temperatures over appreciable timescales. These two properties – the long-range nature of the Coulomb force and the disparity in electron and ion masses – give the physics of plasmas its particular character. A further series of complex phenomena occurs when the plasma is permeated by a large-scale magnetic field; this is particularly relevant for the study of gas accreting on to highly magnetized neutron stars and white dwarfs.

3.2 Charge neutrality, plasma oscillations and the Debye length

Let us begin by examining in more detail the consequences of the long-range character of the Coulomb force between charged particles.

Note first that the number densities of ions and electrons at any point must be approximately equal, and therefore a plasma must always be close to charge neutrality: even a small charge imbalance would result in very large electric fields which would act to move the plasma particles so as to restore neutrality very quickly. As an example, suppose that there is a 1% charge imbalance in a sphere of

Charge neutrality, plasma oscillations, Debye length

radius r in a plasma of number density N; then electrons near the edge of the sphere experience in an electric field E and an acceleration

$$\dot{v} = \frac{e|\mathbf{E}|}{m_e} \simeq \frac{4\pi r^3}{3m_e} \frac{N}{100} \frac{e^2}{[4\pi\varepsilon_0]r^2}$$

where $(-e)$ is the charge on an electron. For a sphere of radius $r = 1$ cm in a typical astrophysical plasma having $N = 10^{10}$ cm^{-3} we find $\dot{v} \sim 10^{17}$ cm s^{-2}, so electrons would move very rapidly to neutralize the charge imbalance in a time of order 3×10^{-9} s. Indeed, they would move so rapidly that they would overshoot, and induce *oscillations* of the plasma. Since all plasmas are subject to small perturbations (e.g. the passage through them of electromagnetic radiation, or the thermal motion of the plasma particles) which tend to disturb charge neutrality, the natural frequency of these oscillations is a fundamental quantity called the *plasma frequency*.

To calculate the plasma frequency, let us consider an otherwise uniform plasma with a small excess of electrons in some (small) region. We assume the ions have on average a charge $Ze \simeq e$, and we take the ion and electron number densities as

$$N_i \simeq N_0, \quad N_e = N_0 + N_1(\mathbf{r}, t),$$

with $N_1 \ll N_0$, and N_1 zero outside our small region. The charge excess N_1 gives rise to an electric field $\mathbf{E}$ through the Maxwell equation

$$\boldsymbol{\nabla} \cdot \mathbf{E} = -\frac{4\pi}{[4\pi\varepsilon_0]} N_1 e. \tag{3.1}$$

This electric field causes motion of the plasma particles. Since $m_i \gg m_e$, we may neglect the motions of the ions which are much slower than those of the electrons. The electrons will move as a fluid with a velocity field $\mathbf{v}_e(\mathbf{r}, t)$ (cf. Chapter 2), where $|\mathbf{v}_e|$ is proportional to N. Since N_1 is assumed small, we can neglect the term $(\mathbf{v}_e \cdot \boldsymbol{\nabla}) \mathbf{v}_e$ in the Euler equation (cf. (2.3)) for the electron fluid. Further, the force density is $-N_e e\mathbf{E}$ and the mass density $N_e m_e$ so that the equation of motion of the electron fluid reduces to that of the individual electrons:

$$m_e \frac{\partial \mathbf{v}_e}{\partial t} = -e\mathbf{E}. \tag{3.2}$$

Finally, there must be an equation ensuring charge conservation for the electrons: as charge is a scalar quantity carried around, like mass, by each electron, the required equation must resemble the continuity equation (2.1):

$$\frac{\partial N_e}{\partial t} + \boldsymbol{\nabla} \cdot (N_e \mathbf{v}_e) = 0 \tag{3.3a}$$

or, neglecting products of small quantities,

$$\frac{\partial N_1}{\partial t} + N_0 \boldsymbol{\nabla} \cdot \mathbf{v}_e = 0. \tag{3.3b}$$

Plasma concepts

We can now eliminate $\mathbf{v}_e$ from these equations by taking the divergence of (3.2) and the time derivative of (3.3b) and subtracting:

$$\frac{1}{N_0}\frac{\partial^2 N_1}{\partial t^2} - \frac{e}{m_e}\nabla\cdot\mathbf{E} = 0.$$

Using (3.1) we get

$$\frac{\partial^2 N_1}{\partial t^2} + \left\{\frac{4\pi}{[4\pi\varepsilon_0]}\frac{N_0 e^2}{m_e}\right\}N_1 = 0.$$

Thus, the charge imbalance N_1 oscillates with the *plasma frequency*

$$\omega_p = \left\{\frac{4\pi}{[4\pi\varepsilon_0]}\frac{N_0 e^2}{m_e}\right\}^{1/2}. \tag{3.4}$$

Numerically, with N_0 measured in cm^{-3}, we have

$$\left.\begin{aligned}\omega_p &= 5.7\times 10^4\, N_0^{1/2}\,\text{rad s}^{-1} \\ \text{and} \\ v_p &= \frac{\omega_p}{2\pi} = 9.0\times 10^3\, N_0^{1/2}\,\text{Hz.}\end{aligned}\right\} \tag{3.5}$$

A plasma is opaque to electromagnetic radiation with frequency $v < v_p$ because the plasma oscillations are more rapid than the variations in the applied electromagnetic fields, and the plasma electrons move to 'short out' the radiation. For the Earth's ionosphere $N_0 \cong 10^6$ cm^{-3}, so that radio waves of frequency less than about 10^7 Hz cannot penetrate it and are reflected.

Associated with the typical timescale v_p^{-1} of charge oscillations must be a typical lengthscale l for the field set up by the charge imbalance N_1. This is given to an order of magnitude by estimating the derivatives as

$$\partial/\partial t \sim \omega_p, \quad \nabla \sim 1/l.$$

From (3.3a) we find

$$l \sim v_e/\omega_p. \tag{3.6}$$

Thus, the field E set up by the charge imbalance is limited in range to $\sim l$ by the shielding effect of the flow of electrons. Since even an unperturbed plasma will be subject to charge fluctuations due to the thermal motion of the electrons, there is a fundamental shielding distance λ_{Deb}, the *Debye length*, given by setting $v_e \sim (kT_e/m_e)^{1/2}$ in (3.6) where T_e is the electron temperature

$$\lambda_{\text{Deb}} = \left\{\frac{[4\pi\varepsilon_0]kT_e}{4\pi N_0 e^2}\right\}^{1/2}. \tag{3.7}$$

Collisions

Numerically, with N_0 measured in cm^{-3},

$$\lambda_{\text{Deb}} \cong 7\, (T_e/N_0)^{1/2} \text{ cm}. \tag{3.8}$$

The importance of λ_{Deb} is that it gives the lengthscale over which there can be appreciable departures from charge neutrality; thus it gives an effective range for Coulomb collisions in a plasma. Clearly, for our description of plasmas to be consistent, we require that a large number of particles are involved in a plasma oscillation, and also that plasma quantities do not vary on lengthscales L smaller than λ_{Deb}: thus

$$N_e \lambda_{\text{Deb}}^3 \gg 1, \quad L \gg \lambda_{\text{Deb}}. \tag{3.9}$$

If these conditions are not satisfied, collective effects will not be important, and we can treat the gas as a system of independent particles.

3.3 Collisions

Let us now consider Coulomb collisions between plasma particles. Because these collisions involve the acceleration of charged particles, electromagnetic radiation must be produced at the expense of the kinetic energy of the two particles. However, it is possible to show (e.g. Jackson (1975)) that the energy loss rates P_{rad}, P_{coll} of a particle of speed v, due to radiation and collisions respectively, are in the ratio

$$\frac{P_{\text{rad}}}{P_{\text{coll}}} \lesssim \frac{e^2}{[4\pi\varepsilon_0]\hbar c}\left(\frac{v}{c}\right)^2 \sim \frac{1}{137}\left(\frac{v}{c}\right)^2 \ll 1, \tag{3.10}$$

where $e^2/[4\pi\varepsilon_0]\hbar c$ is the fine-structure constant and $\hbar = h/2\pi$, the reduced Planck constant. Thus, the radiation energy loss is negligible in a collision and, to a good approximation, we can regard collisions as *elastic*, i.e. energy conserving. Thus, the problem concerns two particles, charges e_1, e_2, interacting via the Coulomb force $e_1 e_2/[4\pi\varepsilon_0]r^2$ at a separation r.

In such problems, it is easiest to study the trajectories of the colliding particles in a reference frame moving with the centre of mass. If the initial velocities of the two particles, in the laboratory frame are $\mathbf{v}_1, \mathbf{v}_2$, and they have masses m_1, m_2, the centre-of-mass frame has velocity

$$\mathbf{v}_{\text{CM}} = \frac{m_1 \mathbf{v}_1 + m_2 \mathbf{v}_2}{m_1 + m_2}. \tag{3.11}$$

In this frame, the incident particles have equal and opposite momenta $\pm m\mathbf{v}$, and their total kinetic energy is $\tfrac{1}{2}mv^2$, where

$$\left.\begin{aligned} m &= m_1 m_2/(m_1 + m_2), \\ \mathbf{v} &= \mathbf{v}_1 - \mathbf{v}_2. \end{aligned}\right\} \tag{3.12}$$

Since the Coulomb force obeys an inverse square law, the trajectories of the colliding particles in the centre-of-mass frame are hyperbolae, which can be characterized by an *impact parameter* b (Fig. 2). Furthermore, the particle speeds,

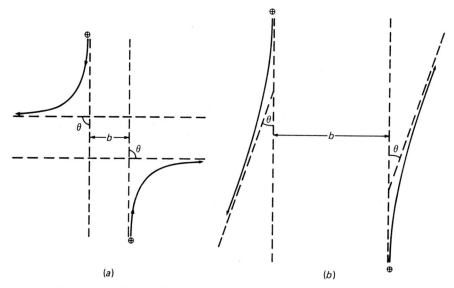

Figure 2. Coulomb collisions of two particles of like (here positive) charge, (a) close collision, (b) distant collision.

and hence energies, in this frame are unaltered by the collisions: the particles are merely deflected. The deflection angle θ will be appreciable, say of order 90°, when b takes a value b_0 such that the kinetic and potential energies of the particles are comparable at closest approach:

$$\frac{e_1 e_2}{[4\pi\varepsilon_0]b_0} \sim \tfrac{1}{2}mv^2 \sim kT \tag{3.13}$$

for thermal particles. For $b \gg b_0$, $\theta \sim e_1 e_2/[4\pi\varepsilon_0]bmv^2$ which is very small. In an atomic or molecular gas, b_0 would measure the size of the gas particles, and give a cross-section

$$\sigma_\perp = \pi b_0^2 = \frac{\pi}{[4\pi\varepsilon_0]^2} \frac{e_1^2 e_2^2}{(kT)^2}. \tag{3.14}$$

For a gas particle density N, the mean free path $\lambda_\perp$ between such collisions would then be

$$\lambda_\perp = \frac{1}{N\sigma_\perp} \cong \frac{[4\pi\varepsilon_0]^2}{\pi N}\left(\frac{kT}{e_1 e_2}\right)^2 \tag{3.15}$$

since, by definition of $\lambda_\perp$ and $\sigma_\perp$, the $N\lambda_\perp$ particles in a volume of unit cross-section and length $\lambda_\perp$ must present a unit area target to an incoming particle. From (3.15), for $e_1 e_2 = e^2$, we obtain

$$\lambda_\perp \cong 7 \times 10^5 \, T^2/N \text{ cm}. \tag{3.16}$$

However, any charged particle must interact electrostatically with all the plasma

Collisions

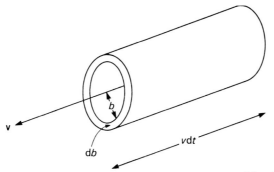

Figure 3. Coulomb encounters of a plasma particle of velocity **v**.

particles inside a sphere of radius λ_{Deb} at any time. These distant collisions produce only small deflections, since

$$\lambda_{\text{Deb}}/b_0 \sim [4\pi\varepsilon_0]kT\lambda_{\text{Deb}}/e^2 \sim 4\pi N_0 \lambda_{\text{Deb}}^3 \gg 1$$

using (3.7) and (3.9). Nonetheless, there are so many particles within a Debye sphere that the cumulative effect of the distant collisions outweighs that of the close encounters with $b \sim b_0$. We can show this most easily in the case where the colliding particles have comparable masses: $m_1 \sim m_2 \sim m$ (e.g. electron–electron or ion–ion encounters). Then each particle has initial speed $\sim v$, and in a distant collision acquires a velocity increment

$$\Delta v_b \sim v\theta \sim \frac{e_1 e_2}{[4\pi\varepsilon_0]mvb}$$

orthogonal to its original direction of motion. For *comparable masses* and *initial* (e.g. *thermal*) *speeds* v_1, v_2, this estimate carries over into the laboratory frame. (Note, however, that in general this will not be so.) The total effect of all the distant collisions in deflecting a given particle is obtained by suitably summing the increments Δv_b over b: since Δv_b can be either positive or negative, depending on initial conditions, it is natural to consider its square:

$$(\Delta v_b)^2 \sim \frac{e_1^2 e_2^2}{[4\pi\varepsilon_0]^2 m^2 v^2 b^2}.$$

In a time dt, the given particle 'collides' at impact parameter b with all the $2\pi b\, db\, v\, dt\, N$ particles in the cylindrical shell (see Fig. 3), where N is the number density of target particles (electrons or ions). Hence, the total rate of change of $(\Delta v)^2$ is

$$\frac{d(\Delta v)^2}{dt} \sim Nv \int (\Delta v_b)^2 2\pi b\, db$$

$$= \frac{2\pi N e_1^2 e_2^2}{[4\pi\varepsilon_0]^2 m^2 v} \ln\left(\frac{b_{\max}}{b_{\min}}\right) \qquad (3.17)$$

where $b_{\max}$ and $b_{\min}$ are the largest and smallest values of b contributing to the integral. We shall assign values to them below.

To establish the relative importance of distant and close collisions, we need to associate a mean free path with the change $(\Delta v)^2$ in (3.17), to compare with $\lambda_\perp$. To this end we define, from (3.17), a *deflection time*:

$$t_d = \frac{v^2}{d(\Delta v)^2/dt} = \frac{[4\pi\varepsilon_0]^2 m^2 v^3}{2\pi N e_1^2 e_2^2 \ln \Lambda} \tag{3.18}$$

where we have used the standard notation $\Lambda = b_{max}/b_{min}$. Clearly, t_d measures the time that elapses before a plasma particle of velocity v is appreciably deflected from its initial trajectory. Associated with t_d is a mean free path or *deflection length*

$$\lambda_d = v t_d = \frac{[4\pi\varepsilon_0]^2 m^2 v^4}{2\pi N e_1^2 e_2^2 \ln \Lambda}. \tag{3.19}$$

Comparing (3.19) and (3.14) and using $mv^2 \sim kT$, we find that

$$\frac{\lambda_\perp}{\lambda_d} \sim \ln \Lambda. \tag{3.20}$$

Thus, small angle distant collisions dominate close encounters if $\ln \Lambda > 1$.

We shall see below that the so-called *Coulomb logarithm*, $\ln \Lambda$, is normally of order 10–20, and the result will be established.

Since $\ln \Lambda$ rather than Λ is required, a fairly rough estimate of Λ will suffice. First, a possible candidate for b_{min} is the classical size b_0 given by (3.13). However, we must not violate the quantum-mechanical uncertainty principle, which asserts that we cannot simultaneously know the position and momentum of a particle with uncertainties less than Δx, Δp, related by

$$\Delta x \, \Delta p \sim \hbar.$$

Thus, we cannot use $b_{min} = b_0$ if this is smaller than

$$b_{min}(\text{QM}) \sim \hbar/\Delta p \sim \hbar/mv \tag{3.21}$$

where we have used $\Delta p \sim mv$. Equation (3.21) gives $b_{min}(\text{QM})$ as the de Broglie wavelength of a particle of momentum mv. In general,

$$b_{min} = \max\{\hbar/mv, 2e_1 e_2/[4\pi\varepsilon_0]mv^2\}, \tag{3.22}$$

so b_{min} is given by the classical value, b_0, unless

$$v \gtrsim \frac{e^2}{[4\pi\varepsilon_0]\hbar c} c = \frac{c}{137}. \tag{3.23}$$

To find b_{max} we recall that the plasma reacts to charge disturbances in a characteristic time $\sim 1/\nu_p$ (cf. (3.5)). Thus, the plasma can 'screen off' the Coulomb field of any particle whose characteristic interaction time $\sim b/v$ is longer than $1/\nu_p$. Hence, we can set

$$b_{max} = \frac{v}{\nu_p}.$$

Thermal plasmas: relaxation times, mean free path 29

Of course, if v is the thermal velocity v_e of an electron, we find

$$b_{\max} = \lambda_{\text{Deb}}$$

as expected. In general in this book we shall be interested in $\ln \Lambda$ for thermal particles with $mv^2/2 \sim kT$ in (3.22), so that

$$\ln \Lambda = \ln(\lambda_{\text{Deb}}/b_0);$$

i.e., using (3.8),

$$\ln \Lambda \cong 10 + 3.45 \log T - 1.15 \log N_e \tag{3.24}$$

with N_e in cm^{-3}. It is easy to verify that $\ln \Lambda$ is in the range 10–20 for all astrophysical plasmas. This finally proves our claim that the effect of many weak, distant collisions in a plasma is more important than the strong, close collisions. As a matter of practical importance, it shows that $\ln \Lambda$ is extremely insensitive to errors in our estimates of $b_{\max}$, $b_{\min}$: if $\ln \Lambda = 15$ an error in $b_{\max}/b_{\min}$ of a factor 100 either way only produces $\ln \Lambda = 10$–20. Thus, (3.24) is likely to be a reasonable estimate, even if (3.23) is satisfied.

3.4 Thermal plasmas: relaxation times and mean free path

Let us apply our basic results (3.18), (3.19) to the case of plasma particles moving with thermal velocities. We shall avoid the problem of summation over the range of velocities of the particles contributing to the deflection by simply putting $mv^2 \sim kT$. This will produce the correct order of magnitude for λ_d in (3.25) below. However, in general, a full kinetic theory analysis of the time evolution of the velocity distribution, given by the Fokker–Planck equation, is required. This is particularly important for the energy exchange time t_E for fast (suprathermal) particles to be considered in Section 3.5. (For further details see, for example, Chapter 10 of Boyd & Sanderson (1969).)

Using $mv^2 \sim kT$ in (3.19), we see that, for colliding particles of comparable mass,

$$\lambda_d \cong \frac{7 \times 10^5}{\ln \Lambda} \frac{T^2}{N} \text{ cm}. \tag{3.25}$$

This is independent of the particle mass, so the average distance travelled before an appreciable deflection occurs is the same for electrons and ions, and λ_d given by (3.25) is the mean free path for both types of particles. Note that the deflection time $t_d(\text{e–e})$ for electrons with electrons is $(m_p/m_e)^{1/2} \sim 43$ times shorter than that for ions with ions, but this is compensated for by the higher electron velocity. For unequal masses, it can be shown by a similar analysis that formulae of the form (3.18), (3.19) hold with m of the order of the smaller mass and v of the order of the greater thermal speed. Thus, for electron–ion encounters, $m \sim m_e$, $v \sim (kT/m_e)^{1/2}$,

$$t_d(\text{e–i}) \sim \frac{[4\pi\varepsilon_0]^2 m_e^{1/2} (kT)^{3/2}}{2\pi N e_1^2 e_2^2 \ln \Lambda} \sim t_d(\text{e–e}). \tag{3.26}$$

Thus, electron–ion encounters occur at a similar rate to electron–electron encounters. One sometimes defines an (electron) collision frequency as

$$\nu_c \sim 1/t_d(\text{e–i}) \cong 2 \ln \Lambda \, Z^2 N_e T^{-3/2} \text{ s}^{-1} \tag{3.27}$$

(N_e in cm^{-3}), where Ze is the average ion charge, and an ion–ion collision frequency as

$$\nu_c(\text{i–i}) \sim 1/t_d(\text{i–i}) \cong 5 \times 10^{-2} \ln \Lambda \, Z^4 N_e T^{-3/2} \text{ s}^{-1}. \tag{3.28}$$

So far, we have considered only the *deflections* of plasma particles: the timescales t_d give an approximate measure of the time for an initially anisotropic velocity distribution to relax to isotropy. Ultimately, any initial distribution of particle velocities must tend to an equilibrium Maxwell–Boltzmann distribution; the timescale for this relaxation to occur is the *energy exchange timescale*:

$$t_E = \frac{E^2}{\text{d}(\Delta E)^2/\text{d}t} \tag{3.29}$$

where E is the energy of a test particle and ΔE the change due to collisions (cf. (3.18)).

Note that there is no *a priori* reason to expect t_E to be similar to t_d; after all, in the centre-of-mass frame there is a deflection θ, but *no* energy exchange between the particles, since, as we remarked in Section 3.3, their speeds remain unaltered. In some important special cases (e.g. particles of comparable masses and similar initial velocities) t_E and t_d *are* very similar, and, indeed, the distinction between them is not always adequately drawn in the astrophysical literature. In general, however, t_E can exceed t_d considerably.

An important example of this difference occurs for electron–ion collisions in a plasma in thermal equilibrium: since $m_e \ll m_i \cong m_p$ and the centre-of-mass frame differs from the laboratory frame only by a velocity $\mathbf{v}_{\text{CM}} \cong (m_e/m_i)^{1/2}\mathbf{v}_i \ll \mathbf{v}_i$ (cf. (3.11)). From the fact that no energy is exchanged in the centre-of-mass frame, we find the maximum energy transfer in the laboratory frame in a head-on collision is of order

$$\Delta E = 2m_i v_{\text{CM}}^2 \cong 2m_e v_i^2,$$

i.e.

$$\frac{\Delta E}{\tfrac{1}{2}m_i v_i^2} \sim \frac{m_e}{m_i} \ll 1. \tag{3.30}$$

Thus, collisions between electrons and ions are very inefficient in transferring energy. Hence, although, as remarked after (3.26), electron–ion encounters and electron–electron encounters occur at similar rates in a plasma, a difference in average energy between electrons and ions will take $\sim m_i/m_e \sim m_p/m_e \sim 1836$ times longer to even out by collisions than a similar difference among the electrons themselves.

For like-particle encounters with comparable velocities, we expect $t_E \sim t_d$, so (3.27, 3.28) now imply that the relaxation times to establish, from an originally non-equilibrium situation, (i) an electron thermal distribution, (ii) an ion thermal distribution, (iii) equipartition between electrons and ions, are in the ratio

$$t_E(\text{e–e}):t_E(\text{i–i}):t_E(\text{e–i}) = 1:(m_p/m_e)^{1/2}:(m_p/m_e). \tag{3.31}$$

Thus, the electrons reach equilibrium first, followed by the ions, followed finally by equipartition. This behaviour is particularly important in the study of shock waves in plasmas (see Sections 3.8 and 6.4).

3.5 The stopping of fast particles by a plasma

In our study of accretion, we shall need to consider how high-velocity infalling matter is incorporated into the material of the accreting star. This process again involves collisions, but with the difference now that the velocities of the infalling particles greatly exceed those of the 'background' particles of the stellar envelope.

Let a test particle (i.e. a particle whose influence on the ambient medium can be neglected) having mass m_1, velocity U, and charge e_1 be incident on a plasma of number density N, temperature T, particle mass m_2, and charge e_2. This will allow us to consider the stopping by the electron and ion plasmas separately. We suppose our test particle to be fast (suprathermal) in the sense that

$$\tfrac{1}{2} m_1 U^2 \gg \tfrac{3}{2} kT = \tfrac{1}{2} m_2 v_{\text{th}}^2;$$

i.e. U greatly exceeds the typical thermal speed of the background particles. A full kinetic theory treatment (e.g. Boyd & Sanderson (1969) Chapter 10) yields a further timescale, characterizing the stopping of the test particle, in addition to t_d and t_E. This is the *slowing-down timescale*:

$$t_s = -\frac{U}{dU/dt}. \tag{3.31a}$$

For suprathermal particles, but assuming non-relativistic velocities $U < c$, one finds

$$t_s \sim \frac{[4\pi\varepsilon_0]^2 m_1^2 U^3}{8\pi N e_1^2 e_2^2 \ln \Lambda} \frac{m_2}{m_1 + m_2},$$

$$t_d \sim \frac{[4\pi\varepsilon_0]^2 m_1^2 U^3}{8\pi N e_1^2 e_2^2 \ln \Lambda},$$

$$t_E \sim \frac{[4\pi\varepsilon_0]^2 m_1^2 U^3}{8\pi N e_1^2 e_2^2 \ln \Lambda} \frac{m_2 U^2}{2kT}. \tag{3.32}$$

Here $\ln \Lambda$ is the Coulomb logarithm, with $b_{\max} = U/v_p$. Of course, since we are again considering collisions, there is a family resemblance between these expressions and our earlier one (3.18) for t_d for thermal particles. Two interesting conclusions follow from the expressions (3.32):

32 Plasma concepts

(i) The factor $m_2/(m_1+m_2)$ in t_s means that suprathermal ions are best slowed down by electrons; this occurs in a time
$$t_s \sim (m_e/m_p)t_d.$$

(ii) Suprathermal electrons are slowed down at similar rates by both electrons and ions; the time scale t_s is the same as the deflection time t_d.

It is often of interest to calculate the average distance $\lambda_s \sim Ut_s$ travelled by a particle before slowing down; this is sometimes referred to as the *stopping length*. For a proton incident on the plasma, we have

$$\lambda_s = Ut_s \sim \frac{[4\pi\varepsilon_0]^2 m_p^2 U^4}{8\pi Ne^4 \ln \Lambda} \frac{m_e}{m_p} \qquad (3.33)$$

since the stopping is mainly by electrons. Writing $\tfrac{1}{2}m_p U^2 = E_p$, the kinetic energy of the proton,

$$\lambda_s = \frac{[4\pi\varepsilon_0]^2 m_e}{2\pi Ne^4 \ln \Lambda m_p} E_p^2. \qquad (3.34)$$

Using the fact that $Nm_p \cong \rho$, the background plasma mass density, we see that the column density $\rho\lambda_s$ of plasma required to slow down an incident proton depends only on the square of the proton energy (neglecting the slow dependence of $\ln \Lambda$ on plasma properties). In fact, the stopping power of a plasma is often expressed in terms of the column density per (keV)^2 of incident particle energy: numerically,

$$\rho\lambda_s = 4.6 \times 10^{-10} \text{ g cm}^{-2}\,(\text{keV})^{-2} \quad \text{for } \ln \Lambda = 15. \qquad (3.35)$$

The energy of a supersonic accretion flow is almost all carried by ions. Therefore, as an example of the use of λ_s let us consider the problem of a proton freely falling on to a star. It is of interest to know whether or not the proton will penetrate the photosphere of the star before being slowed down; if this is the case, most of the accretion energy supplied by a stream of such protons will be released as blackbody radiation at a temperature $\sim T_b$ (equation (1.6)). It is easy to answer this question in the case in which the opacity of the surface layers of the star is due to electron scattering, as, for example, for neutron stars. The condition for penetration is that λ_s should exceed the mean free path λ_{ph} of photons through the surface material. For an electron scattering opacity

$$\lambda_{ph} \cong 1/N\sigma_T$$

where

$$\sigma_T = [4\pi\varepsilon_0]^{-2} \frac{8\pi}{3} \frac{e^4}{m_e^2 c^4} \qquad (3.36)$$

is the Thomson cross-section. Using (3.33), (3.36), we find

$$\frac{\lambda_s}{\lambda_{ph}} = \frac{m_p}{m_e} \frac{1}{3 \ln \Lambda} \left(\frac{U}{c}\right)^4.$$

Hence, for $\ln \Lambda = 15$, the condition for penetration is

$$U \gg 2.6(m_e/m_p)^{1/4} c \sim 0.4c. \tag{3.37}$$

If for U we take the free-fall speed, $(2GM/R_*)^{1/2}$, it is clear that (3.37) implies that penetration by ions is possible only for objects which are *compact* in the sense that

$$R_* \lesssim 6R_{\text{Schw}} \tag{3.38}$$

where R is the gravitational or Schwarzschild radius of the object.

$$R_{\text{Schw}} = \frac{2GM}{c^2} \cong 3 \text{ km} \left(\frac{M}{M_\odot}\right).$$

While the inequality (3.38) obviously holds for black holes ($R_* = R_{\text{Schw}}$), the concept of penetration in this case is not meaningful. Penetration is, in principle, possible for neutron stars ($R_* \cong 3R_{\text{Schw}}$), but not for less compact objects.

3.6 Transport phenomena: viscosity

The fact that the particles of a plasma are constantly in thermal motion, carrying with them their momenta, energies, etc., means that they tend to share out these quantities and try to even out any gradients in the plasma. Thus, the microscopic structure of a plasma, on scales of a mean free path, affects its macroscopic (gas dynamical) properties through *transport processes*, as these effects are known. Examples of transport processes, which we briefly encountered in Chapter 2, are viscosity and thermal conductivity; in these cases the quantities being transferred are various components of momentum and (thermal) energy respectively. As we shall see, transport processes become important when there are large gradients (e.g. of velocity or temperature) in a plasma, a situation which often occurs in accretion problems.

Let us examine how viscosity works in a simple case. We consider a plasma flowing with (bulk) velocity $v(x)$ in the x-direction (Fig. 4). We draw a surface S orthogonal to the gas velocity at a certain point, and consider the transfer of x-momentum across S by thermal motions. This will give us the effect of *bulk* viscosity, which can be important in shock fronts (see Section 3.8). The transfer of the other (y, z) components of momentum give the *shear* viscosity which is generally more important in influencing the gas motion (see Section 4.6). By symmetry, the transfer of x-momentum will be zero unless there is a velocity gradient at S: we write $\partial v/\partial x = v/L$, so that L is the scale length of velocity variation at S. The particles which can carry momentum across S, by virtue of their thermal motions, are just the plasma particles originating from within one deflection length, λ_d (equation (3.19)), of S: since no appreciable deflection occurs over a length shorter than λ_d these particles carry essentially all their momentum across S. Of course, we should add up the momentum transfers due to all particles on each side of S, taking account of their differing values of λ_d. We shall obtain a correct result, in an order-of-magnitude sense, if we assume that most of the momentum transfer is performed

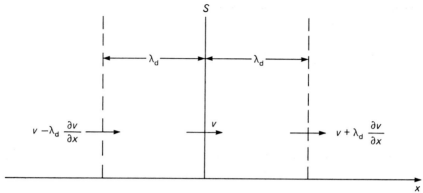

Figure 4. Bulk viscous transport of momentum.

by particles with thermal energy $E \sim kT$: then the appropriate value for λ_d is the thermal mean free path (3.25) for all particles. A second simplification follows from realizing that the systematic momentum (the momentum associated with the bulk motion v) of a plasma flow resides almost totally in the ions because of their greater mass. Finally, we assume that only the velocity v, not ρ, λ_d or the sound speed c_s, changes over the lengthscale λ_d; hence, the net rate at which systematic momentum is transferred across unit area of S is approximately

$$-\rho c_s \left(v + \lambda_d \frac{\partial v}{\partial x}\right) + \rho c_s \left(v - \lambda_d \frac{\partial v}{\partial x}\right) \cong -2\rho c_s \lambda_d \frac{\partial v}{\partial x}. \tag{3.39}$$

Here we have used the facts that the ions of a plasma have thermal velocities of the order of the sound speed c_s (equations (2.16a, b)) and that the change in bulk velocity over a distance λ_d is $\sim \lambda_d \partial v/\partial x$.

Now let us consider a volume element, dV, in the plasma flow, bounded by surfaces S, S', at x, $x + dx$ (Fig. 5), each of area dA; so $dV = dx\,dA$. The momentum transfer rate (3.39) means that there is a force acting in the x-direction on the volume element dV due to viscosity. This force is given by the difference between the expression (3.39) evaluated at x and at $x + dx$:

$$dF_{\text{visc}} = 2\rho c_s \lambda_d \, dA \left(\frac{\partial v}{\partial x} + \frac{\partial^2 v}{\partial x^2} dx\right)$$

$$- 2\rho c_s \lambda_d \, dA \frac{\partial v}{\partial x} dx$$

$$\cong 2\rho c_s \lambda_d \frac{\partial^2 v}{\partial x^2} dV.$$

Hence, there is a (bulk) *viscous force density*

$$f_{\text{visc,bulk}} \cong -2\rho c_s \lambda_d \frac{\partial^2 v}{\partial x^2} \tag{3.40}$$

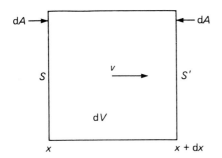

Figure 5. Bulk viscous force density.

acting at each point of the plasma. Note that the force vanishes (correctly) if there is no velocity gradient.

To see when this viscous force is important, we recall the Euler equation (2.3): clearly, we must compare f_{visc} with the pressure gradient $\partial P/\partial x$. If the pressure changes over scale length L, the ratio is

$$\frac{\text{viscous force}}{\text{pressure force}} \sim \rho c_s \lambda_d \frac{v}{L^2} \frac{L}{P}.$$

But, from the equation of state (2.2), we have $P \cong \rho c_s^2$, so the ratio becomes

$$\frac{\text{viscous force}}{\text{pressure force}} \sim \frac{v}{c_s} \frac{\lambda_d}{L}. \qquad (3.41)$$

Equation (3.41) shows that viscosity will be important for supersonic flow ($v > c_s$) with large gradients ($L \sim \lambda_d$) – precisely the conditions which are encountered in shock waves. If gradients are small ($L \gg \lambda_d$) viscosity can generally be neglected, unless the flow is highly supersonic ($v \gg c_s$). This is fortunate, since the viscous force in (3.40) involves second derivatives of **v** and its inclusion considerably complicates the problem of solving the equations of gas dynamics.

In the discussion above we have considered viscous effects only in a one-dimensional gas flow, so-called bulk viscous effects. Clearly, if the velocity field varies in directions orthogonal to the flow (shearing motions) similar (shear) viscous effects will come into play. Further, if there are more violent random motions in the gas involving the bodily displacement of blobs of material (turbulent motion), these can also give rise to (turbulent) viscous effects. We shall be much concerned with turbulent shear viscosity in our discussion of accretion discs (Chapters 5, 6); it is worth bearing in mind that all these viscous effects involve the same basic transport process we have discussed above.

Another similar transport process, referred to at the beginning of this section, is thermal conduction. The main difference here is that, because the quantity being transferred is internal (thermal) energy, of which the electrons have the same amount as the ions, the transport process is dominated by the electrons, since they are quicker by a factor $(m_i/m_e)^{1/2} \sim 43$. With this modification, a calculation very

36   Plasma concepts

similar to the one performed above for bulk viscosity shows that a temperature gradient $\partial T/\partial x$ gives rise to an electron thermal conduction flux:

$$q \sim -\frac{N_e kT}{m_e v_e}\lambda_d k \frac{\partial T}{\partial x} \qquad (3.42)$$

where $v_e = (kT/m_e)^{1/2}$. The formula for the coefficient of thermal conductivity used in (2.6) follows directly from (3.42) and (3.25).

3.7   The effect of strong magnetic fields

Frequently, in accretion problems, we deal with plasmas immersed in strong magnetic fields. For neutron stars, surface fields of 10^{12} G ($= 10^8$ T) are often met with. Many accreting white dwarfs possess magnetic fields of the order of 10^6–10^7 G (10^2–10^3 T). Because plasmas are, by definition, electrically conducting gases, there is a host of effects produced by the field. For example, the transfer of radiation, particularly taking into account its polarization properties, becomes immensely more complicated. In this book we shall touch only rather superficially upon two particular types of magnetic effect: the influence of magnetic fields on gas flows and on transport processes.

In outline, the presence of currents in a moving plasma modifies the magnetic field, while the electromagnetic field acts on the charges to produce currents. Therefore, in general, the influence of a magnetic field on the gas flow is quite complicated. However, we shall find that, if the plasma has sufficiently high electrical conductivity, a condition usually realized in practice, then the plasma and the magnetic field move together in a sense that will be explained precisely below. We shall then determine the conditions under which the gas follows the motion of the magnetic field or, conversely, the magnetic field is dragged around by the gas.

The magnetic field, **B**, is governed by Maxwell's equations:

$$\left. \begin{array}{l} \nabla \cdot \mathbf{B} = 0, \\[6pt] \nabla \wedge \mathbf{B} = \left[\dfrac{1}{c}\right]\dfrac{1}{c}\dfrac{\partial \mathbf{E}}{\partial t} + \left[\dfrac{c\mu_0}{4\pi}\right]\dfrac{4\pi}{c}\mathbf{j}. \end{array} \right\} \qquad (3.43)$$

The first tells us that we can represent **B** by unending lines of force; in the second, the displacement current term ($\propto \partial \mathbf{E}/\partial t$) can usually be neglected in astrophysics, so we shall drop it here. The current density **j** is now related to the electromagnetic field by a form of Ohm's law appropriate to moving media; in the so-called hydromagnetic approximation (see Boyd & Sanderson (1969) Chapter 3), this is

$$\mathbf{j} = \sigma(\mathbf{E} + [c]\mathbf{v} \wedge \mathbf{B}/c) \qquad (3.44)$$

where σ is the electrical conductivity. Essentially, this states that the force due to the fluid velocity in the magnetic field moves charges in addition to the electromotive force. Taking the curl of (3.43) and using (3.44), with σ assumed constant, we obtain

$$\nabla \wedge (\nabla \wedge \mathbf{B}) = \left[\frac{c\mu_0}{4\pi}\right]\frac{4\pi}{c}\sigma(\nabla \wedge \mathbf{E} + [c]\nabla \wedge (\mathbf{v} \wedge \mathbf{B})/c).$$

The effect of strong magnetic fields

On the right hand side Faraday's law,

$$\nabla \wedge \mathbf{E} = -[c]\frac{1}{c}\frac{\partial \mathbf{B}}{\partial t},$$

can be used to eliminate the electric field $\mathbf{E}$. The left hand side is

$$\nabla \wedge (\nabla \wedge \mathbf{B}) = -\nabla^2 \mathbf{B} - \nabla(\nabla \cdot \mathbf{B}) = -\nabla^2 \mathbf{B};$$

so we obtain

$$\frac{\partial \mathbf{B}}{\partial t} = \nabla \wedge (\mathbf{v} \wedge \mathbf{B}) + \left[\frac{4\pi}{c^2 \mu_0}\right]\frac{c^2}{4\pi\sigma}\nabla^2 \mathbf{B}. \tag{3.45}$$

Thus, the time rate of change of the magnetic field is governed by diffusion of the field ($\propto \sigma^{-1}\nabla^2 \mathbf{B}$) and convection ($\nabla \wedge (\mathbf{v} \wedge \mathbf{B})$) of the field by the fluid. We can neglect diffusion if for changes in $\mathbf{B}$ on a lengthscale L

$$\left[\frac{4\pi}{c^2\mu_0}\right]\frac{c^2}{4\pi\sigma}\frac{|\mathbf{B}|}{L^2}\left(\frac{|\mathbf{v}||\mathbf{B}|}{L}\right)^{-1} = \left[\frac{4\pi}{c^2\mu_0}\right]\frac{c^2}{4\pi\sigma|\mathbf{v}|L} = \frac{1}{Rm} \ll 1,$$

i.e. for sufficiently large σ. Rm is known as the magnetic Reynolds number.

To see how the field and fluid move together for $Rm \gg 1$, consider the magnetic flux, Φ, through a surface $S(t)$ moving with the fluid (Fig. 6).
We have

$$\frac{d\Phi}{dt} = \lim_{dt \to 0}\left[\int_{S(t+dt)} \mathbf{B}(t+dt) \cdot d\mathbf{S} - \int_{S(t)} \mathbf{B}(t) \cdot d\mathbf{S}\right]\frac{1}{dt}.$$

But, since the flux through a closed surface is zero, we have, at time $t+dt$,

$$\int_{S(t+dt)} \mathbf{B}(t+dt) \cdot d\mathbf{S} + \int_{\mathscr{S}} \mathbf{B}(t+dt) \cdot d\mathscr{S} - \int_{S(t)} \mathbf{B}(t+dt) \cdot d\mathbf{S} = 0.$$

Therefore:

$$\frac{d\Phi}{dt} = \lim_{dt \to 0}\left[\int_{S(t)} (\mathbf{B}(t+dt) - \mathbf{B}(t)) \cdot d\mathbf{S} - \int_{\mathscr{S}} \mathbf{B}(t+dt) \cdot d\mathbf{r} \wedge \mathbf{v}\,dt\right]\frac{1}{dt},$$

since

$$d\mathscr{S} = d\mathbf{r} \wedge \mathbf{v}\,dt,$$

and

$$\frac{d\Phi}{dt} = \int_{S(t)} \frac{\partial \mathbf{B}}{\partial t}\cdot d\mathbf{S} - \int_{\mathscr{C}} (\mathbf{v} \wedge \mathbf{B})\cdot d\mathbf{r} = \int_{S(t)}\left[\frac{\partial \mathbf{B}}{\partial t} - \nabla \wedge (\mathbf{v} \wedge \mathbf{B})\right]\cdot d\mathbf{S}$$

using Stokes' theorem. Thus, neglecting the diffusion term in (3.45) we obtain

$$\frac{d\Phi}{dt} = 0.$$

Plasma concepts

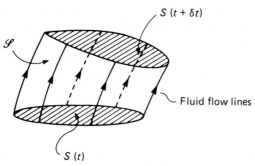

Figure 6.

Therefore, the flux through a fluid element is conserved. In fact, the fluid and magnetic field move exactly together (in the sense that a magnetic field-line consists of the same fluid particles at all times). This is harder to prove (see e.g. Boyd & Sanderson (1969), Chapter 4), but is clearly a plausible consequence of the above result applied to an infinitesimal surface element. The conclusion is expressed by saying that the magnetic field is *frozen* into a perfectly conducting fluid.

Two further consequences of infinite conductivity, to which we shall need to refer in Chapter 9, follow from (3.44). If $\sigma \to \infty$ we must have

$$\mathbf{E} + \frac{[c]}{c}\mathbf{v} \wedge \mathbf{B} = 0, \tag{3.46}$$

called the *perfect magnetohydrodynamic* condition. From (3.46) we then derive

$$\mathbf{E} \cdot \mathbf{B} = 0, \tag{3.47}$$

a condition known as *degeneracy*.

To complete the discussion, consider the magnetic forces on the plasma. Each charge is subject to the Lorentz force $(e/c)\mathbf{v} \wedge \mathbf{B}[c]$, so the magnetic force density is

$$\mathbf{f}_{\text{mag}} = \frac{[c]}{c}\mathbf{j} \wedge \mathbf{B}.$$

Using (3.43), without the displacement current term, to eliminate $\mathbf{j}$ yields

$$\mathbf{f}_{\text{mag}} = \left[\frac{4\pi}{\mu_0}\right]\frac{1}{4\pi}(\nabla \wedge \mathbf{B}) \wedge \mathbf{B}$$

and this is equivalent to

$$\mathbf{f}_{\text{mag}} = \left[\frac{4\pi}{\mu_0}\right]\left\{-\nabla\left(\frac{B^2}{8\pi}\right) + \frac{1}{4\pi}(\mathbf{B} \cdot \nabla)\mathbf{B}\right\}. \tag{3.48}$$

Comparison with the Euler equation (2.3) shows that the first term on the right of (3.48) behaves as a hydrostatic pressure of magnitude

$$P_{\text{mag}} = \left[\frac{4\pi}{\mu_0}\right]\frac{B^2}{8\pi}. \tag{3.49}$$

The effect of strong magnetic fields

The final term in (3.48) is less important, but, by considering special cases, for example $\mathbf{B} = (0, 0, B(z))$, one can show it behaves as a tension of magnitude $[4\pi/\mu_0](B^2/4\pi)$ along the field lines.

An order-of-magnitude analysis of the Euler equation (2.3) now shows that the relative importance of the inertia (or *ram pressure*), gas pressure and magnetic pressure terms is given by

$$\rho v^2 : \rho c_s^2 : \frac{B^2}{8\pi}\left[\frac{4\pi}{\mu_0}\right]. \tag{3.50}$$

It follows that the fluid velocity will be determined by the magnetic field if the magnetic pressure dominates. If the ram pressure is the largest, then the magnetic field is dragged around by the fluid, which moves essentially as if the magnetic forces were absent. In the intermediate case, $\rho c_s^2 \gg B^2/8\pi[4\pi/\mu_0] \gg \rho v^2$, the magnetic field moves the fluid, but is itself confined by gas pressure. The equality of kinetic and magnetic energy densities defines a velocity

$$v_A = \left(\frac{B^2}{4\pi\rho}\left[\frac{4\pi}{\mu_0}\right]\right)^{1/2},$$

the Alfvén velocity, at which magnetic disturbances propagate in the plasma. From (3.50), magnetic forces dominate if $v_A \gg \max(v, c_s)$. We shall meet examples of the influence of strong magnetic fields on accretion flows in Chapter 6.

The importance of magnetic fields in modifying transport processes arises from the effect these fields have on the motions of individual charged particles. It is shown in many books on dynamics and electrodynamics that a non-relativistic charged particle, of mass m and charge q, will *spiral* about a uniform magnetic field $\mathbf{B}$ with angular frequency

$$\Omega = \frac{qB}{mc}[c] \quad \text{rad s}^{-1}, \tag{3.51}$$

variously known as the Larmor frequency, gyro frequency, or cyclotron frequency. The radius of the spiral is given by the particle velocity component $v_\perp$ orthogonal to $\mathbf{B}$; this is the Larmor radius:

$$r_L = v_\perp / \Omega.$$

If $v_\perp$ is given by an average thermal velocity $(kT/m)^{1/2}$ we have

$$r_L = \frac{(mkT)^{1/2}}{qB}\frac{c}{[c]},$$

which for electrons and ions of mass Am_p and charge Ze give

$$\left.\begin{array}{l} r_{Le} \cong 2.2 \times 10^{-2}\, T^{1/2}/B \text{ cm}, \\ r_{Li} \cong AT^{1/2}/ZB \text{ cm} \end{array}\right\} \tag{3.52}$$

respectively (B in Gauss $= 10^{-4}$ Tesla). Clearly, r_L limits the distance which a particle can travel through the plasma in a given direction orthogonal to $\mathbf{B}$. Thus, transport processes across the field-lines will tend to be suppressed if $r_L < \lambda_d$ for the

40   Plasma concepts

relevant species of particle, while transport rates along the field-lines are unaffected. For example, if conditions in the plasma are such that $r_{Le} < \lambda_d$ (equation (3.25)), electron thermal conduction across field-lines will be greatly reduced.

3.8   Shock waves in plasmas

Very often in astrophysics we have to deal with situations where the bulk velocity of a gas flow makes a transition between supersonic and subsonic values. A good example of this is provided by our study of steady adiabatic, spherical, accretion (Section 2.5). The gas velocity will, in general, be close to the free-fall velocity v_{ff} and therefore highly supersonic near to the surface of the accreting star (as noted in number 3 of the conclusions at the end of Section 2.5). Clearly, in order to accrete to the star, the gas must somehow make a transition to a subsonic 'settling' flow, in which the gas velocity becomes very small near the stellar surface.

From the point of view of the gas, i.e. seen in a reference frame moving with the gas velocity v, the surface of the star is pushing into the gas at a supersonic speed $\cong v_{ff}$. To simplify the geometry, let us approximate this to a piston being pushed into a tube of gas (Fig. 7). We assume that the changes induced by the piston are so rapid that the gas behaves adiabatically. As the piston starts from rest with a small displacement, a signal travels into the gas at the sound speed and the gas is slightly compressed. As the piston continues to accelerate into the gas, a further signal travels into the already compressed gas and acts to compress it further; this signal travels at the sound speed of the already compressed gas. Because the sound speed $c_s^{ad} \propto \rho^{1/3}$ (equation (2.16a)) this new sound speed is slightly higher than the original. Hence, arbitrarily dividing the gas in front of the piston into zones labelled 1, 2, 3, ..., we see that the densities $\rho(1), \rho(2), \rho(3), ...$ in these zones obey $\rho(1) > \rho(2) > \rho(3) > ...$, so that the sound speeds obey $c_s^{ad}(1) > c_s^{ad}(2) > c_s^{ad}(3) ...$. Thus, 'news' of the piston travels into the gas at c_s^{ad}, and gas sufficiently far to the right remains at its original density until this 'news' has had time to arrive. Hence, the denser zones to the left continually try to catch up the lower-density zones to the right; the compression of the left hand zones continually increases, and the density profile in front of the piston therefore steepens (Fig. 8).

But the steepening of the density profile cannot continue indefinitely: once the density scale length, $L \cong \rho/|\nabla \rho|$, becomes of the order of the mean free path, λ_d, then since the gas velocity is of the order of the sound speed, c_s, in the higher-density region (and thus supersonic for the undisturbed gas), our estimate (3.41) shows that bulk viscous forces, so far neglected, will become important. The effect of the viscous forces on the gas is to *dissipate* energy: i.e. ordered kinetic energy gets converted by viscous processes into chaotic thermal motions. The precise description of this conversion is, in general, a very difficult problem in kinetic theory, particularly for shock waves in plasmas, where the inefficient energy exchange between ions and electrons complicates matters still further. In many cases; however, a detailed picture of the structure of the shock is not required; since the shock thickness ($\sim \lambda_d$) is much smaller than the lengthscales of gradients in the gas on each side of it, we can approximate the shock as a discontinuity in the gas flow. The connection between the gas density, velocity and pressure (or tem-

Shock waves in plasmas

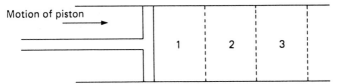

Figure 7. Compression of a gas in a cylinder.

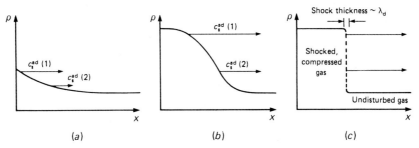

Figure 8. The formation of a shock wave in a gas. (*a*) The denser gas in zone 1 tends to overtake the dilute gas in zone 2 since its adiabatic sound speed is greater; (*b*) as a result, the density gradient, and hence the overtaking tendency, increase, leading (*c*) to a shock wave of thickness of the order of the mean free path of the gas particles.

perature) across this idealized discontinuity can be found by applying conservation laws.

To see how this works, it is convenient to choose a reference frame in which the shock is at rest. Since the shock thickness is so small, we can regard it as locally plane; further, since the gas flows through the shock so quickly, changes in the gas conditions cannot affect the details of the transition across the shock, so we can regard the flow into and out of the shock as steady. Hence, we have the situation depicted in Fig. 9 in which gas flows into the shock with density, velocity and pressure ρ_1, v_1, P_1, and emerges with ρ_2, v_2, P_2. (If the shock is oblique to the flow, v_1, v_2 are just the components of the upstream and downstream flow velocities which are orthogonal to the shock front; the tangential components are continuous. The jump conditions can be modified accordingly.) Let us now apply the conservation laws of mass, momentum and energy across the shock front; this is equivalent to integrating the appropriate forms of the gas dynamics equations, (2.1), (2.3), (2.5), across the discontinuity. If x measures distance from the shock front, we integrate with respect to x from $-dx$ to $+dx$, where $dx \sim \lambda_d/2$, and take the limit $dx \to 0$.

First, we have from the continuity equation (2.1):

$$\frac{d}{dx}(\rho v) = 0,$$

which integrates to

$$\rho_1 v_1 = \rho_2 v_2 = J. \qquad (3.53)$$

Clearly, the quantity $J = \rho v$ which is conserved here is the mass flux.

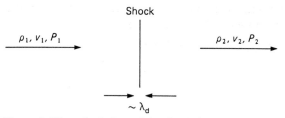

Figure 9. The calculation of the shock jump (Rankine–Hugoniot) conditions.

From the Euler equation (2.3) we get

$$\rho v \frac{dv}{dx} + \frac{dP}{dx} = f_x$$

or

$$\frac{d}{dx}(P + \rho v^2) = f_x$$

where f_x is the force density in the x direction. Integrating across the shock front $x = 0$, we get

$$(P_1 + \rho_1 v_1^2) - (P_2 + \rho_2 v_2^2) = \lim_{dx \to 0} \int_{-dx}^{dx} f_x \, dx = 0,$$

since f_x is finite. Thus,

$$P_1 + \rho_1 v_1^2 = P_2 + \rho_2 v_2^2 = I. \tag{3.54}$$

Here, the conserved quantity $I = P + \rho v^2$ is the momentum flux.

Finally, the energy equation (2.5) implies

$$\frac{d}{dx}[v(\tfrac{1}{2}\rho v^2 + \rho \varepsilon + P)] = f_x v$$

if we assume that radiative losses and thermal conduction, etc., across the shock front are negligible. This is the adiabatic assumption, and the resulting jump conditions are known as the adiabatic shock conditions. The assumption has, of course, to be tested in any individual case. Using $\rho v = $ constant and $E = \tfrac{3}{2} P/\rho$ the energy equation can be written as

$$\rho v \frac{d}{dx}\left[\tfrac{1}{2} v^2 + \tfrac{5}{2} \frac{P}{\rho}\right] = f_x v$$

for a monatonic gas. Integrating over the shock front implies (using 3.53)

$$\tfrac{1}{2} v_1^2 + \tfrac{5}{2} \frac{P_1}{\rho_1} = \tfrac{1}{2} v_2^2 + \tfrac{5}{2} \frac{P_2}{\rho_2} = E, \tag{3.55}$$

since $\int f_x v \, dx$ over the shock vanishes. The quantity $E = \tfrac{1}{2} v^2 + \tfrac{5}{2} P/\rho$ is then the specific total energy, where account is taken of the work done by pressure in compressing the gas across the shock. The three conservation equations (3.53), (3.54), (3.55), relating the upstream and downstream values of ρ, v, P are known as

Shock waves in plasmas

Rankine–Hugoniot conditions; they give three equations to determine the jumps in ρ, v, P. In general, deriving these relations involves a little algebra.

First, let us divide I by Jv: ((3.53), (3.54)) give

$$\frac{I}{Jv} = \frac{P}{\rho v^2} + 1.$$

Defining the Mach number $\mathcal{M}$ by

$$\mathcal{M}^2 = \frac{v^2}{(c_s^{\text{ad}})^2} = \tfrac{3}{5}\frac{\rho v^2}{P}, \qquad (3.56)$$

we get

$$\frac{I}{Jv} = \frac{3}{5\mathcal{M}^2} + 1. \qquad (3.57)$$

Also, using (3.53), (3.54) in (3.55),

$$E = \frac{v^2}{2} + \tfrac{5}{2}\left(\frac{Iv}{J} - v^2\right)$$

or

$$v^2 - \tfrac{5}{4}\frac{I}{J}v + \frac{E}{2} = 0.$$

This is a quadratic equation for v, and therefore has two roots. Since the quadratic was derived by manipulation of the conservation laws, these roots can only be v_1, v_2. Hence, $v_1 + v_2$ is given by the sum of the roots of the quadratic:

$$v_1 + v_2 = \tfrac{5}{4}\frac{I}{J},$$

i.e.

$$1 + \frac{v_2}{v_1} = \tfrac{5}{4}\frac{I}{Jv_1} = \tfrac{5}{4}\left[\frac{3}{5\mathcal{M}_1^2} + 1\right] \qquad (3.58)$$

using (3.57). Here $\mathcal{M}_1 = v_1/c_s^{\text{ad}}(1)$ is the upstream Mach number.

Very often we encounter situations where this quantity is large, i.e. where the flow is highly supersonic. The limit $\mathcal{M}_1 \gg 1$ is the case of a *strong shock*: for such shocks we get, from (3.58),

$$\frac{v_2}{v_1} = \tfrac{1}{4}, \qquad (3.59)$$

i.e. the velocity drops to one quarter of its upstream value across the shock. From (3.53) we get

$$\frac{\rho_2}{\rho_1} = 4, \qquad (3.60)$$

which shows that the gas is compressed by a factor 4 in a strong shock. The values

44 Plasma concepts

$\frac{1}{4}$ and 4 in (3.59) and (3.60) come from taking the adiabatic index $\gamma = \frac{5}{3}$; they can easily be found for a general value of γ.

From (3.54) we have

$$I = \rho_1 v_1^2 \left(1 + \frac{3}{5\mathcal{M}_1^2}\right) \cong \rho_1 v_1^2$$

for a strong shock. Thus, ahead of a strong shock the thermal pressure P_1 is negligible, compared to the ram pressure $\rho_1 v_1^2$. From (3.54) we thus get the pressure behind the shock as

$$P_2 = \rho_1 v_1^2 - \rho_2 v_2^2 = \rho_1 v_1 (v_1 - v_2),$$

using (3.53), so that

$$P_2 = \tfrac{3}{4} \rho_1 v_1^2, \qquad (3.61)$$

from (3.59). Thus, behind the shock the thermal pressure is $\frac{3}{4}$ of the ram pressure. Clearly, what has happened is that the dissipation processes (e.g. viscosity) operating in the shock front have converted most of the kinetic energy of the ordered, supersonic flow into heat – i.e. random motion of the post-shock gas particles. This post-shock gas is indeed subsonic, for its sound speed is

$$c_s(2) = \left(\frac{5 P_2}{3 \rho_2}\right)^{1/2} = \left(\frac{5 \rho_1 v_1^2}{16 \rho_1}\right)^{1/2} = \frac{\sqrt{5}}{4} v_1 > v_2 = \frac{v_1}{4}, \qquad (3.62)$$

using (3.61), (3.60) and (3.59).

Note that we have assumed implicitly throughout that (ρ_1, v_1, P_1) and (ρ_2, v_2, P_2) refer to the pre- and post-shock gas respectively, not the reverse, so that the gas flow goes from supersonic to subsonic. This is justified by the result that ordered motion (pre-shock) gets randomized in the shock (equation (3.61)): it can easily be shown that the entropy of the gas in the state (ρ_2, v_2, P_2) is greater than that of the state (ρ_1, v_1, P_1), confirming that the 'arrow of time' indeed points in the direction we have assumed.

The temperature T_2 of the post-shock gas is given by the perfect gas law (2.2):

$$T_2 = \frac{\mu m_H P_2}{k \rho_2} = \tfrac{3}{16} \frac{\mu m_H}{k} v_1^2. \qquad (3.63)$$

If, as in the example which motivated our discussion of shocks, v_1 is the free-fall velocity, equal to $(2GM/R_*)^{1/2}$, on to an accreting star of mass M and radius R_*, (3.63) gives

$$T_2 = \tfrac{3}{8} \frac{GM \mu m_H}{k R_*}. \qquad (3.64)$$

Comparing (3.64) with (1.7), we see that the temperature estimate T_{th} is quite close to the shock temperature, T_2, since $m_H \sim m_p$, $\mu \sim 1$ and $\frac{3}{8} \sim \frac{1}{3}$. (The value $\frac{3}{8}$ in (3.64) arises from taking $\gamma = \frac{5}{3}$; it can easily be generalized to other values of γ.) Thus, in a strong shock the ordered kinetic energy of the supersonic pre-shock gas is converted to random thermal motions; the post-shock ion thermal velocity ($\cong c_s(2)$) is of the order of the pre-shock bulk velocity (cf. (3.62)). The shock temperature T_2 will only observably characterize the emergent radiation if the post-shock gas is optically thin; otherwise, a blackbody temperature (cf. (1.6)) will be detected. Shock heating of optically thin gas therefore offers a way of generating high temperatures and, consequently, high-energy emission.

Although we have restricted our discussion to strong shocks, $\mathcal{M}_1 \gg 1$, there is no difficulty, in principle, in the extension of these results to weak shocks (finite $\mathcal{M}_1$). Relations similar to (3.59), (3.60) and (3.61) are obtained with coefficients dependent on the strength ($\mathcal{M}_1^2$) of the shock.

4

Accretion in binary systems

4.1 Introduction

The importance of accretion as a power source was first widely recognized in the study of binary systems, especially X-ray binaries. This is still the area where the greatest progress in the understanding of accretion has been made. The reason for this is simply that, by their very nature, binaries reveal more about themselves, notably their masses and dimensions, than do other astronomical objects. This is particularly true in the case of eclipsing binaries, where we get direct information about spatial relations within the source. The importance of accretion is further manifested by the realization that probably a majority of all stars are members of binary systems which, at some stage of their evolution, undergo mass transfer.

The detailed study of interacting binary systems has revealed the importance of angular momentum in accretion. In many cases, the transferred material cannot land on the accreting star until it has rid itself of most of its angular momentum. This leads to the formation of *accretion discs*, which turn out to be efficient machines for extracting gravitational potential energy and converting it into radiation. This property has made accretion discs attractive candidates for the role of the central engine in quasars and active galactic nuclei; the question of whether or not this assignment is correct will be a theme of later chapters. In the context of binaries, however, their existence and importance is well attested through observation; we shall discuss the evidence for this in Chapter 5.

4.2 Interacting binary systems

There are two main reasons many binaries transfer matter at some stage of their evolutionary lifetimes:

(i) In the course of its evolution, one of the stars in a binary system may increase in radius, or the binary separation shrink, to the point where the gravitational pull of the companion can remove the outer layers of its envelope (Roche lobe overflow).

(ii) One of the stars may, at some evolutionary phase, eject much of its mass in the form of a stellar wind; some of this material will be captured gravitationally by the companion (stellar wind accretion).

The situation described in (i) can come about other than by the swelling up of one of the stars in the course of its evolution; for example, the binary separation may shrink because orbital angular momentum is being lost from the system as a consequence of stellar wind mass loss, or gravitational radiation. For the purpose of studying the accretion process, the distinctions between these alternatives are unimportant, although crucial to the study of binary evolution (see Sections 4.4 and 6.8). In our discussion of binary accretion we shall have much to say about (i) and comparatively little about (ii), because the former is considerably better understood.

4.3   Roche lobe overflow

The situation described in (i) above, in which one star orbits another so closely that its outer layers are disrupted by the companion, has close parallels in other branches of astronomy. The problem was first studied by the 19th century French mathematician Edouard Roche in connection with the destruction or survival of planetary satellites, and is usually associated with his name. The essence of the Roche approach is to consider the orbit of a test particle in the gravitational potential due to two massive bodies orbiting each other under the influence of their mutual gravitational attractions. These bodies (in our case the two stars of the binary system) are assumed to be so massive that the test particle does not perturb their orbits (because of this the problem is sometimes known as the restricted three-body problem). Thus, the two stars execute Kepler orbits about each other in a plane: the Roche problem assumes these orbits to be circular. This is usually a good approximation for binary systems, since tidal effects tend to circularize originally eccentric orbits on timescales short compared to the time over which mass transfer occurs, although there are important examples where this is not so. A further restriction on the Roche problem, in addition to circular orbits, is the assumption that the two stars are 'centrally condensed' in the sense that they can be regarded as point masses for dynamical purposes. Again, this is normally an excellent approximation. Thus, we have the configuration shown in Fig. 10.

It is convenient to write the masses of the two stars as $M_1 M_\odot$ and $M_2 M_\odot$ since M_1 and M_2 then lie in the range 0.1–100 for all types of star. The binary separation, a, is then given in terms of the fundamental observational quantity, the binary period, P, through Kepler's law:

$$4\pi^2 a^3 = G(M_1 + M_2) M_\odot P^2. \tag{4.1}$$

For binary periods of the order of years, days, or hours, a can be conveniently expressed in the alternative forms:

$$a = \begin{cases} 1.5 \times 10^{13} M_1^{1/3} (1+q)^{1/3} P_{\text{yr}}^{2/3} \text{ cm}, \\ 2.9 \times 10^{11} M_1^{1/3} (1+q)^{1/3} P_{\text{day}}^{2/3} \text{ cm}, \\ 3.5 \times 10^{10} M_1^{1/3} (1+q)^{1/3} P_{\text{hr}}^{2/3} \text{ cm}, \end{cases} \tag{4.2}$$

giving the typical sizes of systems in terms of the periods ($P_{\text{yr}} = P$ in years, etc.) and the *mass ratio*

$$q = \frac{M_2}{M_1}. \tag{4.3}$$

Any gas flow between the two stars is governed by the Euler equation (2.3). It is

Accretion in binary systems

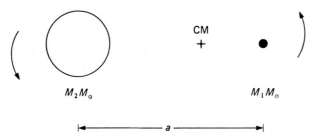

Figure 10. A binary system with a compact star of mass $M_1 M_\odot$ and 'normal' star of mass $M_2 M_\odot$ orbiting their common centre of mass (CM) with separation a.

convenient to write this in a frame of reference rotating with the binary system, with angular velocity ω relative to an inertial frame, since in the rotating frame the two stars are fixed. This introduces extra terms in the Euler equation to take account of centrifugal and Coriolis forces. With the assumptions made for the Roche problem, the Euler equation takes the form

$$\frac{\partial \mathbf{v}}{\partial t}+(\mathbf{v}\cdot\nabla)\mathbf{v} = -\nabla\Phi_R - 2\omega \wedge \mathbf{v} - \frac{1}{\rho}\nabla P, \tag{4.4}$$

with the angular velocity of the binary, ω, given in terms of a unit vector, $\mathbf{e}$, normal to the orbital plane by

$$\omega = \left[\frac{G(M_1+M_2)M_\odot}{a^3}\right]^{1/2} \mathbf{e}.$$

The term $-2\omega \wedge \mathbf{v}$ is the Coriolis force per unit mass; $-\nabla\Phi_R$ includes the effects of both gravitation and centrifugal force. Φ_R is known as the Roche potential and is given by

$$\Phi_R(\mathbf{r}) = -\frac{GM_1 M_\odot}{|\mathbf{r}-\mathbf{r}_1|} - \frac{GM_2 M_\odot}{|\mathbf{r}-\mathbf{r}_2|} - \tfrac{1}{2}(\omega \wedge \mathbf{r})^2 \tag{4.5}$$

where $\mathbf{r}_1, \mathbf{r}_2$ are the position vectors of the centres of the two stars. We gain considerable insight into accretion problems by plotting the equipotential surfaces of Φ_R and, in particular, their sections in the orbital plane (Fig. 11). When doing this, we must be careful to remember that some of the forces, in particular the Coriolis forces, acting on the accreting gas are not represented by Φ_R.

The shape of the equipotentials is governed entirely by the mass ratio q, while the overall scale is given by the binary separation a. Fig. 11 is drawn for the case $q = 0.25$, but its qualitative features apply for any mass ratio. Matter orbiting at large distances ($r \gg a$) from the system sees it as a point mass concentrated at the centre of mass (CM). Thus, the equipotentials of Φ_R at large distances are just those of a point mass viewed in a rotating frame. Similarly, there are circular equipotential sections around the centres of each of the two stars ($\mathbf{r}_1, \mathbf{r}_2$); the motion of matter here is dominated by the gravitational pull of the nearer star. Hence, the potential Φ_R has two deep valleys centred on $\mathbf{r}_1, \mathbf{r}_2$. The most interesting and important feature of Fig. 11 is the figure-of-eight area (the heavy line), which shows how these

Roche lobe overflow

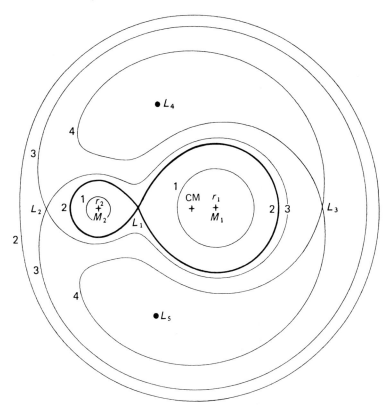

Figure 11. Sections in the orbital plane of the Roche equipotentials $\Phi_R =$ constant, for a binary system with mass ratio $q = M_2/M_1 = 0.25$. Shown are the centre of mass (CM) and Lagrange points L_1–L_5. The equipotentials are labelled 1–4 in order of *increasing* Φ_R. Thus the saddle point L_1 (the inner Lagrange point) forms a 'pass' between the two 'Roche lobes', the two parts of the figure-of-eight equipotential 2. The Roche lobes are roughly surfaces of revolution about the line of centres $M_1 M_2$. L_4 and L_5 (the 'Trojan asteroid' points) are local maxima of Φ_R, but Coriolis forces stabilize synchronous orbits of test bodies at these points.

two valleys are connected. In three dimensions, this 'critical surface' has a dumbbell shape; the part surrounding each star is known as its *Roche lobe*. The lobes join at the *inner Lagrangian point* L_1, which is a saddle point of Φ_R; to continue the analogy, L_1 is like a high mountain pass between two valleys. This means that material inside one of the lobes in the vicinity of L_1 finds it much easier to pass through L_1 into the other lobe than to escape the critical surface altogether.

Now let us examine the consequences of this picture for mass transfer in binaries. Let us suppose initially that both stars are considerably smaller than their Roche lobes and that the rotation of each star on its axis is synchronous with the orbital motion, with axes orthogonal to the binary plane. Synchronism is usually a good assumption, as tidal forces tend to bring it about on a timescale similar to that for circularization of the binary orbit. In this case, the surface of each star will conform to one of the approximately circular equipotential sections in Fig. 11 (or the

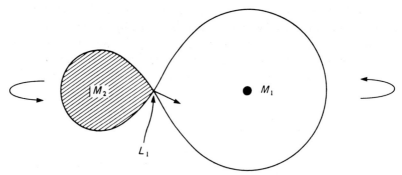

Figure 12. A binary system with the secondary star filling the Roche lobe and transferring mass through L_1 into the lobe of the compact primary.

qualitatively identical version of this figure for the appropriate q). This follows from the Euler equation (4.4) since $\mathbf{v} = 0$ and P is constant on the surface. There is then no tendency for material to be pulled off either of the stars by the gravitational attraction of the other; the binary is said to be *detached*, and mass transfer could proceed only via the wind mechanism ((ii) of Section 4.2). Suppose now that for some reason (e.g. stellar evolution), one of the stars swells up so that its surface, which must lie on an equipotential of Φ_R, eventually fills its Roche lobe. We shall usually label the stars so that this is star 2. Thus, we have the situation shown in Fig. 12. Henceforth, we shall often refer to star 2 which fills the Roche lobe as the *secondary* and star 1 as the *primary*. Note that this does not necessarily agree with the designation of primary and secondary on the basis of, say, the optical luminosity.

We note that, particularly for a mass ratio $q \ll 1$, the surface of the secondary is very distorted from the spherical. This distortion can sometimes be detected as *ellipsoidal variation* in the light curves of close binaries, sometimes even when they are detached. More important, however, is the fact that part of the envelope of the secondary lies very close to the inner Lagrangian point L_1. Any perturbation of this material will push it over the L_1 point into the Roche lobe of the primary, where it must, eventually, be captured by the star. Such perturbations are always present, provided, for example, by pressure forces (see (4.4)). A system like this is said to be *semi-detached*, and will efficiently transfer mass from star 2 to star 1 for as long as the secondary remains in contact with its Roche lobe: we will touch briefly on this problem in the next section. This type of mass transfer is called Roche lobe overflow. A further interesting situation (although not one of much relevance for this book) comes about if *both* stars for some reason fill their Roche lobes simultaneously. Systems of this type (e.g. the W UMa stars) are called *contact binaries*.

4.4 Roche geometry and binary evolution

To treat Roche lobe overflow in a quantitative manner, we need some idea of the geometry of the critical surface; in particular, its dependence on the mass ratio q and binary separation a. The main quantities we shall need are measures of

Roche geometry and binary evolution

the sizes of the Roche lobes and of the distances of the L_1 point from either star. Since the lobes are not spherical we need some average radius to characterize them; a suitable measure is the radius of a sphere having the same volume as the lobe. The complicated form (4.5) of Φ_R means that we must resort to numerical calculation to find this; but a good fit to these results for all values of q is given by the approximate analytic formula (Eggleton (1983))

$$\frac{R_2}{a} = \frac{0.49 q^{2/3}}{0.6 q^{2/3} + \ln(1+q^{1/3})}. \tag{4.6a}$$

Obviously the lobe radius R_1 of the primary is given by replacing q by q^{-1}. For $0.1 \lesssim q \lesssim 0.8$ it is often convenient to use a simple form due to Paczyński:

$$\frac{R_2}{a} = \frac{2}{3^{4/3}}\left(\frac{q}{1+q}\right)^{1/3} = 0.462\left(\frac{M_2}{M_1+M_2}\right)^{1/3}. \tag{4.6b}$$

The distance b_1 of the L_1 point from the centre of the primary (mass M_1) is given to good accuracy by a fitted formula of Plavec and Kratochvil:

$$\frac{b_1}{a} = 0.500 - 0.227 \log q. \tag{4.7}$$

An interesting consequence of the form (4.6b) for R_2/a for $q \lesssim 0.8$ is that the mean density $\bar{\rho}$ of a lobe-filling star is determined solely by the binary period P:

$$\bar{\rho} = \frac{3M_2 M_\odot}{4\pi R_2^3} \cong 115 P_{hr}^{-2} \text{ g cm}^{-3} \tag{4.8}$$

where we have used (4.1) to eliminate a. Equation (4.8) shows that, for binary periods of a few hours, stars with mean densities typical of the lower main sequence ($\bar{\rho} \sim 1-100$ g cm^{-3}) can fill their Roche lobes. If one assumes a structure for the lobe-filling star, and thus a relation $R_2(M_2)$, (4.8) fixes its properties uniquely for a given period. For example, if we assume that the lobe-filling star is close to the lower main sequence we know that its radius and mass are approximately equal in solar units, i.e.

$$M_2 \cong R_2/R_\odot$$

(e.g. Kippenhahn & Weigert (1990)). Thus

$$\bar{\rho} = \frac{3M_2 M_\odot}{4\pi R_2^3} = \frac{3M_\odot}{4\pi R_\odot^3}\frac{1}{M_2^2} = \frac{1.4}{M_2^2} \text{ g cm}^{-3}$$

where we have used the solar mean density $\bar{\rho}_\odot = 1.4$ g cm^{-3}. This relation and (4.8) now give a *period–mass* relation

$$M_2 \cong 0.11 P_{hr} \tag{4.9}$$

and a *period–radius* relation

$$R_2 \cong 0.11 P_h = 7.9 \times 10^9 P_{hr} \text{ cm}. \tag{4.10}$$

Note that these very simple relations come about because of our assumption that the secondary was close to the lower main sequence: however, since the star is in a binary and losing mass to the companion it is not obvious that its structure will be the same as an isolated star of the same mass. Thus we cannot in general assume a main-sequence (or any other) structure without checking carefully the conditions for this to hold (see below).

The vigilant reader will have realized that the process of mass transfer in a binary will change its mass ratio q. Equally true, if not quite so obvious, is that the period P and separation a will be altered by the same process, because of the redistribution of angular momentum within the system. Since the Roche geometry is determined by a and q, it is important to ask whether the effect of these changes is to shrink or swell the Roche lobe of the mass-losing star. In the former case, the lobe-overflow process will be self-sustaining at least for some time; in the latter case the mass transfer will switch off unless some effect, such as the nuclear evolution of the mass-losing star, can increase its radius at a sufficient rate. The determining quantity for these questions is the orbital angular momentum J. Writing ω for the binary's angular velocity $(= 2\pi/P)$, this is

$$J = (M_1 a_1^2 + M_2 a_2^2) M_\odot \omega,$$

where

$$a_1 = \left(\frac{M_2}{M}\right) a, \quad a_2 = \left(\frac{M_1}{M}\right) a$$

are the distances of the two stars from the centre of mass, and $M = M_1 + M_2$. Substituting for a_1, a_2 above and using (4.1) gives

$$J = M_1 M_2 \left(\frac{Ga}{M}\right)^{1/2} M_\odot^{3/2}. \tag{4.11}$$

Usually it is a good approximation to assume that all the mass lost by the secondary star is accreted by the primary, so that $\dot{M}_1 + \dot{M}_2 = 0$, $\dot{M}_2 < 0$ (the treatment can be extended to cover cases where this does not hold). Then logarithmically differentiating (4.11) with respect to time gives

$$\frac{\dot{a}}{a} = \frac{2\dot{J}}{J} + \frac{2(-\dot{M}_2)}{M_2}\left(1 - \frac{M_2}{M_1}\right). \tag{4.12}$$

Conservative mass transfer is characterized by constant binary mass and angular momentum; setting $\dot{J} = 0$ in (4.12) and remembering that $\dot{M}_2 < 0$, we see that the binary expands ($\dot{a} > 0$) if conservative transfer takes place from the less massive to the more massive star: more matter is placed near the centre of mass, so the remaining mass M_2 must move in a wider orbit to conserve angular momentum. Conversely, transfer from the more massive to the less massive star shrinks the binary separation. The Roche lobe size is affected by the change in mass ratio as well as separation. Logarithmic differentiation of (4.6b) gives

$$\frac{\dot{R}_2}{R_2} = \frac{\dot{a}}{a} + \frac{\dot{M}_2}{3M_2},$$

so combining with (4.12) yields

$$\frac{\dot{R}_2}{R_2} = \frac{2\dot{J}}{J} + \frac{2(-\dot{M}_2)}{M_2}\left(\frac{5}{6} - \frac{M_2}{M_1}\right). \tag{4.13}$$

Clearly there are two cases, depending on whether q is larger or smaller than $\frac{5}{6}$. For $q > \frac{5}{6}$ conservative mass transfer shrinks the Roche lobe down on the mass-losing star, and any angular momentum loss ($\dot{J} < 0$) will accentuate this. Unless the star can contract rapidly enough to keep its radius smaller than R_2, the overflow process will become very violent, proceeding on a dynamical or thermal timescale depending on whether the star's envelope is convective or radiative. The overflow stops once the mass ratio is reversed. The possibility that the star could shrink makes the critical value of q for this type of unstable mass transfer depend on its mass–radius relation, so it may differ slightly from $\frac{5}{6}$. All interacting binaries must undergo unstable mass transfer when the more massive star expands as it evolves off the main sequence, and other episodes can occur. However, the process is probably unobservable; it is certainly too violent and short-lived to be typical of the mass-exchanging binaries we do observe, which must therefore correspond to the case $q \lesssim \frac{5}{6}$. Equation (4.13) shows that conservative mass transfer will expand the Roche lobe in this case. Mass transfer therefore only continues if either (i) the star expands, or (ii) the binary loses angular momentum. Case (i) occurs when the secondary star evolves off the main sequence. Then its radius will expand on a nuclear timescale t_{nuc} determined by hydrogen shell burning. The star can never greatly exceed its Roche lobe as mass is then lost on a dynamical timescale, so the lobe too will expand on the timescale t_{nuc}. Thus $\dot{R}_2/R_2 \sim t_{\mathrm{nuc}}^{-1}$, and (4.13) fixes the transfer rate in terms of t_{nuc}:

$$-\dot{M}_2 = \frac{M_2}{(5/3 - 2q)t_{\mathrm{nuc}}}. \tag{4.14}$$

Combining this with (4.12) and (4.1) shows that a and P increase on similar timescales:

$$\frac{\dot{a}}{a} = \frac{2\dot{P}}{3P} = \frac{(1-q)}{(5/3 - 2q)t_{\mathrm{nuc}}}.$$

Case (i) describes mass transfer driven by evolutionary expansion of the secondary star. In many binaries (cataclysmic variables are an example) the secondary has such a low mass that its main-sequence lifetime exceeds the Hubble time, and this mechanism cannot be operating. Mass transfer must be driven by angular momentum loss (case ii) in such systems. A variety of effects can give $\dot{J} < 0$: in short-period systems gravitational radiation is quite efficient, and any mechanism which spins down the secondary star will ultimately remove angular momentum from the binary, as tides force the secondary to corotate. Angular momentum loss in a wind magnetically linked to the secondary is a currently favoured possibility.

The nature of the transfer process in case (ii) depends on the secondary's reaction to the mass loss, and hence on the behaviour of its radius as M_2 is reduced. If mass loss is sufficiently gentle the star will stay close to thermal equilibrium, so that a secondary on the main sequence before mass transfer started will remain close to it, keeping R_2 approximately proportional to M_2. Setting $\dot{R}_2/R_2 = \dot{M}_2/M_2$ in (4.13) gives

$$\frac{-\dot{M}_2}{M_2} = \frac{-\dot{J}/J}{4/3 - M_2/M_1}. \tag{4.15}$$

Mass transfer therefore proceeds on the timescale for angular momentum loss, and (4.12) and (4.1) now show that a and P decrease on similar timescales:

$$\frac{\dot{a}}{a} = 2\frac{\dot{P}}{3P} = \frac{2\dot{J}/3J}{4/3 - q}. \tag{4.16}$$

It is important to realise that the transfer rates (4.14) and (4.15) are the *average* rates driven by evolutionary expansion or angular momentum loss respectively. There can be wide deviations from these rates for epochs short compared with the evolution time (i.e. the time for the Roche lobe to move through one stellar scaleheight) which can be $\gtrsim 10^5$ yr. In various types of binary system, (4.14) and (4.15) do account successfully for the average observed accretion rates $\sim 10^{15} - 10^{18}$ g s^{-1} ($\sim 10^{-11} - 10^{-8}\, M_\odot$ yr^{-1}).

4.5 Disc formation

The considerations of the last two sections have shown how mass transfer in close binary systems can occur via Roche lobe overflow. A consequence of this process is that the transferring material has rather high specific angular momentum in many cases, so that it cannot accrete directly on to the mass-capturing star. This is illustrated in Fig. 13. Note that matter must pass from the Roche lobe of the secondary to that of the primary through the L_1 point; as far as the primary is concerned, it is as though material were being squirted at it from a nozzle which rotates around it in the binary plane. Unless the binary period is quite long, this nozzle rotates so rapidly that the gas stream appears to the primary to move almost orthogonally to the line of centres as it emerges through L_1. For let $v_\parallel$, $v_\perp$ be the components of the stream velocity in a *non*-rotating frame along and at right-angles to the instantaneous line of centres. We have

$$v_\perp \sim b_1 \omega$$

while

$$v_\parallel \lesssim c_s,$$

where c_s is the sound speed in the envelope of the secondary, since the gas is presumably pushed through L_1 by pressure forces. From (4.7) we have $b_1 \gtrsim 0.5a$, unless q is very large; using $\omega = 2\pi/P$ and (4.2) for a we obtain

$$v_\perp \sim 100 M_1^{1/3}(1+q)^{1/3}\, P_{\text{day}}^{-1/3}\text{ km s}^{-1}.$$

Disc formation

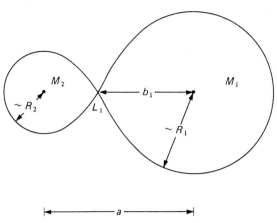

Figure 13. Roche geometry.

By contrast, (2.17) shows that $v_\parallel \sim c_s$ cannot be much more than about 10 km s^{-1} for normal stellar envelope temperatures ($< 10^5$ K).

The discussion above also shows that the gas stream issuing through L_1 is supersonic. This will hold *a fortiori* as the stream is accelerated by the primary's gravitational attraction within its Roche lobe. This fact enormously simplifies the calculation of the stream trajectory, for the work of Chapter 2 shows that pressure forces can be neglected: the stream will follow a ballistic trajectory determined by the Roche potential, as though it consisted simply of a set of test particles. Further, since $v_\parallel \sim c_s$ is much smaller than the velocities acquired during free fall towards the primary, the initial conditions near L_1 have little effect on the trajectory, which is therefore essentially unique (on energetic grounds one can show that the width of the stream issuing through L_1 is $\sim 0.1 P c_s$, much smaller than the radius of the secondary for example). To a good approximation we can take the stream trajectory as the orbit of a test particle released from rest at L_1 (and thus with a given angular momentum), falling in the gravitational field of the primary alone. (Note that in Fig. 11 the equipotential curves are approximately circular within the lobes.) This would give an elliptical orbit lying in the binary plane: the presence of the secondary causes this to precess slowly. A continuous stream trying to follow this orbit will therefore intersect itself, resulting in dissipation of energy via shocks. On the other hand, the gas has little opportunity to rid itself of the angular momentum it had on leaving L_1, so it will tend to the orbit of lowest energy for a given angular momentum, i.e. a circular orbit. We thus expect the gas initially to orbit the primary in the binary plane at a radius R_{circ} such that the Kepler orbit at R_{circ} has the same specific angular momentum as the transferring gas had on passing through L_1. Thus the gas will have circular velocity

$$v_\phi(R_{\text{circ}}) = \left(\frac{GM_1 M_\odot}{R_{\text{circ}}}\right)^{1/2}$$

where

$$R_{\text{circ}} v_\phi(R_{\text{circ}}) = b_1^2 \omega.$$

Using $\omega = 2\pi/P$ where P is the binary period these can be combined to give

$$R_{\text{circ}}/a = (4\pi^2/GM_1 M_\odot P^2)a^3(b_1/a)^4$$

which, by (4.1), becomes

$$R_{\text{circ}}/a = (1+q)(b_1/a)^4.$$

Finally, we use (4.7) to write

$$R_{\text{circ}}/a = (1+q)[0.500-0.227 \log q]^4. \qquad (4.17)$$

The radius R_{circ} is often called the *circularization radius*. The particular form of (4.17) depends on the assumption that the accretion occurs via Roche lobe overflow (see Section 4.8 for R_{circ} under a different assumption).

It is easy to show that R_{circ} is always smaller than the lobe radius R_L of the primary. Typically it is a factor 2–3 smaller, except for very small q: for example, for $q = 1$, $R_{\text{circ}} = 0.125a$, $R_L = 0.38a$. Hence, left to itself, the captured material would orbit the primary well inside the Roche lobe. This will be prevented, however, if the primary already occupies this space, i.e. if its radius $R_* > R_{\text{circ}}$. Using (4.2) to rewrite (4.17) as

$$R_{\text{circ}} \cong 4(1+q)^{4/3}[0.500-0.227 \log q]^4 P_{\text{day}}^{2/3} R_\odot \qquad (4.18)$$

we find that R_{circ} can be of the order $R_\odot$ or less for typical parameters. For example,

$$R_{\text{circ}} \cong 1.2 P_{\text{day}}^{2/3} R_\odot \quad \text{for } q = 0.3$$

and

$$R_{\text{circ}} \cong 0.6 P_{\text{day}}^{2/3} R_\odot \quad \text{for } q \geqslant 0.5.$$

Thus in systems where the accreting star is extended, for example a main-sequence star, it is quite possible that $R_{\text{circ}} < R_*$ if the period is less than about 100 days. In this case, the gas stream from the secondary must crash obliquely into the primary. This must presumably happen in some of the Algol systems which have periods of a few days. The details of what happens in the region where the gas stream hits the star will be quite complex, so it is fortunate that this is not the case of greatest interest for accreting binaries: as we have pointed out in Chapter 1, accretion is more efficient as a power source the more *compact* the accreting object is. Any stellar mass compact object will have a radius R_* less than R_{circ} for realistic binary parameters, since (4.17) and (4.2) imply

$$R_{\text{circ}} \gtrsim 3.5 \times 10^9 P_{\text{hr}}^{2/3} \text{ cm} \qquad (4.19)$$

while R_* cannot be larger than the radius of a low-mass white dwarf:

$$R_* \lesssim 10^9 \text{ cm}. \qquad (4.20)$$

The implications of the inequality between R_* and R_{circ} given by (4.19) and (4.20) are profound. If we were considering a single test particle we would deduce that it would simply orbit the primary in an ellipse of characteristic dimensions $R \sim R_{\text{circ}}$. But for the continuous stream of gas captured from the secondary which is

Disc formation

Figure 14. Accretion disc.

envisaged here, the corresponding configuration is a ring of matter at $R = R_{\text{circ}}$. Clearly, within such a ring, there will be dissipative processes (e.g. collisions of gas elements, shocks, viscous dissipation, etc.) which will convert some of the energy of the ordered bulk orbital motion about the primary into internal (i.e. heat) energy. Eventually, of course, some of this energy is radiated and therefore lost to the gas. The only way in which the gas can meet this drain of energy is by sinking deeper into the gravitational potential of the primary, that is, by orbiting it more closely; this in turn requires it to lose angular momentum. The timescale on which the orbiting gas can redistribute its angular momentum is normally much longer than both the timescale over which it loses energy by radiative cooling, t_{rad}, and the dynamical (or orbital) timescale, t_{dyn}. Therefore, the gas will lose as much energy as it can for a given angular momentum, which itself decreases more slowly. Since a circular orbit has the least energy for a given angular momentum, we might expect most of the gas to spiral slowly inwards towards the primary through a series of approximately circular orbits in the binary orbital plane, a configuration known as an accretion disc (Fig. 14). This spiralling-in process entails a loss of angular momentum: in the absence of external torques, this can only occur by transfer of angular momentum outwards through the disc by internal torques. Thus the outer parts of the disc will gain angular momentum and will spiral outwards. The original ring of matter at $R = R_{\text{circ}}$ will spread to both smaller and larger radii by this process.

In most cases, the total mass of gas in the disc is so small, and its mean density so much lower than that of the primary, that we may neglect the self-gravity of the disc. The circular orbits are then Keplerian, with angular velocity

$$\Omega_{\text{K}}(R) = (GM_1 M_\odot / R^3)^{1/2}. \tag{4.21}$$

The binding energy of a gas element of mass m in the Kepler orbit which just grazes the surface of the primary is $\tfrac{1}{2}GMm/R_*$. Since the gas elements start at large distances from the star with negligible binding energy, the total disc luminosity in a steady state must be

$$L_{\text{disc}} = \frac{GM\dot{M}}{2R_*} = \tfrac{1}{2}L_{\text{acc}} \tag{4.22}$$

where $\dot{M}$ is the accretion rate and L_{acc} the accretion luminosity (1.3). The other half of L_{acc} has still to be released very close to the star (see Section 6.2).

Thus half of the available energy is radiated while the matter in the disc spirals slowly inwards. On the other hand, since the specific angular momentum $R^2\Omega(R) \propto R^{1/2}$ and $R_* \ll R_{\text{circ}}$, the material must somehow get rid of almost all its original angular momentum. Clearly, the dissipation processes which cause the conversion of ordered kinetic energy into heat must also exert torques on the spiralling material. The effect of these is to transport angular momentum outward through the disc (Section 4.6). Near the outer edge of the disc some other process must finally remove this angular momentum. It is likely that angular momentum is fed back into the binary orbit through tides exerted on the outer disc by the secondary (see Section 5.9).

An important consequence of this angular momentum transport is that the outer edge of the disc will in general be at some radius R_{out} *exceeding* R_{circ} given by (4.17); R_{circ} is often referred to as the minimum disc radius.

4.6 Viscous torques

From our discussion of the last section we see that an accretion disc can be an efficient 'machine' for slowly lowering material in the gravitational potential of an accreting object and extracting the energy as radiation. A vital part of this machinery is the dissipation process which converts the orbital kinetic energy into heat. One of the main unsolved problems of disc structure is the precise nature of this dissipation: it is surprising (and fortunate) that despite this lack of knowledge considerable progress has been achieved in some areas.

The Keplerian rotation law (4.21) implies *differential rotation*; i.e. material at neighbouring radii moves with different angular velocity Ω. This will also be true in cases other than Keplerian: whenever the rotation law departs from $\Omega = $ const or *solid body rotation*, fluid elements on neighbouring streamlines will slide past each other. Because of the ever present chaotic thermal motions of fluid molecules or turbulent motions of fluid elements, viscous stresses are generated just as in the case of bulk viscosity discussed in Section 3.6. The only difference here is that we consider transport of momentum, etc. *orthogonal* to the gas motion, rather than along it. This type of transport process is known as *shear viscosity*, as the effect must vanish (by symmetry) if there is no sliding (shearing) of adjacent gas streams past one another. Shearing is absent in two particularly simple states of fluid motion: uniform translational motion with $v = $ const and solid body (uniform) rotation. In both cases the relative distances between any pairs of fluid elements remain constant in time, or put another way, the fluid moves without internal distortions.

Given a typical scale λ and speed $\tilde{v}$ for the chaotic motions, the force density and torques exerted by shear viscosity can be calculated by a method similar to that used to obtain the bulk viscous force (3.40). In that calculation, we took λ to be of the order of the mean free path λ_d, and $\tilde{v}$ to be of the order of the sound speed c_s, but we shall leave λ and $\tilde{v}$ free here, in order to treat the viscous effect of other chaotic motions, especially turbulent ones. With the application of accretion discs in mind, we consider the situation shown in Fig. 15, where the gas flow lies between the planes $z = 0$ and $z = H$ of cylindrical polar coordinates (R, ϕ, z). We assume azimuthal symmetry, with gas flowing in the ϕ direction with angular velocity $\Omega(R)$ about the origin.

Let us consider two neighbouring rings of width λ on both sides of a surface $R = $ const. Because of the chaotic motions, gas elements such as A and B are constantly exchanged across this surface, with speeds $\sim \tilde{v}$. A typical element travels a distance $\sim \lambda$ before interacting on the other side of the surface $R = $ const. As

Viscous torques

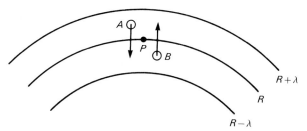

Figure 15. Viscous angular momentum transport in a shearing medium.

these elements of fluid are exchanged, they carry slightly different amounts of angular momentum: elements such as A will *on average* carry an angular momentum corresponding to the location $R-\lambda/2$, while elements such as B will be representative of a radial location $R+\lambda/2$. Because the chaotic motion takes place in an equilibrium flow, this process cannot result in the net transfer of any matter between the two rings. Therefore, mass crosses the surface $R = $ const at equal rates in both directions, of the order $H\rho\tilde{v}$, per unit arc length, where $\rho(R)$ is the mass density. Since the two mass fluxes nonetheless carry different angular momenta, there is a transport of angular momentum due to the chaotic process, i.e. a *viscous torque* exerted on the outer stream by the inner stream. To calculate its magnitude we must estimate the net angular momentum flux outwards as seen by a corotating observer at generic point P on the surface $R = $ const. As seen by an observer corotating with the fluid at P with angular velocity $\Omega(R)$, the fluid at $R-\lambda/2$ will appear to move with velocity $(R-\lambda/2)\Omega(R-\lambda/2)+\Omega(R)\lambda/2$. Thus the average angular momentum flux per unit arc length through $R = $ const in the outward direction is

$$\rho\tilde{v}H(R-\lambda/2)[(R-\lambda/2)\Omega(R-\lambda/2)+\Omega(R)\lambda/2].$$

An analogous expression, changing the sign of λ, gives the average inward angular momentum flux per unit arc length. The torque exerted on the outer ring by the inner ring is given by the *net* outward angular momentum flux. Since the mass flux due to chaotic motions is the same in both directions, one obtains to first order in λ the torque per unit arc length as

$$-\rho\tilde{v}H\lambda R^2\Omega',$$

where $\Omega' = d\Omega/dr$, and we have assumed that the angular velocity changes slowly over the lengthscale of chaotic motions. For the case of circular rings of gas (as in an accretion disc), we get the total torque simply by multiplying by the length $2\pi R$ of the circular boundary. Setting $\rho H = \Sigma$, the *surface density*, we can write the torque exerted by the *outer* ring on the *inner* ($= -$ the torque of the inner on the outer) as

$$G(R) = 2\pi R\nu\Sigma R^2\Omega' \tag{4.23}$$

where we have defined

$$\nu \sim \lambda\tilde{v} \tag{4.24}$$

as the coefficient of kinematic viscosity.

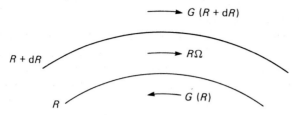

Figure 16. Differential viscous torque.

Let us check that the form (4.23) for $G(R)$ agrees with our expectations. First, the torque vanishes for the case of *rigid* rotation $\Omega' = 0$. This is reasonable since there is no shearing of gas elements in this case. Second, the sign of (4.23) means that for a rotation law in which $\Omega(R)$ decreases outwards (e.g. $\Omega =$ the Keplerian value given by (4.21)) $G(R)$ is negative; hence, in this case, the inner rings lose angular momentum to the outer ones and the gas slowly spirals in, as envisaged above.

Now let us consider the net torque on a ring of gas between R and $R+\mathrm{d}R$. As this has both an inner and an outer edge, it is subject to competing torques (Fig. 16); the net torque (trying to speed it up) is

$$G(R+\mathrm{d}R) - G(R) = \frac{\partial G}{\partial R}\mathrm{d}R.$$

Because this torque is acting in the sense of angular velocity $\Omega(R)$, there is a rate of working

$$\Omega \frac{\partial G}{\partial R}\mathrm{d}R = \left[\frac{\partial}{\partial R}(G\Omega) - G\Omega'\right]\mathrm{d}R \qquad (4.25)$$

by the torque. But the term

$$\frac{\partial}{\partial R}(G\Omega)\,\mathrm{d}R$$

is just the rate of 'convection' of rotational energy through the gas by the torques; if we were to integrate (4.25) over the whole disc, this term would give a contribution $[G\Omega]_{\text{inner edge}}^{\text{outer edge}}$ determined solely by conditions at the edges of the disc. The term $-G\Omega'\,\mathrm{d}R$, on the other hand, represents a local rate of loss of mechanical energy to the gas. This lost energy must go into internal (heat) energy. The viscous torques therefore cause *dissipation* within the gas at a rate $G\Omega'\,\mathrm{d}R$ per ring width $\mathrm{d}R$. Ultimately, this energy will be radiated over the upper and lower faces of the disc; we are therefore interested in the dissipation rate per unit plane surface area $D(R)$. Remembering that each ring has two plane faces and thus a plane area $4\pi R\,\mathrm{d}R$ we find that

$$D(R) = \frac{G\Omega'}{4\pi R} = \tfrac{1}{2}\nu\Sigma(R\Omega')^2 \qquad (4.26)$$

using (4.23). Note that $D(R) \geqslant 0$, vanishing only for the case of rigid rotation.

4.7 The magnitude of viscosity

So far, our treatment of viscosity has been rather formal in that we have not given any physical prescription for the lengthscale λ and speed $\tilde{v}$. It might be wondered why we did not simply set $\lambda \sim \lambda_{\rm d}$ and $\tilde{v} \sim c_{\rm s}$ as in the case of the plasma bulk viscosity discussed in Section 3.6. The reason is connected with the magnitude of the viscous force terms. In Section 4.6 we calculated only viscous *torques* in the accretion disc; had we calculated a force density, as in Section 3.6, we would have found that shear viscosity gives a force density in the ϕ direction which is dimensionally identical to the bulk viscous force density (3.40):

$$f_{\rm visc,\,shear} \sim \rho\lambda\tilde{v}\frac{\partial^2 v_\phi}{\partial R^2} \sim \rho\lambda\tilde{v}\frac{v_\phi}{R^2}.$$

Let us compare this with the inertia terms $\rho(\partial \mathbf{v}/\partial t + (\mathbf{v}\cdot\nabla)\mathbf{v})$ in the Euler equation (2.3). The *Reynolds number*

$$Re = \frac{\text{inertia}}{\text{viscous}} \sim \frac{v_\phi^2/R}{\lambda\tilde{v}v_\phi/R^2} = \frac{Rv_\phi}{\lambda\tilde{v}} \qquad (4.27)$$

measures the importance of viscosity: if $Re \ll 1$ viscous forces dominate the flow; if $Re \gg 1$ the viscosity associated with the given λ and $\tilde{v}$ is dynamically unimportant. This is certainly the case for the standard 'molecular' viscosity given by $\lambda \sim \lambda_{\rm d}$, $\tilde{v} \sim c_{\rm s}$, because using (3.25), (2.17) and (4.21) we find that

$$Re_{\rm mol} \sim 0.2\,NM_1^{1/2}R_{10}^{1/2}T_4^{-5/2}. \qquad (4.28)$$

Here R_{10} is the distance from the centre of the accreting star in units of 10^{10} cm, T_4 is the gas temperature in units of 10^4 K and N (cm^{-3}) is the gas density. In typical regions of an accretion disc (see Chapter 5), we shall have $M_1^{1/2} \sim R_{10}^{1/2} \sim T_4^{-5/2} \sim 1$, while N usually comfortably exceeds 10^{15} cm^{-3}, so that

$$Re_{\rm mol} > 10^{14}. \qquad (4.29)$$

Even in regions where extreme conditions hold, it is hard to reduce $Re_{\rm mol}$ very much below this estimate. Hence, molecular viscosity is far too weak to bring about the dissipation and angular momentum transport we require.

Curiously, a clue to the right kind of viscosity is provided by the very large estimate (4.29); it is known from laboratory experiments that, if the (molecular) Reynolds number of a fluid flow is gradually increased, there is a *critical* Reynolds number at which *turbulence* sets in: i.e. the fluid velocity $\mathbf{v}$ suddenly begins to exhibit large and chaotic variations on an arbitrarily short time and lengthscale. For most laboratory fluids, the critical Reynolds number is of the order $10-10^3$. Because of the size of the estimate (4.29), we may conjecture that the gas flow in accretion discs is also turbulent, although there is, as yet, no proof that this is so. If this is the case, the flow will be characterized by the size $\lambda_{\rm turb}$ and turnover velocity $v_{\rm turb}$ of the largest turbulent eddies (see e.g. Landau & Lifshitz (1959)). Since the turbulent motion is completely chaotic about the mean gas velocity, our simple viscosity calculations apply: there is a turbulent viscosity $v_{\rm turb} \sim \lambda_{\rm turb} v_{\rm turb}$. Although this result has a neat appearance, it is here that our troubles with viscosity really begin.

62   Accretion in binary systems

Turbulence is one of the major uncharted areas of classical physics and we do not understand the onset of turbulence, still less the physical mechanisms involved and how they determine the lengthscale, λ_{turb}, and turnover velocity v_{turb}. The most we can do with present knowledge is to place plausible limits on λ_{turb} and v_{turb}.

First, the typical size of the largest turbulent eddies cannot exceed the disc thickness H, so $\lambda_{\text{turb}} \lesssim H$. Second, it is unlikely that the turnover velocity v_{turb} is supersonic, for, in this case, the turbulent motions would probably be thermalized by shocks (see Section 3.8). Thus, we can write

$$v = \alpha c_s H \tag{4.30}$$

and expect $\alpha \lesssim 1$. This is the famous *α-prescription* of Shakura and Sunyaev. It is important to realize that (4.30) is merely a parametrization: all our ignorance about the viscosity mechanism has been isolated in α rather than in v and apart from the expectation $\alpha \lesssim 1$ we have gained nothing. There is certainly no cogent reason, for example, for believing α to be constant throughout a disc structure; even $\alpha \lesssim 1$ might be violated in small regions of the disc where some physical input continually feeds the supersonic turbulence. Nonetheless, the α-prescription has proved a useful parametrization of our ignorance and has encouraged a semi-empirical approach to the viscosity problem, which seeks to estimate the magnitude of α by a comparison of theory and observation; there is, for example, some reason to believe that $\alpha \sim 1$ in accretion discs in cataclysmic variables, at least some of the time (see Section 5.8). Moreover, the α-prescription can be formulated in an identical fashion for viscosities due to the winding-up of chaotic magnetic fields in the disc; here, too, one expects $\alpha \lesssim 1$ on dimensional grounds.

4.8   Accretion in close binaries: other possibilities

In this chapter we have concentrated almost entirely on accretion via Roche lobe overflow, since this is the best understood of the various possibilities and clearly demonstrates the necessity of disc formation in many cases. We should not lose sight of the fact that in many close binaries accretion proceeds via different mechanisms. Indeed, we mentioned in Section 4.2 the possibility of accretion from a stellar wind. This is particularly relevant for systems containing an early-type (O or B) star and a neutron star (or a black hole) in a close orbit. Often the neutron star spins asynchronously and gives rise to regular X-ray pulses. Such systems can be very luminous and many of the first galactic X-ray sources to be discovered (e.g. Cen X-3, SMC X-1, Vela X-1) probably fall into this category. Cyg X-1 is a non-pulsed source where the compact object is thought to be a black hole accreting from a stellar wind. The general situation is depicted in Fig. 17.

The stellar wind of the early-type companion is both intense, with mass loss rates 10^{-6}–$10^{-5} M_\odot$ yr^{-1}, and highly supersonic, with velocity

$$v_w(r) \sim v_{\text{esc}}(R_E) = \left(\frac{2GM_E}{R_E}\right)^{1/2} \tag{4.31}$$

at a distance r from the centre of the early-type star. Here, M_E and R_E are the mass and radius of this star and v_{esc} is the escape velocity at its surface. For typical

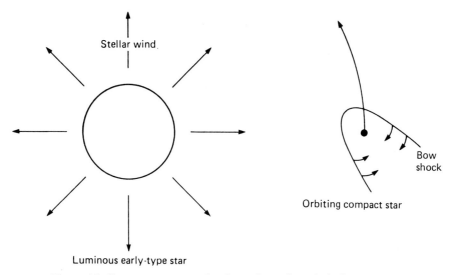

Figure 17. Compact star accreting from the stellar wind of an early-type companion.

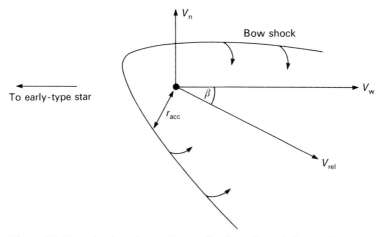

Figure 18. Bow shock and accretion radius in stellar wind accretion.

parameters v_w is a few thousand km s^{-1} which greatly exceeds the sound speed $c_s \sim 10$ km s^{-1} (equation (2.17)). If the orbital velocity of the compact star (which we henceforth assume to be a neutron star) about its companion is v_n, the wind sweeps past the neutron star at an angle $\beta \cong \tan^{-1}(v_n/v_w)$ to the line of centres (Fig. 18) with speed

$$v_{\rm rel} \cong (v_n^2 + v_w^2)^{1/2}. \tag{4.32}$$

Since the gas flow relative to the neutron star is highly supersonic, we can neglect the gas pressure and think of the flow as a collection of test particles. Then those wind particles which pass so close to the neutron star that their kinetic energy is less than the gravitational potential energy will be captured and ultimately accreted by it. Thus capture will occur within a cylindrical region with axis along the relative wind direction (v_{rel}) and radius

$$r_{acc} \sim 2GM_n/v_{rel}^2. \tag{4.33}$$

Note the strong resemblance between this equation and (2.33) for capture from the interstellar medium: the basic argument is the same. Again, we call r_{acc} the accretion radius: but note from the discussion of Section 2.5 that it is a rather ill-defined quantity. Because the wind flow is highly supersonic there will be a strong 'bow' shock at about r_{acc} from the neutron star (see Fig. 18). Of more concern to us here is roughly how much wind material gets captured, and whether or not this has enough angular momentum to form an accretion disc.

For simplicity, let us consider these questions for the case $v_w \gg v_n$, so that $\beta = 0$ and $v_{rel} = v_w$. This is usually quite a good approximation, as one can see from (4.2), which implies $v_n = 200(M_E/M_\odot)^{1/3}(1+q)^{1/3} P_{day}^{-1/3}$ km s^{-1} with $q = M_n/M_E$. Then the fraction of the stellar wind captured by the neutron star is given by comparing the mass flux into the accretion cylinder of radius r_{acc} with the total mass loss rate $-\dot{M}_w$ of the star: using (4.33),

$$\frac{\dot{M}}{-\dot{M}_w} \cong \frac{\pi r_{acc}^2 v_w(a)}{4\pi a^2 v_w(a)} = \frac{G^2 M_n^2}{a^2 v_w^4(a)} \tag{4.34}$$

where a is the binary separation. From (4.31) we find

$$\frac{\dot{M}}{-\dot{M}_w} \cong \tfrac{1}{4}\left(\frac{M_n}{M_E}\right)^2 \left(\frac{R_E}{a}\right)^2. \tag{4.35}$$

For parameters typical of the X-ray binaries, (4.35) implies accretion rates of order 10^{-4}–10^{-3} of the mass-loss rate $-\dot{M}_w$. Thus accretion from a stellar wind is a very inefficient process by comparison with Roche lobe overflow, where almost all the material lost by one star is accreted by the other. It is only because the mass-loss rates are themselves so large (10^{-6}–10^{-5} $M_\odot$ yr^{-1}) that sources powered in this way are observable; from (1.4b) we have

$$L_{acc} \sim 10^{37} \left(\frac{\dot{M}}{-10^{-4} \dot{M}_w}\right)\left(\frac{-\dot{M}_w}{10^{-5} M_\odot \text{ yr}^{-1}}\right) \text{erg s}^{-1}$$

for a neutron star accreting from a stellar wind. Although the luminosity of the early-type companion may well exceed this, the fact that L_{acc} is emitted mainly as X-rays makes it readily detectable.

Now let us ask if this accreted material has sufficient angular momentum for disc formation to occur. If we were to neglect completely the neutron star's orbital motion about the companion, the accreted angular momentum would be zero by symmetry. However, since the whole accretion cylinder rotates about the companion with the orbital angular velocity $\omega(=v_n/a)$ the specific angular momentum l of the captured material is very approximately that of a flat circular disc of radius r_{acc} rotating rigidly about a diameter:

$$l \sim \tfrac{1}{4} r_{acc}^2 \omega. \tag{4.36}$$

This is smaller by a factor $\sim (r_{acc}/a)^2$ than the corresponding value for Roche lobe accretion; thus, we expect that the chances of disc formation are smaller in this case. To make this more precise we again define a circularization radius R_{circ} given by $l = (GM_n R_{circ})^{1/2}$. From (4.36) and (4.33), we get

$$R_{circ} \cong \frac{G^3 M_n^3 \omega^2}{v_w^8}. \tag{4.37}$$

Setting $v_w^2 = \lambda(r) v_{esc}^2(R_E)$ with $\lambda(r) \sim 1$ (cf. (4.31)), we find from (4.31) and Kepler's law (4.1)

$$\frac{R_{circ}}{a} = \frac{M_n^3(M_n+M_E)}{16\lambda^4(a)M_E^4}\left(\frac{R_E}{a}\right)^4 \tag{4.38}$$

which can be compared with (4.17) for the Roche lobe case.

The first point to note about R_{circ} given by (4.37) and (4.38) is that it is very uncertain: theory and observations of stellar winds are not yet in such good accord that one can be confident about any wind law $\lambda(r)$, not to mention the problem of defining r_{acc} mentioned above. Since v_w appears to the eighth power ($\lambda^4(a)$) in the formulae for R_{circ}, our lack of knowledge of λ can have a serious effect on estimates of R_{circ}. As a numerical example, let us take $R = 0.5a$ and $M_E = 10M_n$. Then

$$R_{circ} = \frac{13}{\lambda^4(a)}\left(\frac{M_E}{M_\odot}\right)^{1/3} P_{day}^{1/3} \text{ km}$$

where we have used Kepler's law (4.2) for a. If, say, $M_E = 10M_\odot$ and $P = 10$ days, this comfortably exceeds the radius (~ 10 km) of a neutron star for $\lambda(a) = 1$. But a change in $\lambda(a)$ by a factor of less than two completely alters this result. Thus, it is no easy matter to decide if disc formation occurs in stellar wind accretion: if the neutron star has a strong magnetic field, as is often the case, the problem is further complicated, since R_{circ} must exceed the radius at which the magnetic field begins to control the flow if a disc is to form (see Section 6.3). We recall, however, that provided the necessary condition on R_{circ} is satisfied, the resulting disc can be much larger than R_{circ} because of angular momentum transport by viscous torques, as noted at the end of Section 4.5. This possibility also means that the accretion process can, in principle, extract angular momentum and so 'spin down' the neutron star, even though accretion is taking place. (See Section 6.3 for a further discussion.)

Three further possible modes of mass transfer in close binary systems are worth a mention. The first is that of non-synchronous rotation: if the 'normal' star in a close binary rotates asynchronously, its surface need not lie on a Roche equipotential; indeed, it can lose matter to its companion when smaller than its Roche lobe. Asynchronism of this kind is expected to be a rather short-lived phenomenon, as tidal effects tend to enforce corotation. A similarly short-lived situation occurs when the orbits of the stars are eccentric. (Indeed, the circularization and synchronization timescales are comparable.) This can come about, for example, as the result of the supernova which precedes the formation of a neutron star: if this involves a sufficiently large fraction of the total mass ($\lesssim 0.5$) the system may become almost unbound, with a very eccentric orbit. If the periastron separation of the neutron star and the companion is sufficiently small, matter can be pulled off the envelope of the companion, giving a burst of accretion each time the neutron star reaches periastron, even though for most of the orbit the stars are too far apart for accretion to occur. Eccentric orbits for the neutron stars in some X-ray binaries are observed, as shown by Doppler analysis of the X-ray pulses (see Section 6.3). Finally, it is possible that in some X-ray sources the mass transfer process is actually stimulated by the X-ray emission itself: heating of the part of the companion star facing the accreting compact component via absorption of some of the X-rays can cause the companion to lose mass at an accelerated rate.

5
Accretion discs

5.1 Introduction

In the previous chapter we have seen that in many binary systems undergoing mass transfer the accreting material will have sufficient angular momentum to form an accretion disc. In this chapter we shall study these discs in detail. In many cases the disc flow is confined so closely to the orbital plane that to a first approximation one can regard the disc as a two-dimensional gas flow. This *thin disc approximation*, which will be made quantitative in the course of our discussion, has proved very successful, and allows quite an elaborate theory to be developed. Much of the present interest in accretion discs stems from the encouraging results of comparison between this theory and observations of close binary systems.

5.2 Radial disc structure

Let us consider the dynamics of a thin disc. We assume that in cylindrical polar coordinates (R, ϕ, z) the matter lies very close to the plane $z = 0$. We assume that the matter moves with angular velocity Ω in circles about the accreting star which has mass $M = M_1 M_\odot$ and radius R_*. Often the angular velocity will have the Keplerian value

$$\Omega = \Omega_K(R) = \left(\frac{GM}{R^3}\right)^{1/2} \tag{5.1}$$

so that the circular velocity is then

$$v_\phi = R\Omega_K(R). \tag{5.2}$$

In addition to v_ϕ, the gas is assumed to possess a small radial 'drift' velocity v_R, which is negative near the central star, so that matter is being accreted. In general, v_R will be a function of both radius R and time t, because we shall want to be able to treat time-varying situations. The disc is then characterized by its *surface density* $\Sigma(R, t)$, which is the mass per unit surface area of the disc, given by integrating the gas density ρ in the z-direction (see Section 4.6).

We now exploit these assumptions to write conservation equations for the mass and angular momentum transport in the disc due to the radial drift motions. An annulus of the disc material lying between R and $R + \Delta R$ has total mass $2\pi R \Delta R \Sigma$

and total angular momentum $2\pi R\Delta R\Sigma R^2\Omega$. The rate of change of both of these quantities is given by the net flow from the neighbouring annuli. For the mass of the annulus,

$$\frac{\partial}{\partial t}(2\pi R\Delta R\Sigma) = v_R(R, t)2\pi R\Sigma(R, t) - v_R(R+\Delta R, t)$$

$$\times 2\pi(R+\Delta R)\Sigma(R+\Delta R, t)$$

$$\cong -2\pi\Delta R \frac{\partial}{\partial R}(R\Sigma v_R).$$

Thus, in the limit $\Delta R \to 0$ we get the mass conservation equation

$$R\frac{\partial \Sigma}{\partial t} + \frac{\partial}{\partial R}(R\Sigma v_R) = 0. \tag{5.3}$$

The conservation equation for angular momentum is similar, except that we must include the transport due to the net effects of the viscous torques $G(R, t)$ (equation (4.23)). Thus,

$$\frac{\partial}{\partial t}(2\pi R\Delta R\Sigma R^2\Omega) = v_R(R, t)2\pi R\Sigma(R, t)R^2\Omega(R)$$

$$-v_R(R+\Delta R, t)2\pi(R+\Delta R)\Sigma(R+\Delta R, t)$$

$$\times (R+\Delta R)^2\Omega(R+\Delta R) + \frac{\partial G}{\partial R}\Delta R$$

$$\cong -2\pi\Delta R\frac{\partial}{\partial R}(R\Sigma v_R R^2\Omega) + \frac{\partial G}{\partial R}\Delta R.$$

Again proceeding to the limit $\Delta R \to 0$, we get

$$R\frac{\partial}{\partial t}(\Sigma R^2\Omega) + \frac{\partial}{\partial R}(R\Sigma v_R R^2\Omega) = \frac{1}{2\pi}\frac{\partial G}{\partial R}. \tag{5.4}$$

Equations (5.3) and (5.4), together with the relation from Section 4.6

$$G(R, t) = 2\pi R v \Sigma R^2 \Omega' \tag{4.23}$$

defining the torque, and a relation for v in terms of the other variables, determine the disc structure in the radial direction. Using (5.3), (5.4) can be simplified to give

$$R\Sigma v_R(R^2\Omega)' = \frac{1}{2\pi}\frac{\partial G}{\partial R}, \tag{5.5}$$

where we have assumed $\partial\Omega/\partial t = 0$. This will hold for orbits in a fixed gravitational potential. Combining equations (5.3) and (5.5) to eliminate v_R yields

$$R\frac{\partial \Sigma}{\partial t} = -\frac{\partial}{\partial R}(R\Sigma v_R) = -\frac{\partial}{\partial R}\left[\frac{1}{2\pi(R^2\Omega)'}\frac{\partial G}{\partial R}\right].$$

Radial disc structure

Only at the next step do we use the assumption of Keplerian, rather than merely circular, orbits. Substituting for G from (4.23) and using (5.1), we get finally

$$\frac{\partial \Sigma}{\partial t} = \frac{3}{R}\frac{\partial}{\partial R}\left\{R^{1/2}\frac{\partial}{\partial R}[\nu\Sigma R^{1/2}]\right\}. \tag{5.6}$$

This is the basic equation governing the time evolution of surface density in a Keplerian disc. In general, it is a *nonlinear* diffusion equation for Σ, because ν may be a function of local variables in the disc, i.e. Σ, R and t. Given a solution of (5.6), v_R follows using (5.5) and (4.23):

$$v_R = -\frac{3}{\Sigma R^{1/2}}\frac{\partial}{\partial R}[\nu\Sigma R^{1/2}]. \tag{5.7}$$

Clearly, to make further progress we need some prescription for ν. As we stressed in the last chapter, there is no particularly cogent choice here. We gain some insight into the evolution of discs by choosing forms for ν which render (5.6) tractable; but we should remember that this is in general just a mathematical exercise designed to show how the disc 'machinery' works, rather than a physically based prediction of the theory. First, we note that (5.6) is linear in Σ if ν is independent of Σ. Moreover, if ν varies as a power of R, (5.6) can be solved by separation of variables.

As an example, suppose $\nu = $ constant. Then (5.6) can be rewritten as

$$\frac{\partial}{\partial t}(R^{1/2}\Sigma) = \frac{3\nu}{R}\left(R^{1/2}\frac{\partial}{\partial R}\right)^2 (R^{1/2}\Sigma).$$

Thus, setting $s = 2R^{1/2}$, the equation becomes

$$\frac{\partial}{\partial t}(R^{1/2}\Sigma) = \frac{12\nu}{s^2}\frac{\partial^2}{\partial s^2}(R^{1/2}\Sigma).$$

Hence we can write $R^{1/2}\Sigma = T(t)S(s)$, finding

$$\frac{T'}{T} = \frac{12\nu}{s^2}\frac{S''}{S} = \text{constant} = -\lambda^2.$$

The separated functions T and S are therefore exponentials and Bessel functions respectively. It is interesting to find the Green function, which is, by definition, the solution for $\Sigma(R, t)$ taking as the initial matter distribution a ring of mass m at $R = R_0$:

$$\Sigma(R, t=0) = \frac{m}{2\pi R_0}\delta(R-R_0)$$

where $\delta(R-R_0)$ is the Dirac delta function. Standard methods give

$$\Sigma(x, \tau) = \frac{m}{\pi R_0^2}\tau^{-1}x^{-1/4}\exp\left\{-\frac{(1+x^2)}{\tau}\right\}I_{1/4}(2x/\tau) \tag{5.8}$$

where $I_{1/4}(z)$ is a modified Bessel function and we have used the dimensionless radius and time variables $x = R/R_0$, $\tau = 12\nu t R_0^{-2}$. Fig. 19 shows $\Sigma(x, \tau)$ as a function of x for various values of τ. We can see from the figure that viscosity has the effect of spreading the original ring in radius on a typical timescale

$$t_{\text{visc}} \sim R^2/\nu \tag{5.9}$$

at each radius R, obtained by setting $(1+x^2)\tau^{-1} \sim x^2\tau^{-1} \sim 1$ in the argument of the exponential in (5.8). Note that $\tau \sim t/t_{\text{visc}}(R_0)$. From (5.7),

$$v_R \sim \nu/R; \tag{5.10}$$

therefore (5.9) may be re-expressed as

$$t_{\text{visc}} \sim R/v_R; \tag{5.11}$$

t_{visc} is known as the *viscous* or *radial drift timescale*, since in the form (5.11) it gives an estimate of the timescale for a disc annulus to move a radial distance R.

From the diffusion equation (5.6) we see that if in some region of the disc Σ has spatial gradients characterized by a lengthscale $l \neq R$ in general, t_{visc} will be given by $\sim l^2/\nu$. In particular, density enhancements involving sharp spatial gradients (small l) diffuse more quickly than smoother density distributions.

From (5.7) and (5.8), with $\nu = $ constant, we get

$$v_R = -3\nu \frac{\partial}{\partial R} \ln(R^{1/2}\Sigma)$$

$$= -\frac{3\nu}{R_0}\frac{\partial}{\partial x} \ln(x^{1/2}\Sigma)$$

$$= -\frac{3\nu}{R_0}\frac{\partial}{\partial x}\left\{\tfrac{1}{4}\ln x - \frac{(1+x^2)}{\tau} + \ln I_{1/4}\left(\frac{2x}{\tau}\right)\right\}.$$

The asymptotic behaviour

$$I_{1/4}(z) \propto z^{-1/2} e^z, \quad z \gg 1$$
$$\propto z^{1/4}, \quad z \ll 1$$

now shows that

$$v_R \sim \frac{3\nu}{R_0}\left\{\frac{1}{4x} + \frac{2x}{\tau} - \frac{2}{\tau}\right\} > 0 \quad \text{for } 2x \gg \tau$$

and

$$v_R \sim \frac{-3\nu}{R_0}\left\{\frac{1}{2x} - \frac{2x}{\tau}\right\} < 0 \quad \text{for } 2x \ll \tau.$$

Hence the outer parts of the matter distribution ($2x \gg \tau$) move outwards, taking away the angular momentum of the inner parts, which move inwards towards the

Steady thin discs

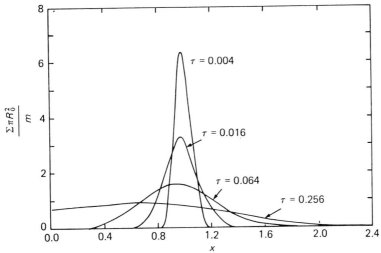

Figure 19. A ring of matter of mass m placed in a Kepler orbit at $R = R_0$ spreads out under the action of viscous torques. The surface density Σ is shown as a function of $x = R/R_0$ and the dimensionless time variable $\tau = 12\nu t R_0^{-2}$, with ν the viscosity. (Reproduced, with permission, from J. E. Pringle, *Ann Rev. Astron. Astrophys.*, Vol. 19, © 1981 by Annual Reviews Inc.)

accreting star. Moreover, the radius at which v_R changes sign moves outwards itself, for if we choose some point at which x is initially much greater than τ, after a sufficient time this same point will have x much less than τ. Thus parts of the matter distribution which are at radii $R > R_0$ just after the initial release of the ring ($t \sim 0$) at first move to larger radii, but later begin themselves to lose angular momentum to parts of the disc at still larger radii and thus drift inwards. At very long times ($\tau \gg 1$) after the initial release, almost all of the original mass m has accreted on to the central star ($R \sim 0$); all of the original angular momentum has been carried to very large radii by a very small fraction of the mass.

5.3 Steady thin discs

We have seen in the previous section that changes in radial structure in a thin disc occur on timescales $\sim t_{\text{visc}} \sim R^2/\nu$. In many systems external conditions (e.g. mass transfer rate) change on timescales rather longer than t_{visc}. In this case, the disc will settle to a steady-state structure, which we can examine by setting $\partial/\partial t = 0$ in the conservation equations (5.3) and (5.4).

From (5.3) we get, with $\partial/\partial t = 0$,

$$R\Sigma v_R = \text{constant}.$$

Clearly, as an integral of the mass conservation equation, this represents the constant inflow of mass through each point of the disc. Since $v_R < 0$ we write

$$\dot{M} = 2\pi R \Sigma (-v_R) \tag{5.12}$$

where $\dot{M}$ (g s^{-1}) is the accretion rate.

72   Accretion discs

From the angular momentum equation (5.4) we get, with $\partial/\partial t = 0$,

$$R\Sigma v_R R^2 \Omega = \frac{G}{2\pi} + \frac{C}{2\pi}$$

with C a constant. Using (4.23) for G we get

$$-\nu\Sigma\Omega' = \Sigma(-v_R)\Omega + C/(2\pi R^3). \tag{5.13}$$

The constant C is related to the rate at which angular momentum flows into the compact star, or equivalently, the couple exerted by the star on the inner edge of the disc. For example, let us suppose that our disc extends all the way down to the surface $R = R_*$ of the central star. In a realistic situation, the star must rotate more slowly than break-up speed at its equator, i.e. with angular velocity

$$\Omega_* < \Omega_K(R_*). \tag{5.14}$$

In this case the angular velocity of the disc material remains Keplerian and thus increases inwards, until it begins to decrease to the value Ω_* in a 'boundary layer' of radial extent b. Hence, there exists a radius $R = R_* + b$ at which $\Omega' = 0$. We shall show (Section 6.2) that $b \ll R_*$, and thus that Ω is very close to its Keplerian value at the point where $\Omega' = 0$, i.e.

$$\Omega(R_* + b) = \left(\frac{GM}{R_*^3}\right)^{1/2} [1 + O(b/R_*)]. \tag{5.15}$$

If b is instead comparable to R_*, the thin disc approximations all break down at $R = R_* + b$: this is no longer a boundary layer, but some kind of thick disc. At $R = R_* + b$, (5.13) becomes

$$C = 2\pi R_*^3 \Sigma v_R \Omega(R_* + b)|_{R_*+b}$$

which, using (5.15) and (5.12), implies

$$C = -\dot{M}(GMR_*)^{1/2} \tag{5.16}$$

to terms of order b/R_*. Substituting this into (5.13) and setting $\Omega = \Omega_K$ (equation (5.1)), we find

$$\nu\Sigma = \frac{\dot{M}}{3\pi}\left[1 - \left(\frac{R_*}{R}\right)^{1/2}\right]. \tag{5.17}$$

Although this is the usual result, there are cases where it is not valid, i.e. where C does not have the value (5.16). For example, if the central star has a sufficiently strong magnetic field that it can control the flow of disc material out to radii R such that

$$v_\phi = R\Omega_* > v_\phi(\text{Kepler}) = R\Omega_K(R)$$

then (5.17) is invalid; in this case, the disc would be pulled around by the star, and indeed would extract rotational energy and angular momentum from it.

Steady thin discs

The expression (5.17) for a steady-state disc with a slowly rotating central star (or the equivalent one for a disc with a central torque) has an important and elegant result. If we set $\Omega = \Omega_K$ in the expression (4.26) for the viscous dissipation per unit disc face area $D(R)$ and use (5.17), we obtain

$$D(R) = \frac{3GM\dot{M}}{8\pi R^3}\left[1-\left(\frac{R_*}{R}\right)^{1/2}\right]. \tag{5.18}$$

Hence, the energy flux through the faces of a steady thin disc is independent of viscosity. This is an important result: $D(R)$ is a quantity of prime observational significance, and (5.18) shows that its dependence on $\dot{M}$, R, etc. is known, even though we are at present quite ignorant as to the physical nature of the viscosity v. The result relies on the implicit assumption that v can adjust itself to give the required $\dot{M}$. The independence of $D(R)$ from v has come about because we were able to use conservation laws to eliminate v; clearly the other disc properties (e.g. Σ, v_R, etc.) do depend on v.

Let us use (5.18) to find the luminosity produced by the disc between radii R_1 and R_2. This is

$$L(R_1, R_2) = 2\int_{R_1}^{R_2} D(R) 2\pi R\,dR,$$

the factor 2 coming from the two sides of the disc. From (5.18) we get

$$L(R_1, R_2) = \frac{3GM\dot{M}}{2}\int_{R_1}^{R_2}\left[1-\left(\frac{R_*}{R}\right)^{1/2}\right]\frac{dR}{R^2}$$

which can be evaluated by putting $y = R_*/R$, with the result

$$L(R_1, R_2) = \frac{3GM\dot{M}}{2}\left\{\frac{1}{R_1}\left[1-\tfrac{2}{3}\left(\frac{R_*}{R_1}\right)^{1/2}\right]-\frac{1}{R_2}\left[1-\tfrac{2}{3}\left(\frac{R_*}{R_2}\right)^{1/2}\right]\right\}. \tag{5.19}$$

Letting $R_1 = R_*$ and $R_2 \to \infty$, we obtain the luminosity of the whole disc:

$$L_{\text{disc}} = \frac{GM\dot{M}}{2R_*} = \tfrac{1}{2}L_{\text{acc}} \tag{5.20}$$

which is our old result (4.22). As we explained in Section 4.5, the fact that matter at R_* still retains as kinetic energy half of the potential energy it has lost in spiralling in, means that half of L_{acc} is still available to be radiated from the boundary layer itself, which is therefore just as important as the disc for the total emission. We shall discuss the boundary layer in the next chapter.

Before investigating further the structure of steady discs we should consider the dissipation rate (5.18) a little more closely. The total rate at which energy is dissipated in a ring between R and $R+dR$ is

$$2 \times 2\pi R\,dR D(R) = \frac{3GM\dot{M}}{2R^2}\left[1-\left(\frac{R_*}{R}\right)^{1/2}\right]dR.$$

Of this total, $GM\dot{M}\,dR/2R^2$ comes from the rate of release of gravitational binding energy between $R+dR$ and R. The remainder

$$\frac{GM\dot{M}}{R^2}\left[1-\tfrac{3}{2}\left(\frac{R_*}{R}\right)^{1/2}\right]dR \qquad (5.21)$$

is convected into the ring from smaller R. Thus, for radii $R > 9R_*/4$ the rate of energy release exceeds that due to the release of binding energy; for $R \gg R_*$ this viscous transport of energy released lower down in the potential well is twice as important as the local binding energy loss. The viscous transport only redistributes the energy release within the disc; it cannot change the total rate. To compensate for the higher rate in the outer parts of the disc the energy released in the region $R_* \leqslant R < 9R_*/4$ is *less* than the binding energy loss, because the expression (5.21) is negative.

We are now in a position to check some of the assumptions we have made in constructing our steady-state disc. In particular, we may see if the azimuthal velocity v_ϕ conforms to the Keplerian value (5.2), and whether the assumption that the disc is thin is self-consistent.

We consider first the structure of the disc in the z-(vertical) direction. As there is essentially no flow in this direction, hydrostatic equilibrium must hold:

$$\frac{1}{\rho}\frac{\partial P}{\partial z} = \frac{\partial}{\partial z}\left[\frac{GM}{(R^2+z^2)^{1/2}}\right].$$

(This is the z-component of the Euler equation (2.3) with all the velocity terms neglected.) For a thin disc, $z \ll R$ and this becomes

$$\frac{1}{\rho}\frac{\partial P}{\partial z} = -\frac{GMz}{R^3}. \qquad (5.22)$$

If the typical scale-height of the disc in the z-direction is H, we may set $\partial P/\partial z \sim P/H$ and $z \sim H$; the 'thin' assumption amounts then to requiring $H \ll R$. With $P \sim \rho c_s^2$, where c_s is the sound speed, we find

$$H \cong c_s \left(\frac{R}{GM}\right)^{1/2} R \qquad (5.23)$$

so that we demand

$$c_s \ll \left(\frac{GM}{R}\right)^{1/2}, \qquad (5.24)$$

i.e. *for a thin disc we require that the local Kepler velocity should be highly supersonic*. This is clearly a condition on the temperature of the disc and thus, ultimately, on the cooling mechanism, which must be checked at each point of the disc and in many cases is satisfied. It need not hold throughout a disc, however, so that we must

The local structure of thin discs

consider cases where the thin disc approximation breaks down. In Section 5.6 we shall show that this occurs in the central regions of some discs. In Chapter 10, we discuss disc structures for which the 'thin disc' requirement is relaxed completely.

We can now show that *if the thin disc condition (5.24) holds, the circular matter velocity v_ϕ will be very close to the Keplerian value.* The radial component of the Euler equation (2.3) is

$$v_R \frac{\partial v_R}{\partial R} - \frac{v_\phi^2}{R} + \frac{1}{\rho}\frac{\partial P}{\partial R} + \frac{GM}{R^2} = 0. \tag{5.25}$$

Because of (5.24) we may neglect the pressure term $(1/\rho)\partial P/\partial R \sim c_s^2/R$ by comparison with the gravity term. To evaluate the $v_R \partial v_R/\partial R$ term we use (5.12) and (5.17) to obtain

$$v_R = -\frac{3\nu}{2R}\left[1-\left(\frac{R_*}{R}\right)^{1/2}\right]^{-1}. \tag{5.26}$$

Note that this is indeed of order ν/R as anticipated on dimensional grounds (5.10). For any reasonable viscosity, v_R is highly subsonic, since, adopting the α-parametrization (4.30), we have

$$v_R \sim \frac{\nu}{R} \sim \alpha c_s \frac{H}{R} \ll c_s. \tag{5.27}$$

Hence the $v_R \partial v_R/\partial R$ term in (5.25) is even smaller than the pressure gradient term. Defining the *Mach number* $\mathcal{M}$ (cf. (3.56)),

$$\mathcal{M} = v_\phi/c_s, \tag{5.28}$$

(5.25) implies

$$v_\phi = \left(\frac{GM}{R}\right)^{1/2}[1+O(\mathcal{M}^{-2})] \tag{5.29}$$

while (5.23) and (5.27) can be rewritten as

$$\left. \begin{array}{l} H \sim \mathcal{M}^{-1}R, \\ v_R \sim \alpha\mathcal{M}^{-1}c_s. \end{array} \right\} \tag{5.30}$$

Thus in a thin disc the circular velocity v_ϕ is Keplerian and highly supersonic; the radial drift velocity and vertical scale-height are self-consistently small.

5.4 The local structure of thin discs

If the thin disc approximation holds, the task of computing the detailed disc structure is enormously simplified. Both the pressure and temperature gradients are essentially vertical, so that the vertical and radial structures are largely decoupled. Thus we can treat the vertical disc structure at a given radius as if it were a one-dimensional version of stellar structure. As in stellar structure (see e.g. Kippenhahn & Weigert (1990)) we have the equations of hydrostatic equilibrium and energy transport to solve, with the radial disc structure only entering the calculation in the fixing of the local energy generation rate $D(R)$.

For an isothermal z-structure, the hydrostatic equation (5.22) would give

$$\rho(R, z) = \rho_c(R) \exp(-z^2/2H^2) \quad (5.31)$$

with H given by (5.23); here $\rho_c(R)$ is the density on the central disc plane $z = 0$. Thus in general we can define a central disc density approximately by

$$\left. \begin{array}{l} \rho = \Sigma/H, \\ H = Rc_s/v_\phi. \end{array} \right\} \quad (5.32)$$

The sound speed c_s is given by

$$c_s^2 = P/\rho \quad (5.33)$$

where in general the pressure P is the sum of gas and radiation pressures:

$$P = \frac{\rho \kappa T_c}{\mu m_p} + \frac{4\sigma}{3c} T_c^4. \quad (5.34)$$

Here σ is the Stefan–Boltzmann constant, and we have assumed that the temperature $T(R, z)$ is close to the central temperature $T_c(R) = T(R, 0)$.

The temperature T_c must itself be given by an energy equation relating the energy flux in the vertical direction to the rate of generation of energy by viscous dissipation. As in stars, the vertical energy transport mechanism may be either radiative or convective, depending on whether or not the temperature gradient required for radiative transport is smaller or greater than the gradient given by the adiabatic assumption (2.10). We shall touch on this question later (Section 5.8), and for the time being assume that the transport is radiative: this is indeed true in many important cases. Because of the thin disc approximation, the disc medium is essentially 'plane-parallel' at each radius, so that the temperature gradient is effectively in the z-direction, as we pointed out above. Under these circumstances, the flux of radiant energy through a surface z-constant is given by

$$F(z) = \frac{-16\sigma T^3}{3\kappa_R \rho} \frac{\partial T}{\partial z} \quad (5.35)$$

where κ_R is the Rosseland mean opacity. It is implicitly assuming in writing (5.35) that the disc is optically thick in the sense that

$$\tau = \rho H \kappa_R(\rho, T_c) = \Sigma \kappa_R \quad (5.36)$$

is $\gg 1$, so that the radiation field is locally very close to the blackbody value: once τ given by (5.36) becomes $\lesssim 1$ the expression (5.35) breaks down as the radiation can escape directly. The energy balance equation is

$$\partial F/\partial z = Q^+$$

The emitted spectrum

where Q^+ is the *volume* rate of energy production by viscous dissipation. Integrating, we have

$$F(H) - F(0) = \int_0^H Q^+(z) \, dz = D(R) \tag{5.37}$$

since the total dissipation rate through one half of the vertical structure must give the dissipation rate per unit face area $D(R)$ (equation (5.18)). From (5.35) we have

$$F(z) \sim (4\sigma/3\tau)T^4(z)$$

using (5.36), so that provided the central temperature greatly exceeds the surface temperature (i.e. $T_c^4 \gg T^4(H)$) (5.37) becomes approximately

$$\frac{4\sigma}{3\tau} T_c^4 = D(R) \tag{5.38}$$

which is our required energy equation.

To complete the set of equations we need a relation $\kappa_R = \kappa_R(\rho, T_c)$ for the opacity, equation (5.17) relating v and Σ to $\dot{M}$, and some relation (such as the α-prescription (4.30)) for the viscosity v. We collect these together:

$$\left. \begin{array}{l} 1. \ \rho = \Sigma/H; \\[4pt] 2. \ H = c_s R^{3/2}/(GM)^{1/2}; \\[4pt] 3. \ c_s^2 = P/\rho; \\[4pt] 4. \ P = \dfrac{\rho k T_c}{\mu m_p} + \dfrac{4\sigma}{3c} T_c^4; \\[4pt] 5. \ \dfrac{4\sigma T_c^4}{3\tau} = \dfrac{3GM\dot{M}}{8\pi R^3}\left[1 - \left(\dfrac{R_*}{R}\right)^{1/2}\right]; \\[4pt] 6. \ \tau = \Sigma \kappa_R(\rho, T_c) = \tau(\Sigma, \rho, T_c); \\[4pt] 7. \ v\Sigma = \dfrac{\dot{M}}{3\pi}\left[1 - \left(\dfrac{R_*}{R}\right)^{1/2}\right]; \\[4pt] 8. \ v = v(\rho, T_c, \Sigma, \alpha, \ldots). \end{array} \right\} \tag{5.39}$$

These eight equations can now be solved for the eight unknowns $\rho, \Sigma, H, c_s, P, T_c, \tau$ and v as functions of $\dot{M}, M, R$ and any parameter (e.g. α) appearing in the viscosity prescription 8. Once the equations (5.39) are solved, the radial drift velocity v_R can be read off from (5.26).

5.5 The emitted spectrum

Before proceeding to solve the disc structure equations (5.39) for a particular case (Section 5.6), let us note an extremely important consequence of the assumption that the disc is optically thick in the z-direction. If this holds, each element of the disc face radiates roughly as a blackbody with a temperature $T(R)$ given by equating the dissipation rate $D(R)$ per unit face area to the blackbody flux:

$$\sigma T^4(R) = D(R). \tag{5.40}$$

Using (5.18), we get

$$T(R) = \left\{\frac{3GM\dot{M}}{8\pi R^3 \sigma}\left[1 - \left(\frac{R_*}{R}\right)^{1/2}\right]\right\}^{1/4}. \qquad (5.41)$$

For $R \gg R_*$,

$$T = T_*(R/R_*)^{-3/4}$$

where

$$\left.\begin{aligned} T_* &= \left(\frac{3GM\dot{M}}{8\pi R_*^3 \sigma}\right)^{1/4} \\ &= 4.1 \times 10^4 \, \dot{M}_{16}^{1/4} M_1^{1/4} R_9^{-3/4} \text{ K} \\ &= 1.3 \times 10^7 \, \dot{M}_{17}^{1/4} M_1^{1/4} R_6^{-3/4} \text{ K.} \end{aligned}\right\} \qquad (5.42)$$

In (5.42) we have used $\dot{M}_{16} = \dot{M}/10^{16}$ g s^{-1}, $M_1 = M/M_\odot$, $R_9 = R_*/10^9$ cm, etc. to express T_* numerically for discs around white dwarfs ($M_1 \sim R_9 \sim \dot{M}_{16} \sim 1$) and neutron stars ($M_1 \sim R_6 \sim \dot{M}_{17} \sim 1$). Note that T_* is of the order of the blackbody temperature defined in Chapter 1. As discussed there, the order of magnitude of this temperature shows that we should expect accretion discs around white dwarfs and neutron stars to be ultraviolet and X-ray sources respectively, since $T(R)$ attains a maximum value $0.488 T_*$ at $R = (49/36) R_*$.

The temperature $T(R)$ plays an analogous role to the effective temperature of a star: very crudely, we can approximate the spectrum emitted by each element of area of the disc as

$$I_\nu = B_\nu[T(R)] = \frac{2h\nu^3}{c^2(e^{h\nu/kT(R)} - 1)} \text{ (erg s}^{-1}\text{ cm}^{-2}\text{ Hz}^{-1}\text{ sr}^{-1}).$$

The approximation involved here neglects the effect of the atmosphere of the disc (i.e. that part of the disc material at optical depths $\tau \lesssim 1$ from infinity) in redistributing the radiation over frequency ν. Hence, this prescription may not represent the detailed spectrum of the disc very well in particular frequency domains, especially those where the atmospheric opacity is a rapidly varying function of frequency. For an observer at a distance D whose line of sight makes an angle i to the normal to the disc plane (i.e for binary inclination i) the flux at frequency ν from the disc is

$$F_\nu = \frac{2\pi \cos i}{D^2} \int_{R_*}^{R_{\text{out}}} I_\nu R \, dR$$

where R_{out} is the outer radius of the disc, since a ring between radii R and $R + dR$ subtends solid angle $2\pi R \, dR \cos i / D^2$. With the blackbody assumption, we get

$$F_\nu = \frac{4\pi h \cos i \, \nu^3}{c^2 D^2} \int_{R_*}^{R_{\text{out}}} \frac{R \, dR}{e^{h\nu/kT(R)} - 1}. \qquad (5.43)$$

An important feature of this result is that F_ν is independent of the disc viscosity. This is a consequence of both the steady and blackbody assumptions. Since both of these are likely to be at least roughly valid for some systems, we should expect the

The emitted spectrum

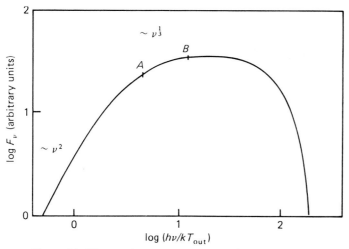

Figure 20. The continuum spectrum F_ν of a steady optically thick accretion disc radiating locally as a black body. Here the outer radius $R_{\text{out}} = 250 R_*$, and $T_{\text{out}} = T(R_{\text{out}})$. Points A and B have 'spectral index' $d(\ln F_\nu)/d(\ln \nu)$ equal to 0.38 and 0.28 respectively, showing the relative shortness of the '$\nu^{1/3}$' section of the continuum.

spectrum specified by (5.43) to give at least a crude representation of the observed spectrum for some systems, in the same way that stellar spectra are to a crude approximation blackbody: any serious discrepancy cannot be due to our ignorance of viscosity.

The spectrum given by (5.43) is shown in Fig. 20. The shape of this spectrum is easy to deduce from (5.43). First, for frequencies $\nu \ll kT(R_{\text{out}})/h$ the Planck function B_ν takes the Rayleigh–Jeans form $2kT\nu^2/c^2$; hence (5.43) gives $F_\nu \propto \nu^2$. For $\nu \gg kT_*/h$ each Planck function B_ν assumes the Wien form $2h\nu^3 c^{-2} e^{-h\nu/kT}$: the integral in (5.43) is dominated by the hottest parts of the disc ($T \sim T_*$) and the integrated spectrum is exponential. For intermediate frequencies ν such that $kT(R_{\text{out}})/h \ll \nu \ll kT_*/h$ we let $x = h\nu/kT(R) \cong (h\nu/kT_*)(R/R_*)^{3/4}$. Then (5.43) becomes approximately

$$F_\nu \propto \nu^{1/3} \int_0^\infty \frac{x^{5/3}}{e^x - 1} dx \propto \nu^{1/3}$$

since the upper limit in the integral is $h\nu/kT(R_{\text{out}}) \gg 1$ and the lower limit is $h\nu/kT_* \ll 1$. Thus the integrated spectrum F_ν (Fig. 20) is a stretched-out blackbody; the 'flat' part $F_\nu \propto \nu^{1/3}$ is sometimes considered a characteristic disc spectrum. However, unless $T_{\text{out}} = T(R_{\text{out}})$ is appreciably smaller than T_* this part of the curve may be quite short and the spectrum is not very different from a blackbody. In Section 5.7 we shall see that the disc continuum is likely to differ considerably in detail from this simple picture.

5.6 The structure of steady α-discs (the 'standard model')

We shall now solve the steady disc equations (5.39) for a simple case. This will allow us to check if the blackbody and thin disc approximations are valid for this particular case, and will illustrate the general procedure for solving these equations. The disc structure we consider was first investigated by Shakura and Sunyaev and is associated with their names. To specify a disc model using the system (5.39) we need give only a viscosity prescription and a relation for the opacity – equations 8 and 6 of (5.39). We take the α-prescription (4.30):

$$v = \alpha c_s H, \tag{4.30}$$

and assume that ρ and T_c are such that the Rosseland mean opacity is well approximated by Kramers' law:

$$\kappa_R = 6.6 \times 10^{22} \rho T_c^{-7/2} \text{ cm}^2 \text{ g}^{-1}. \tag{5.44}$$

In addition, we shall drop the radiation pressure term $(4\sigma/3c)T_c^4$ from the equation of state, equation 4 in (5.39): we can use the solution to check where this assumption is valid.

With our choice for v and κ_R, the system (5.39) is algebraic and can be solved straightforwardly. To show how this is done without the complication of inessential details we omit numerical factors, put $f^4 = 1 - (R_*/R)^{1/2}$ and write D for the right hand side of 5. Thus from 5 (using 6, (5.44) and 2) we get

$$T_c^8 \propto \Sigma^2 D M^{1/2} R^{-3/2}$$

where we have used 3 and 4 to put $c_s \propto T_c^{1/2}$. Combining this with 7 and 8 now gives

$$\Sigma \propto \alpha^{-4/5} \dot{M}^{7/10} M^{1/4} R^{-3/4} f^{14/5}$$

which is one of our required solutions: the others now follow easily and we can use (5.26) to give v_R. To get an idea of the typical sizes of the disc quantities, we express the solutions in terms of $R_{10} = R/(10^{10} \text{ cm})$, $M_1 = M/M_\odot$ and $\dot{M}_{16} = \dot{M}/(10^{16} \text{ g s}^{-1})$. (We take $\mu = 0.615$, appropriate to a fully ionized 'cosmic' mixture of gases.) The Shakura–Sunyaev disc solution is

$$\left.\begin{aligned}
\Sigma &= 5.2 \alpha^{-4/5} \dot{M}_{16}^{7/10} M_1^{1/4} R_{10}^{-3/4} f^{14/5} \text{ g cm}^{-2}, \\
H &= 1.7 \times 10^8 \alpha^{-1/10} \dot{M}_{16}^{3/20} M_1^{-3/8} R_{10}^{9/8} f^{3/5} \text{ cm}, \\
\rho &= 3.1 \times 10^{-8} \alpha^{-7/10} \dot{M}_{16}^{11/20} M_1^{5/8} R_{10}^{-15/8} f^{11/5} \text{ g cm}^{-3}, \\
T_c &= 1.4 \times 10^4 \alpha^{-1/5} \dot{M}_{16}^{3/10} M_1^{1/4} R_{10}^{-3/4} f^{6/5} \text{ K}, \\
\tau &= 33 \alpha^{-4/5} \dot{M}_{16}^{1/5} f^{4/5}, \\
v &= 1.8 \times 10^{14} \alpha^{4/5} \dot{M}_{16}^{3/10} M_1^{-1/4} R_{10}^{3/4} f^{6/5} \text{ cm}^2 \text{ s}^{-1}, \\
v_R &= 2.7 \times 10^4 \alpha^{4/5} \dot{M}_{16}^{3/10} M_1^{-1/4} R_{10}^{-1/4} f^{-14/5} \text{ cm s}^{-1}, \\
\text{with } f &= \left[1 - \left(\frac{R_*}{R}\right)^{1/2}\right]^{1/4}.
\end{aligned}\right\} \tag{5.45}$$

Before checking the validity of the assumptions made in deriving this solution, a number of remarks are in order. First, we note that the unknown α-parameter does not enter any of the expressions for the disc quantities with a high power. This means that the very reasonable-looking orders of magnitudes of these quantities are not particularly sensitive to the actual value of α, which is encouraging. On the other hand, it also means that we cannot expect to discover the typical size of α by direct comparison of steady-state disc theory with observations. Second, we see from (5.45) that

$$H/R = 1.7 \times 10^{-2} \alpha^{-1/10} \dot{M}_{16}^{3/10} M_1^{-3/8} R_{10}^{1/8} f^{3/5} \qquad (5.46)$$

so the disc is indeed thin as long as (5.45) holds. In agreement with this, the radial drift velocity $v_R \sim 0.3$ km s^{-1} is highly subsonic ($c_s \sim 10$ km s^{-1}) while the Kepler velocity $v_\phi \sim 1000$ km s^{-1} is very supersonic. From the expression for τ we see that the disc is certainly optically thick for any reasonable accretion rate, provided that the solution (5.45) is valid. Moreover, the central and surface temperatures T_c and $T(R)$ are in the ratio $\sim \tau^{1/4} \sim 2$, so the disc is roughly uniform in the vertical direction. Thus, unless the solution breaks down (in particular, the Kramers law (5.44) may become invalid), the disc can extend out to quite large radii, of the order of the Roche lobe of the accreting star. Even for a very large disc, extending out to $R \sim 10^{11}$ cm (cf. (4.2)), the solution (5.45) implies that the mass in the disc at any one time is

$$M_{\rm disc} = 2\pi \int_{R_*}^{R_{\rm out}} \Sigma R \, {\rm d}R \lesssim (10^{-10} M_\odot) \alpha^{-4/5} \dot{M}_{16}^{7/10}. \qquad (5.47)$$

Hence unless α is very small ($\sim 10^{-10}$) the disc mass is negligible compared to that of the accreting star. Similarly, the neglect of self-gravity of the disc is fully justified: at the disc surface the z-component of gravitational force due to the central star is GMH/R^3 at radius R, which compares to the self-gravity contribution $\sim G\rho H^3/H^2 \sim G\rho H$, treating the disc as an infinite uniform plane. Thus the condition for the neglect of self-gravitation (and therefore for stability against breaking up into self-gravitating clumps) is

$$\rho \ll M/R^3 \qquad (5.48)$$

which we see from (5.45) is amply satisfied for all disc radii unless again α is very small ($\sim 10^{-10}$).

In spite of this list of desirable properties, it should not be forgotten that the α-disc solutions are based on the entirely *ad hoc* viscosity prescription (4.30). As (4.30) is merely a parametrization, α can in general be a function of $\dot{M}$, M, R, and z. The z-dependence can be ignored if we interpret the α appearing in our solutions (5.45) as a 'vertically averaged' α, because the disc equations (5.39) essentially treat the vertical structure as uniform (in particular, $T(R, z) \cong T_c(R)$ throughout.) Since the structure equations (5.39) are algebraic, the dependence of α on $\dot{M}$, M and R need not be constrained in obtaining the solution (5.45). This means that (5.45) is scarcely a solution at all in a strict mathematical sense: we do not know how α depends on $\dot{M}$, M or R, so the dependence of all the other disc quantities is unknown. This type of situation is of common occurrence in astrophysics; often, we parametrize some unknown element of a theory in order to see how sensitive our conclusions are to our ignorance. Thus the real importance of the solution (5.45) is that for $\alpha \lesssim 1$ we get believable orders of magnitude for the disc quantities, and this is reinforced by the comparatively weak dependence on α noted above. What we cannot necessarily expect the solution (5.45) to do is to predict correctly the way the disc quantities change when $\dot{M}$, M or R are varied.

As an example, consider the ratio of disc height to radius (5.46). For $R \gg R_*$ we have

$$H/R \propto \alpha^{-1/10} R^{1/8}.$$

Hence H/R will be an increasing function of R provided α increases with R more slowly than $\sim R^{5/4}$; the disc will then be concave (Fig. 21). In this case 'hard' radiation from the central regions $R \sim R_*$ of the disc can heat the disc surfaces and cause potentially observable effects. However, we have no *a priori* guarantee that the 'real' α does not increase rapidly with R, so that the disc might be convex in some regions. What we *can* be reasonably confident about is that H/R is always numerically small whenever (5.46) holds.

In this spirit, we can now check where we expect the assumptions (Kramers' opacity and the neglect of radiation pressure) involved in deriving (5.45) to break down. From (5.45) we get

$$\kappa_R \text{ (Kramers)} = \tau/\Sigma = 6.3 \dot{M}_{16}^{-1/2} M_1^{-1/4} R_{10}^{3/4} f^{-2}, \tag{5.49}$$

independent of α, which we must compare with other opacity sources to see where Kramers opacity is dominant. For an ionized gas at $T \gtrsim 10^4$ K the major competitive opacity is electron scattering:

$$\kappa_R \text{(electron scattering)} \cong \sigma_T/m_p \cong 0.4 \text{ cm}^2 \text{ g}^{-1} \tag{5.50}$$

so that Kramers' opacity dominates for

$$R \gtrsim 2.5 \times 10^8 \dot{M}_{16}^{2/3} M_1^{1/3} f^{8/3} \text{ cm}. \tag{5.51}$$

Except for very high accretion rates this is smaller than the radius of a white dwarf, so for the accretion discs in cataclysmic variables we expect Kramers' opacity to dominate in most of the disc. Near the outer edge of the disc ($R_{10} \sim 1$) T_c may drop significantly below the temperature ($\gtrsim 10^4$ K) for which Kramers' opacity is valid, as a result of hydrogen recombining at such temperatures. A more careful consideration of the opacity structure is required in such outer regions. From (5.45) the ratio of radiation to gas pressure is

$$\frac{P_r}{P_g} = 2.8 \times 10^{-3} \alpha^{1/10} \dot{M}_{16}^{7/10} R_{10}^{-3/8} f^{7/5} \tag{5.52}$$

which is small throughout the region where (5.51) holds. Hence the solution (5.45) is valid in this region.

Within the radius R given by (5.51) the dominant opacity is electron scattering, although we still have $P_r \ll P_g$. Since the opacity mechanism no longer involves the microscopic inverse of the processes emitting the radiation (free–free and bound–free emission) the emergent radiation need not be precisely blackbody, even

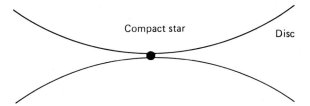

Figure 21. Section, ϕ = constant, (not to scale) through a thin accretion disc with α = constant, showing that the disc faces are concave in this case.

for quite large optical depths τ through the disc. As shown by Shakura and Sunyaev, the integrated disc spectrum in this case is somewhat flatter than in the Kramers region (Section 5.5).

From (5.52) we see that there is a tendency for the importance of radiation pressure to increase relative to that of gas pressure at smaller radii: Shakura and Sunyaev show that this tendency continues, and indeed intensifies, in the region where electron scattering dominates the opacity. Eventually, for radii

$$R \lesssim 24\alpha^{2/21} \dot{M}_{16}^{16/21} M_1^{-3/21} f^{4/21} \text{ km}, \tag{5.53}$$

radiation pressure exceeds gas pressure. Eventually such regions can exist only in discs around neutron stars and black holes, and then only for accretion rates $\dot{M}_{16} \gtrsim 1$. As the α-dependences in (5.51), (5.52) and (5.53) are so weak we can have some confidence in the general picture of the disc structure summarized in Fig. 22.

We can derive a simple and very important result for the scale-height H of any disc region dominated by radiation pressure from the disc structure equations (5.39). Combining equations 3, 4 and 5 of (5.39) and neglecting gas pressure, we get

$$c_s^2 = \frac{3GM\dot{M}\tau}{8\pi R^3 \rho c}\left[1 - \left(\frac{R_*}{R}\right)^{1/2}\right].$$

Since the opacity in such a region must be due to electron scattering, we have from 6 and 1

$$\tau = \Sigma \kappa_R(\text{e.s.}) \cong \rho H \sigma_T / m_p.$$

Thus,

$$c_s^2 \cong \frac{3GM\dot{M}\sigma_T H}{8\pi R^3 m_p c}\left[1 - \left(\frac{R_*}{R}\right)^{1/2}\right]$$

so that using 2 to eliminate c_s gives

$$H \cong \frac{3\sigma_T \dot{M}}{8\pi m_p c}\left[1 - \left(\frac{R_*}{R}\right)^{1/2}\right]. \tag{5.54}$$

Hence, the scale-height of radiation-pressure supported disc regions is essentially independent of R: this really results from the fact that the radiation pressure force $\sim T_c^4 \sim M\dot{M}R^{-3}$ is balancing the z-component of gravity $\propto MHR^{-3}$.

We can put (5.54) into a more instructive form by recalling from Chapter 1 that

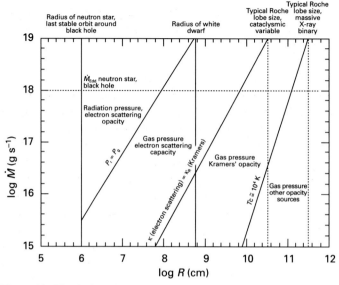

Figure 22. Physical regimes in steady α-discs around compact objects.

radiation pressure limits the accretion luminosity of any object to the Eddington value

$$L_{\text{Edd}} = \frac{4\pi G M m_p c}{\sigma_T}. \tag{1.2a}$$

By (1.5a), this means that there is a *critical accretion rate* $\dot{M}_{\text{crit}}$ which will supply this luminosity:

$$\dot{M}_{\text{crit}} = \frac{L_{\text{Edd}} R_*}{2\eta G M} = 1.5 \times 10^{18} \left(\frac{R_*}{3 \text{ km}}\right) \left(\frac{\eta}{0.1}\right)^{-1} \text{ g s}^{-1} \tag{5.55}$$

where η is the efficiency: we recall that $\eta \sim 0.1$ for neutron stars and black holes. Using (1.2a) we can write

$$\dot{M}_{\text{crit}} = 2\pi R_* m_p c / \eta \sigma_T$$

so that (5.54) becomes

$$H \cong \frac{3 R_*}{4\eta} \frac{\dot{M}}{\dot{M}_{\text{crit}}} \left[1 - \left(\frac{R_*}{R}\right)^{1/2}\right]. \tag{5.56}$$

This equation shows that *at accretion rates $\dot{M}$ approaching $\dot{M}_{\text{crit}}$ the thin disc approximation must break down near the central object*. For $\eta = 0.1$ and $\dot{M} = \dot{M}_{\text{crit}}$ we find that $H/R = 1$ for $R \cong 2 R_*$. Hence, for near-critical accretion rates, radiation pressure acts to make the innermost parts of the disc quasi-spherical. We note that this result is independent of viscosity, since only equations 1–6 of the set (5.39) were used in its derivation. This is a specific example of the breakdown of the thin disc approximation. It is easy to see from equations 2, 3 and 4 of (5.39) that the approximation will also fail if T_c becomes very large – for example, when the cooling mechanism is inefficient, so that the temperature in the disc greatly exceeds the blackbody temperature $T(R)$.

5.7 Steady discs: confrontation with observation

We must now see to what extent the picture of disc accretion presented above is supported by observation. The accretion flow in the immediate vicinity of the compact star (e.g. boundary layers, accretion columns), is likely to be different in character and will therefore be considered separately (Chapter 6). Hence the disc regions under discussion here have characteristic temperatures $T \lesssim 10^6$ K (equations (5.41) and (5.42)) and therefore radiate predominantly in the ultraviolet, optical and infrared. To study these regions observationally, we require binary systems in which the light is dominated by the disc contribution in one or more of these parts of the spectrum. X-ray binaries containing giant or supergiant companions, despite being amongst the first objects for which accretion was shown to be a plausible power supply, largely fail this test. The accretion luminosity, typically up to 10^{38}–10^{39} erg s^{-1}, in these systems is radiated almost entirely as hard X-rays, where it gives information about the accretion process very close to the compact object, but very little about any possible disc flow. The companion O or B giant or supergiant stars produce comparable luminosities (up to $\sim 10^{39}$ erg s^{-1}), but radiate them all in the ultraviolet and optical regions, completely swamping any disc contribution in most cases. Clearly, we have a much better chance of observing discs in systems where the companion (mass-donating) star is small and faint. There are two main classes with this property: the low-mass X-ray binaries (including the 'galactic bulge sources') and the cataclysmic variables (a portmanteau designation including the dwarf novae, nova-like variables, old novae, AM Her and intermediate polar systems). In both classes the companion star is thought to be of low mass ($\sim 0.1\,M_\odot$ to $\sim 1\,M_\odot$) and close to the lower main sequence. This is fairly well established for the cataclysmic variables and rather less so for the low-mass X-ray binaries, where the main difficulty lies in detecting the companion at all. The two classes of system differ in that the compact star is believed to be a neutron star for the low-mass X-ray binaries and a white dwarf in the case of the cataclysmic variables. Nonetheless, in the optical region of the spectrum the two types of system should have similar luminosities (cf. (5.41)) if our picture of disc accretion applies to both. Observationally, this is not the case: the low-mass X-ray binaries are ~ 100 times brighter optically than the cataclysmics. The extra luminosity must presumably be due to re-absorption of some of the X-ray emission, and suggests that these systems will not straightforwardly conform to the picture presented above. We are left then with only the cataclysmic variables as good candidates for testing our ideas about disc accretion. It is fortunate that these systems do indeed give manifold opportunities for such confrontations with observation; it is the successful outcome of these comparisons which is the basis for much of the present interest in accretion disc theory. In the main, therefore, the observations we shall discuss in this section are of cataclysmics: evidence from other types of system is helpful in showing how the standard picture may need modification in some cases.

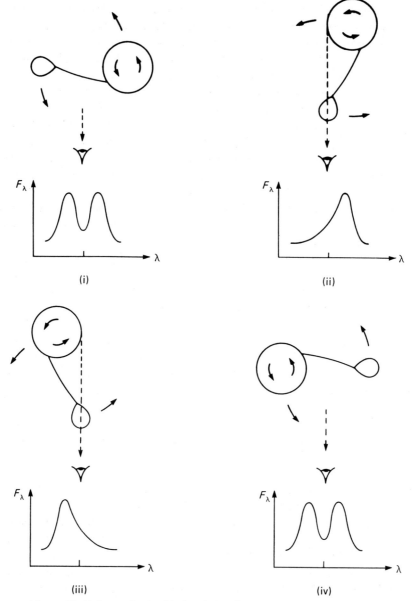

Figure 23. Eclipse of a doubled emission line in a cataclysmic variable. The advancing side of the disc is eclipsed first, (ii), leading to the disappearance of the blueward component of the line. As the eclipse proceeds this side of the disc re-emerges and the receding side is eclipsed, (iii) leading to the re-appearance of the blueward component and loss of the redward component. Outside eclipse ((i) and (iv)) the line appears double because of circular motion in the white dwarf's Roche lobe. Shifts of the line due to the rotation of the white dwarf about the centre of mass are not shown.

Let us begin with the evidence for perhaps the most basic requirement for accretion discs, namely circular motion of material around the accreting star. Conveniently for this purpose, cataclysmic variables show strong emission lines of hydrogen and helium, which because of the phasing of their periodic red- and blueshifts can be firmly identified as coming from the vicinity of the accreting component of the binary system. In many cataclysmics, particularly those where there is reason to believe that the orbital inclination is high, from, for example, the occurrence of eclipses, the emission lines are found to be double peaked with extensive wings. This is exactly what is expected for line emission from a rotating disc of optically thin gas: the velocity-splitting of the line peaks gives the projected circular velocity $v \sin i$ of the outer edge of the disc (and thus increases with inclination i), while the broad wings are easily explicable if the gas motion is Keplerian, and therefore quite rapid near the centre of the gaseous disc. In eclipsing systems where the eclipse is sufficiently long that good spectra can be obtained during it, the evidence for circular motion is still stronger. The double-peaked emission lines lose their blueward peaks near the beginning of the eclipse (ingress) and gradually regain them, losing the redward wing near the end of the eclipse (egress). This shows that the line-emitting gas around the white dwarf orbits in the same sense as the secondary star (see Fig. 23). The only reasonable explanation for these observations is that the gas is in roughly circular motion about the white dwarf. Interpreting this motion as Keplerian, in systems where we can estimate the inclination and masses of the two stars, leads to the conclusion that the emitting gas extends from quite close to the white dwarf out to near its Roche lobe. This good qualitative and quantitative agreement means that we can be fairly confident that disc flow of some kind occurs in these systems.

What is the evidence for *optically thick* disc accretion? There are two ways of trying to find such evidence: we may hope to account for the observed continuum spectrum of an object in terms of a theoretical disc spectrum, or, in eclipsing systems, try to discover the surface brightness distributions across a disc at various wavelengths and compare them with those predicted by theory. In both cases it is sensible to choose binaries where the disc light dominates all other system components at the relevant wavelengths. In the optical this means choosing cataclysmics with short orbital periods: for $P \sim 100$ min the period–mass relation (4.9) gives such a low mass ($\lesssim 0.2 M_\odot$) that the secondary makes almost no contribution at wavelengths shorter than the infrared. However there is still a possible extraneous component, due to the 'bright spot' where the gas stream from the secondary strikes the edge of the disc (Fig. 24). Because this spot presents a varying aspect through the binary cycle, it produces a light variation of its own. In some cataclysmics this variation dominates all the others at optical wavelengths (Fig. 24(*b*)). In systems with high mass transfer rates however, such as novalike variables and dwarf novae in outburst (see Section 5.9) the disc is relatively much brighter, and we do not need to consider the bright spot.

In attempting to compare predicted disc spectra with observation, we can use the property that in the ultraviolet, only the disc is an important emitter. Since cataclysmics are relatively nearby systems (~ 100 pc), interstellar reddening is

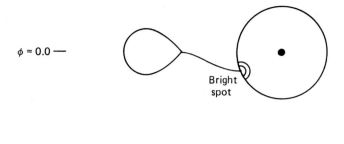

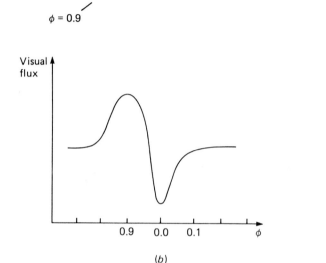

Figure 24. The bright spot in cataclysmic variables; (a) orbital geometry, (b) light curve of an eclipsing system.

usually negligible, and good-quality UV spectra of several systems have been obtained by the IUE satellite, often simultaneously with optical spectra. This is vitally important for the spectral-fitting approach, for one needs as wide a wavelength coverage as possible to constrain the fit to theory. The main problem with this approach comes in modelling the disc spectrum theoretically. The blackbody–disc spectrum (5.43) (Fig. 20) is inadequate for anything but the very crudest comparison. This is hardly surprising, as the disc is very unlikely to radiate locally as a blackbody: a star certainly does not, for example.

The simplest refinement of this picture is given by assuming that the local radiation spectra are given by a standard stellar atmosphere. Such atmospheres are uniquely characterized by the effective temperature T_{eff} and surface gravity g ($=GM/R^2$) of the star; in building up a disc spectrum one therefore uses tabulated atmospheres corresponding to $T_{\text{eff}} = T(R)$ and $g = GMH/R^3$ for each disc annulus, and integrates over R. This more elaborate procedure does not give dramatically better results; in particular the predicted atmospheric absorptions (e.g. the Balmer

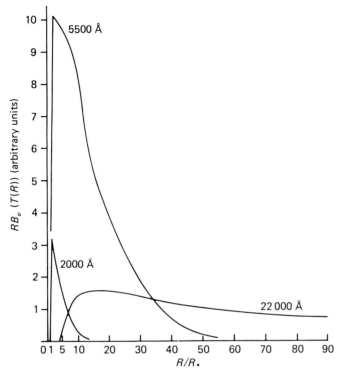

Figure 25. Surface brightness $RB_\nu(T(R))$ versus R/R_* for a steady, optically thick disc with $T_* = 7 \times 10^4$ K. At short wavelengths (here 2000 Å) the light is strongly concentrated towards the central disc regions, while for long wavelengths (here 22 000 Å) the brightness distribution is almost uniform outside the central regions.

jump) are generally rather deeper than those observed. This is not unexpected, as stellar atmosphere models are constructed under the assumption that the outward flux of radiation is all generated deep in the star, at effectively infinite optical depth; there are no energy sources within the atmosphere and the total radiative flux integrated over frequency remains constant, although the flux is redistributed between frequencies. In a disc, even if we implicitly assume that most of the radiation is first produced on the mid-plane (cf (5.38)), the optical depth τ is sufficiently small (5.45) for us to expect deviations from the stellar case. Worse, there is no good reason to expect dissipation to be absent from the surface layers at all. At low optical depths matter will have to heat up in order to radiate the dissipated energy, so the vertical temperature gradient can be flattened or even inverted. The details of the vertical structure now enter the calculation of the emergent spectrum with a vengeance, different viscosity prescriptions for example leading to entirely different spectra. Much remains to be done in this interesting area; it is far from clear for instance, that an increase in $\dot{M}$ always causes F_ν to rise at any given wavelength, although this is almost always implicitly assumed. But there seems little prospect of unambiguous theoretical predictions of the emitted

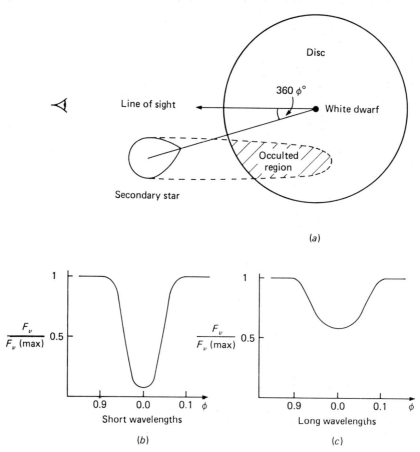

Figure 26. (a) Eclipse geometry for a binary system with an accretion disc. Because of the very different surface brightness distributions at short and long wavelengths (cf. Fig. 25) the light curves predicted are very different: deep and narrow at short wavelengths (b), shallow and broad at long wavelengths (c).

spectrum which could be used to test the basic disc structure. One important point which will probably survive is that at very low optical depths dissipation heats the disc enough to create a corona or mass loss in an outflowing wind.

The second method of checking for the presence of discs, by using eclipses to probe the surface brightness distribution, has proved more successful. The basis of the method is illustrated in Figs. 25 and 26. At short wavelengths, the disc light is concentrated towards its centre, while at longer wavelengths the distribution is more even. Hence if we observe a cataclysmic variable with a sufficiently high orbital inclination that the companion star eclipses the central regions of the disc, there should be a deep and sharp eclipse at short wavelengths, and a shallower, broader one at long wavelengths.

In practice this method can only be used at optical wavelengths, as present ultraviolet satellites do not give enough time resolution to obtain useful light curves. In the most successful technique, a large number of UBV light curves are

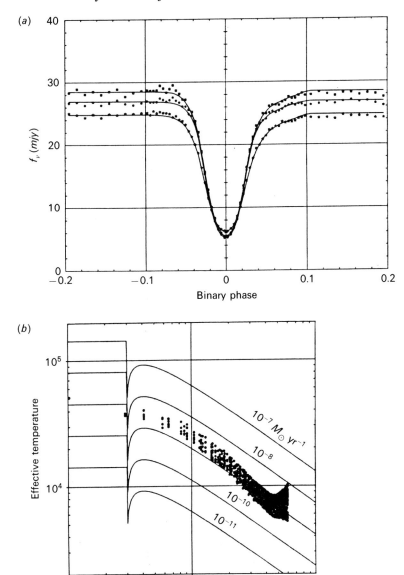

Figure 27. The eclipse mapping technique used to find the surface brightness distribution of the accretion disc in the dwarf nova Z Cha. The observations were made in an outburst, so that the disc dominates the optical light. (a) Eclipse light curves (B, U, V from top). (b) Effective temperature distribution given by maximum-entropy deconvolution, compared with equation (5.41) for various values of $\dot{M}$. (Reproduced from K. Horne, T. R. Marsh, *Physics of Accretion onto Compact Objects* (1986) p. 5. Springer-Verlag.)

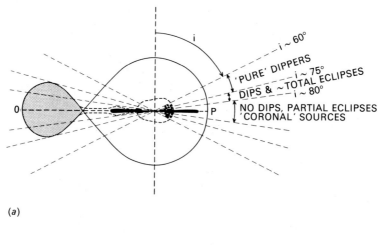

(a)

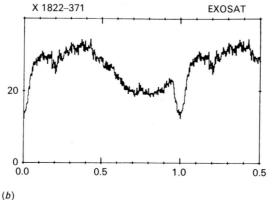

(b)

Figure 28(a, b). For legend see opposite.

averaged so as to smooth out short-term fluctuations. A maximum-entropy method is then used to find the most axisymmetric image of the disc in each waveband which is consistent with the data. The radial flux distributions can then be used to find the dependence $T(R)$ of effective temperature on radius, provided we assume something about the way the disc radiates locally. This of course is precisely the area of greatest difficulty in modelling the integrated disc spectrum; fortunately, the predicted run of $T(R)$ is not greatly affected by what is assumed at this point. Some results of this procedure are shown in Fig. 27. As can be seen, they do point to a dependence of the form $T(R) \propto R^{-3/4}$ in this particular case. This is reassuring, as one can see from Section 5.5 that it is very difficult to avoid such a dependence in a steady-state disc. Further, the estimated accretion rate ($\sim 10^{-9} M_\odot$ yr^{-1}) is very reasonable for a dwarf nova in outburst. Not all systems give such good agreement with simple theory, but this can usually be accounted for by deviations from a steady state.

For systems other than cataclysmic variables the evidence for discs is rather less clearcut, basically for the reasons outlined at the beginning of this section. The low

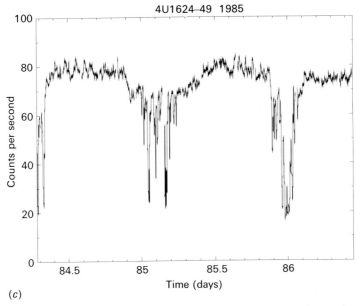

Figure 28. (a) The light curves of low-mass X-ray binaries. (Reproduced from J. Frank, A. R. King & J. P. Lasota (1987) *Astron. Astrophys.*, **178**, 137.) At high inclination only X-rays scattered from an extended corona around the accreting neutron star (or black hole) are seen, and eclipses by the companion star are partial (Fig. 28b). (Reproduced from A. N. Parmar, N. E. White, P. Giommi & M. Gottwald (1986), *Astrophys. J.*, **308**, 199.) At lower inclinations the central point X-ray source is seen, and the X-rays may be modulated by dips (Fig. 28c) (M. G. Watson, R. Willingale, A. R. King, J. E. Grindlay & J. Halpern (1985) unpublished) and occasionally total eclipses by the companion.

mass X-ray binaries have similar periods and secondary stars to those of cataclysmics, but differ from them in having a neutron star (or occasionally a black hole, see Section 6.7) as the accreting component, rather than a white dwarf. Naively one might have expected them to show straightforward evidence of discs. However the brightest of these systems show a marked lack of binary signatures, particularly eclipses, a fact hindering early identification of the class as binaries at all. This is evidently the result of a selection effect: the systems are much brighter in X-rays when viewed at low orbital inclination than when seen at inclinations giving eclipses by the companion star. As fainter systems were discovered, some of them were found to show periodic features in their X-ray light curves; generally these systems also have a lower ratio of X-ray to optical flux. A simple interpretation is that the X-rays from the accreting neutron star are scattered by a tenuous corona above the central regions of the accretion disc (Fig. 28(a)). This corona probably results from the effects of viscous dissipation in the optically thin outer layers of the disc (see above); in low-mass X-ray binaries the extra heating caused by X-ray irradiation of the disc surface intensifies the effect. The scattering corona acts as a weak, extended X-ray source in its own right. If we assume that at high inclinations the disc is able to obscure completely the X-ray source near the neutron star surface, we are left only with this extended source. Eclipses by the companion will be partial and gradual (Fig. 28(*b*)), just like optical eclipses of one star by another, rather than the abrupt total eclipses which would occur if the X-rays came only from the immediate vicinity of the neutron star. At slightly lower inclinations the pointlike X-ray source is visible, and we can have a near-total eclipse (the corona providing the residual flux) or 'dipping' activity (Fig. 28(*c*)) confined to a restricted phase interval. In systems where the orbital phasing can be established, usually from optical data, the dips are found to occur preferentially somewhat before superior conjunction of the X-ray source (i.e. the phase when the companion star is most in front of it). This is reminiscent of the phase of maximum light of the bright spot in cataclysmics (see Fig. 24). The dependence of the dips on X-ray energy shows that they are generally caused by increased photoelectric absorption. However the matter causing the absorption must either be substantially underabundant in heavy elements compared with the solar case, or have most of these elements highly ionized so that they do not absorb. Because all systems which show an eclipse by the companion also have dips whereas the converse is not true, the structure producing the dips must subtend a greater angle at the X-ray source than the secondary star.

Two main lines of explanation for these orbital modulations have been considered. In one, it is assumed that the outer edge of the disc is able to occult the X-ray source in the central regions of the disc. At high inclinations, where only the weak extended X-ray source is seen, a continuous modulation results if the disc edge causes obscuration all the way around the binary orbit. At slightly lower inclinations the pointlike X-ray source near the neutron star is visible, and dipping activity is assumed to result from the obscuration of this by the most vertically extended part of the disc edge.

It is clear that this model implies a substantial deviation from the standard picture. The repeating nature of the continuous modulation demands a disc rim with a stable vertical structure, which must extend to considerable heights from the disc plane. Even for small mass ratios $q \sim 0.1$ the secondary subtends angles $\sin^{-1} R_2/a \gtrsim 10°$ at the X-ray source, implying that the rim structure has a vertical extent H' obeying $H'/R_{\text{out}} \gtrsim 0.2$. This is much larger than the standard thin disc scale height (5.46); the disc rim thus either involves material much hotter than the thin disc estimates for $T_{\text{c}}(R_{\text{out}})$ (5.45), or not in vertical hydrostatic balance at all. In the first case a thin disc description is evidently not appropriate; it is then unclear why changing the central object from a white dwarf to a neutron star should have such a radical effect on the structure of very distant regions of the disc. If we relax the requirement of hydrostatic balance for the obscuring matter, it could be following a ballistic orbit about the neutron star, or escaping it entirely in the form of a wind. The first type of structure might be caused by the impact of the gas stream from the secondary on the disc, or spiral shocks within the disc (see Section 5.10 below), while winds can be driven off the disc by dissipation in surface layers, as we have remarked above. There are difficulties in both cases. Any ballistic orbit will lie in a plane, and intersect the binary plane at two points which are essentially symmetrically disposed about the neutron star. Thus an orbit leaving the disc rim will intersect it again on the opposite side, with the highest point roughly halfway between. Clearly only rather smoothly-varying azimuthal structure is possible above the disc plane, rather than the fairly jagged profiles that would be required to give the X-ray variation: 'bulges' with an azimuthal extent much less than about 180° are impossible. (The same problem applies also to hydrostatically-supported vertical structure, as shown by the equality of the vertical and azimuthal dynamical timescales t_ϕ, t_z defined in equations (5.57) and (5.58) below. It is worth noting that this argument does not depend on details such as cooling times: t_z is the *minimum* time for the matter to return to the orbital plane.) Further, an orbit producing the periodic X-ray dips in lower inclination systems would have to leave the disc plane at a point corresponding to binary phase ~ 0.55 in order to reach its greatest height above the disc at phase ~ 0.8, where the dips are observed. None of the candidate mechanisms for producing the extended vertical structure readily drive such orbits. Obscuration in a wind from the disc surface is possible, although quantitative modelling suggests that the scattering optical depth through such a wind is likely to be too low to give the observed modulations. The major problem in this case is to explain the marked azimuthal structure required for such a wind to give the observed periodic modulation. Note that in the case of a wind the disc's own (optical) photosphere could not share this azimuthal dependence, although periodic optical variations are observed in some systems. Finally, all models placing the obscuring matter near the disc edge require large underabundances (factors $\sim 10^{-3}$) of heavy elements to explain the energy dependence of the X-ray dips: at such distances the effects of photoionization by the central X-ray source are far too weak.

As can be seen, most of the difficulties outlined above would disappear if the obscuring matter were located closer in to the neutron star than the disc edge. One then needs a way of supplying knowledge of the orbital phase to this material; the simplest possibilities are that part of the gas stream from the secondary star either penetrates the disc or skims over its upper and lower faces. The latter is particularly plausible: the estimate $w \sim 0.1 \, P c_s$ (section 4.5) for the stream radius gives $w \sim 10^9$ cm for $P = 5$ h, while H as given by (5.45) is at most a few $\times 10^8$ cm for reasonable parameters. Collisions of the stream either with itself or the disc material then inevitably occur near the neutron star; the resulting strong shocks (involving velocities comparable with the local Kepler value) will automatically put matter at substantial heights above the disc plane, since the shocked gas will have $c_s \lesssim v_\phi$. By (5.30) this gas subtends large angles at the neutron star; this is clearly a promising arrangement for producing X-ray modulations at the orbital period, provided the gas has the required absorption properties. In this type of model the ionization parameter $\Xi = P_{\text{ion}}/P_{\text{gas}}$, where P_{ion} is the pressure of the ionizing radiation and P_{gas} is the gas pressure, always turns out to lie in the range 0.1–10 which implies a two-phase instability (see Fig. 61). Most of this gas condenses into cool ($\sim 10^4$ K) clouds scattered through a low-density hot ($\sim 10^7$ K) medium. This happens independently of the accretion rate because both P_{ion} and P_{gas} are proportional to $\dot{M}$ so that it cancels in the expression for Ξ. We thus have material cool enough to produce substantial photoelectric absorption subtending a large angle at the X-ray source. This is evidently qualitatively just what is needed to produce the various types of X-ray light curves; in particular the fine structure in X-ray dips can result from the passage of individual cool clouds across the line of sight to the X-ray source. Similar structure does not result in cataclysmics, as the ionization parameter is always lower by a factor $R_{\rm wd}/R_{\rm ns} \sim 10^3$, leaving the matter in the cool state.

This generic type of model offers physical explanations for the main features of the X-ray behaviour without requiring us to abandon the thin-disc picture for low-mass X-ray binaries (LMXB). To explain the light curves in detail we must specify the kinematics of the shocked gas stream more closely. The possibility so far explored in most detail is that the gas stream skims over the disc surfaces and forms a ring at the circularization radius $R_{\rm circ}$. The newly-arriving gas then collides with this ring, shocking and rising out of the disc plane at the impact point. The two-phase instability forms cool clouds which give rise to the dips (Fig. 29). The behaviour with orbital phase predicted by this model agrees with most, but not all, of the observations (note that in the thick-rim picture there is no physical constraint at all on the various phases). Another possibility is that no ring forms at $R_{\rm circ}$ at all since the stream pinches off vertically near to the distance of closest approach to the neutron star, colliding with itself and the disc material. This too would give a two-phase instability, although details of the resulting behaviour have still to be worked out.

5.8 Time dependence and stability

In the last five sections, we have dealt in some detail with the theory and observations of steady thin accretion discs. There are several reasons for extending this to the study of the time-dependent behaviour of discs. One is that we must check that the steady-state models are stable against small perturbations: if not, we have probably made some assumption in the course of constructing these discs which is not compatible with the further assumption of steadiness. Another reason is that the observable properties of the steady-state, optically thick discs we have considered above are largely independent of viscosity; this is a fortunate occurrence for the purpose of showing that such discs do indeed exist, but it means that observations of steady discs are unlikely to give much information about the viscosity. The time dependence of disc flow is on the other hand controlled by the size of the viscosity (cf. Section 5.2). Hence observations of time-dependent disc behaviour offer one of the few sources of quantitative information about disc viscosity. In view of our present ignorance of the basic physical processes involved, such a semi-empirical approach to the problem seems the most reasonable. The main arena for the study of time dependence is the problem of the outbursts of dwarf novae, which also represent an area of prime interest in their own right.

We begin by identifying the typical timescales on which the disc structure may vary. We have already encountered the viscous timescale

$$t_{\rm visc} \sim \frac{R^2}{v} \sim \frac{R}{v_{\rm R}} \tag{5.9}$$

which gives the timescale on which matter diffuses through the disc under the effect of the viscous torques. Note that (5.9) assumes that the typical lengthscale for

Time dependence and stability

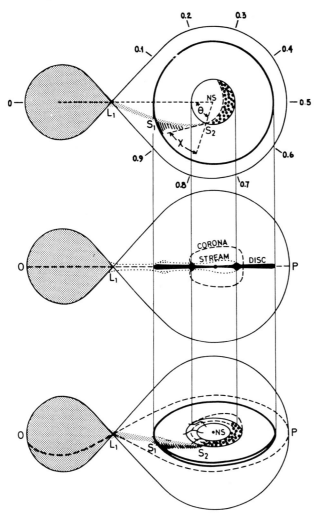

Figure 29. Three different views of a model LMXB in the penetrating gas-stream picture. The situation in the orbital plane with line of sight positions at 0.1 phase intervals is depicted at the top. The mass transfer stream hits the edge of the accretion disc at S_1, and the ring of radius R_{circ} at S_2. A two-phase medium of cold clouds and hot intercloud gas forms after impact at S_2 due to ionization instabilities and splashes off the disc plane. The thick dots represent the cold clouds. The central diagram shows a cut through the binary axis normal to the orbital plane (OP). The disc and ring at R_{circ} appear in cross section. The extent of the stream and corona are schematically indicated. The bottom diagram is a perspective view of the model showing the splitting of the stream at S_1, the ring and the bulge with the cold clouds. (Reproduced from J. Frank, A. R. King & J. P. Lasota (1987) *Astron. Astrophys.*, **178**, 137.)

surface density gradients in the disc is $\sim R$ (see Section 5.2 above). The shortest characteristic timescale of the disc is the *dynamical timescale*

$$t_\phi \sim \frac{R}{v_\phi} \sim \Omega_K^{-1}. \tag{5.57}$$

Clearly, inhomogeneities, such as flares on the disc surface, would cause quasi-periodic variations in the disc light at some wavelengths with this timescale. Deviations from hydrostatic equilibrium in the z-direction are smoothed out on a timescale

$$t_z = H/c_s. \tag{5.58}$$

But from (5.30) we see that

$$t_z \sim R/\mathcal{M}c_s = R/v_\phi \sim t_\phi$$

so that the dynamical timescale also measures the speed with which hydrostatic equilibrium in the vertical direction is established. In addition to t_ϕ, t_z and t_{visc}, we can define the *thermal timescale*:

$$t_{\text{th}} = \frac{\text{heat content per unit disc area}}{\text{dissipation rate per unit disc area}},$$

which gives the timescale for re-adjustment to thermal equilibrium, if, say, the dissipation rate is altered. Since the heat content per unit volume of a gas is $\sim \rho k T/\mu m_p \sim \rho c_s^2$, this gives

$$t_{\text{th}} \sim \Sigma c_s^2/D(R). \tag{5.59}$$

The relation (4.25) between $D(R)$ and $\nu\Sigma$ means that t_{th} can be re-expressed in terms of the viscous timescale: for a Keplerian disc, we have ($\Omega = \Omega_K$)

$$t_{\text{th}} \sim \frac{R^3 c_s^2}{GM\nu} \sim \frac{c_s^2 R^2}{v_\phi^2 \nu} \sim \mathcal{M}^{-2} t_{\text{visc}} \tag{5.60}$$

using (5.28). Using the α-parametrization (4.30) we can also write

$$t_{\text{visc}} \sim \frac{R^2}{\nu} \sim \frac{1}{\alpha}\frac{R}{H}\frac{R}{c_s} = \frac{1}{\alpha}\frac{R}{H}\frac{R}{v_\phi} \cdot \frac{v_\phi}{c_s}$$

which, for a thin disc, gives, on using (5.28), (5.30) and (5.57),

$$t_{\text{visc}} \sim \alpha^{-1} \mathcal{M}^2 t_\phi. \tag{5.61}$$

Collecting results we have

$$t_\phi \sim t_z \sim \alpha t_{\text{th}} \sim \alpha (H/R)^2 t_{\text{visc}}. \tag{5.62}$$

Time dependence and stability

If we assume $\alpha \lesssim 1$ there is thus a well-defined hierarchy of timescales $t_\phi \sim t_z \lesssim t_{\rm th} \ll t_{\rm visc}$. Numerically, we have, for the α-disc solutions (5.45),

$$\begin{aligned} t_\phi \sim t_z \sim \alpha t_{\rm th} &\sim 100 M_1^{-1/2} R_{10}^{3/2}\ {\rm s}, \\ t_{\rm visc} &\sim 3 \times 10^5 \alpha^{-4/5} \dot{M}_{16}^{-3/10} M_1^{1/4} R_{10}^{5/4}\ {\rm s}. \end{aligned} \tag{5.63}$$

Thus the dynamical and thermal timescales are of the order of minutes, and the viscous timescale of the order of days to weeks for typical parameters.

Suppose now that a small perturbation is made to a putative equilibrium solution and that this perturbation continues to grow rather than being damped. Then the supposed steady solution is said to be unstable and cannot occur in reality. The sharp difference we have found in the various timescales means we can distinguish different types of instability. Suppose, for example, that the energy balance is disturbed; any instability will grow on a timescale $t_{\rm th}$, which is much less than $t_{\rm visc}$. Since $t_{\rm visc}$ is the timescale for significant changes in the surface density Σ to occur, we can assume that Σ is fixed during the growth time $t_{\rm th}$. We refer to this as a *thermal instability*. For $\alpha < 1$ we also have $t_{\rm th} > t_\phi \sim t_z$, so the vertical structure of the disc can respond rapidly, on a timescale t_z, to changes due to the thermal instability, and keep the vertical structure close to hydrostatic equilibrium. Hence we can impose the conditions

$$\begin{aligned} \Sigma &\sim {\rm constant}, \\ H &\sim c_{\rm s}(T_{\rm c})(R^3/GM)^{1/2} \end{aligned} \tag{5.64}$$

(from (5.23)) in examining the growth of any thermal instabilities.

Such instabilities arise when the local (volume) cooling rate Q^- (erg s^{-1} cm^{-3}) within the disc can no longer cope with the (volume) heating rate $Q^+ \sim D/H$ due to viscous dissipation. In equilibrium we must have $Q^+ = Q^-$. But if, when the central temperature $T_{\rm c}$ is increased by a small perturbation $\Delta T_{\rm c}$, Q^+ increases faster than Q^-, $T_{\rm c}$ will rise further because the cooling rate is inadequate. In other words, a steady state is impossible in a parameter regime where the instability would grow, despite the fact that formally an 'equilibrium' solution can be found.

To make these ideas clearer, let us consider an explicit example. Suppose the local cooling rate is just the optically thin emissivity $4\pi j\,(\rho, T_{\rm c})$ (erg s^{-1} cm^{-3}); i.e. we assume that the disc density is so low that the vertical optical depth $\tau \ll 1$. Generally, such a gas cools by two-body emission processes so the emissivity depends on the square of the density, and has the form $4\pi j \propto \rho^2 \Lambda(T_{\rm c})$. $\Lambda(T_{\rm c})$ is a function called the *cooling curve*, which depends principally on how the ionization balance of the gas is maintained – for example by photoionization, or by collisions. It is the behaviour of Λ as a function of $T_{\rm c}$ which largely determines whether the instability grows or not. By (5.64) we can write that during the perturbation

$$Q^- = 4\pi j \propto \rho^2 \Lambda \sim (\Sigma/H)^2 \Lambda \sim c_{\rm s}^{-2}\Lambda \sim T_{\rm c}^{-1}\Lambda,$$

where we have assumed negligible radiation pressure in the last step. Similarly, we have

$$Q^+ \sim D/H \sim v/H \sim \alpha c_{\rm s} \sim \alpha T_{\rm c}^{1/2}$$

where we have used the α-parametrization (4.30). Comparing these two expressions, we see that the thermal instability will grow if Q^- increases less rapidly with T_c than does Q^+, i.e. if Λ/α grows more slowly with T_c than $T_c^{3/2}$. The customary way of writing the instability criterion is

$$\frac{d \ln (\Lambda/\alpha)}{d \ln T_c} < \tfrac{3}{2}, \tag{5.65}$$

or, for the general case, in the alternative forms,

$$\left.\begin{array}{c} \dfrac{dQ^-}{dT_c} < \dfrac{dQ^+}{dT_c} \\[2mm] \dfrac{d \ln Q^-}{d \ln T_c} < \dfrac{d \ln Q^+}{d \ln T_c} \end{array}\right\} \Rightarrow \text{unstable.} \tag{5.66}$$

From the form of $\Lambda(T_c)$, which decreases for $T_c \gtrsim 10^4$ K, and with $\alpha \sim$ constant, we expect optically thin regions of the disc to be thermally unstable for $T_c \gtrsim 10^4$ K. Since $\Lambda(T_c)$ for other cases has a similar character, this result is of general validity. It can also be extended to the case where radiation pressure dominates. Note however that it is sensitive to the functional form of α: for $\alpha \sim T_c^{-2}$, say, our conclusion would be reversed. As we have stressed in connection with the Shakura–Sunyaev solution (5.45), we have no knowledge of the form, as opposed to the magnitude, of α. Because we expect $\alpha \lesssim 1$ it is rather unlikely that α could conspire to behave as, say, $\alpha \sim T_c^{-2}$ for a wide range of T_c, so the result above probably holds for most values of $T_c \gtrsim 10^4$ K; but we cannot rule out *a priori* the possibility that isolated stable regimes (for restricted ranges of T_c) exist.

In contrast to the optically thin case, for optically thick discs the *effective* volume cooling rate (taking account of the local re-absorption of radiant energy) is

$$Q^- = \frac{dF}{dz} \sim \frac{\sigma T_c^4}{\kappa_R \rho H^2}$$

by (5.37) and (5.35). For Kramers' opacity (5.44) this can be rewritten as

$$Q^- \sim \frac{T_c^{15/2}}{\rho^2 H^2} = \frac{T_c^{15/2}}{\Sigma^2} \sim T_c^{15/2}$$

omitting constraints and using (5.64). As before, we have $Q^+ \sim \alpha T_c^{1/2}$ if gas pressure dominates, so, unless α decreases more rapidly than T_c^{-7}, optically thick parts of the disc will be thermally stable. Similar conclusions follow when other opacities and radiation pressure dominate.

Let us now consider changes in the disc structure which take place on the viscous timescale. This includes viscous instabilities and the evolution of discs in response to changes in external conditions, such as the mass transfer rate (cf. Sections 5.2). As $t_{\text{visc}} \gg t_{\text{th}} \gtrsim t_z$ we may assume that the disc adjusts so rapidly that it always maintains both thermal and hydrostatic equilibrium. Hence some of equations (5.39) describing steady discs apply also to time-dependent discs: equations 1, 2, 3,

Time dependence and stability

4, 6 and 8 survive unaltered. Equations 5 and 7 explicitly used the assumption of steadiness and must be changed. The energy transport equation 5 is replaced by the combination of (5.38) and (4.26) with $\Omega = \Omega_K$:

$$\frac{4\sigma}{3\tau} T_c^4 = \tfrac{9}{8}\nu\Sigma \frac{GM}{R^3}.$$

Here we have assumed that t_{visc} is much longer than the timescale for radiation to diffuse out of the disc; this is amply satisfied by the resulting disc solutions. Equation 7 is an integral of the mass and angular momentum equations (5.3) and (5.4). In the time-dependent case we can replace it by the evolution equation (5.6). Defining the mass transfer rate $\dot{M}(R, t)$ at each radius by equation (5.12), in which Σ and v_R are now allowed to be functions of t, we can rewrite the mass conservation equation (5.3) as

$$2\pi R \frac{\partial \Sigma}{\partial t} = \frac{\partial \dot{M}}{\partial R}. \tag{5.67}$$

For reference we collect these equations together:

$$\left.\begin{aligned}
&1.\ \rho = \Sigma/H, \\
&2.\ H = c_s R^{3/2}/(GM)^{1/2}, \\
&3.\ c_s^2 = P/\rho, \\
&4.\ P = \frac{\rho k T_c}{\mu m_p} + \frac{4\sigma}{3c} T_c^4, \\
&5.\ \frac{4\sigma}{3c} T_c^4 = \tfrac{9}{8}\nu\Sigma \frac{GM}{R^3}, \\
&6.\ \tau = \Sigma \kappa_R(\rho, T_c) = \tau(\Sigma, \rho, T_c), \\
&7.\ \frac{\partial \Sigma}{\partial t} = \frac{3}{R}\frac{\partial}{\partial R}\left\{R^{1/2}\frac{\partial}{\partial R}(\nu\Sigma R^{1/2})\right\}, \\
&8.\ \nu = \nu(\rho, T_c, \Sigma, \alpha, \ldots).
\end{aligned}\right\} \tag{5.68}$$

Thus again we have eight equations for the eight unknowns ρ, Σ, H, c_s, P, T_c, τ and ν as functions now of R, t and any parameters (e.g. α) in the viscosity prescription 8. Because 7 is a diffusion equation we must also provide appropriate boundary conditions. Once the set (5.68) is solved, $\dot{M}$ may be found from (5.67) and boundary conditions, and v_R read off from (5.12). As before, we can simplify these equations and reduce the time-dependent disc problem to one involving fewer variables.

Since the time dependence enters only through equation 7, it is sensible to try to express the other quantities in terms of Σ as far as possible, as they depend on t only through Σ and not explicitly. In particular, equations 1–4 can be combined to give

$$\frac{GM\Sigma H}{R^3} = \frac{kT_c \Sigma}{\mu m_p H} + \frac{4\sigma}{3c} T_c^4. \tag{5.69}$$

Accretion discs

From 5 to 6 we have

$$T_c^4 = \frac{27}{32\sigma}\Sigma^2 \kappa_R v \frac{GM}{R^3}. \tag{5.70}$$

Since κ_R and v are both functions of Σ/H, T_c and R, we can in principle eliminate κ_R, v and T_c, say, to get H as a function of R and Σ, with similar results for all the other quantities. As an example, if κ_R is given by the Kramers' form (5.44) and v specified by the α-prescription (4.30), (5.70) assumes the form

$$T_c^4 \propto \alpha\Sigma^3 HT_c^{-7/2}(GM/R^3)^{1/2}.$$

Using this equation, T_c can be eliminated from (5.69) to give H as a function of R and Σ. Similarly we get from the α-prescription

$$v = \alpha c_s H = \alpha(kT_c/\mu m_p)^{1/2} H = v(R, \Sigma)$$

and so on. This rather tedious, if straightforward, algebraic manoeuvring illustrates an important result: *the set (5.68) fixes v as a function of R and Σ, so that equation 7 is a nonlinear diffusion equation for $\Sigma(R, t)$.* An exception to this result would arise if v were not a function of local conditions within the disc, but somehow determined from outside, so that it involved an explicit t-dependence.

Clearly to solve the nonlinear equation 7 and subsequently find all the other disc variables is in general a formidable task which must be tackled numerically. We shall consider shortly the results of some such calculations. However, we can draw some important conclusions from the result above and the form of equation 7 alone. Suppose that the surface density in a steady disc is perturbed axisymmetrically at each R, so that

$$\Sigma = \Sigma_0 + \Delta\Sigma$$

where $\Sigma_0(R)$ is the steady-state distribution. Writing $\mu = v\Sigma$ there will be a corresponding perturbation $\Delta\mu$. Since $v = v(R, \Sigma)$ by our result above we have $\mu = \mu(R, \Sigma)$, so that $\Delta\mu = (\partial\mu/\partial\Sigma)\Delta\Sigma$. Because of equation 7 of (5.68) $\Delta\Sigma$ and $\Delta\mu$ are related further by

$$\frac{\partial}{\partial t}(\Delta\Sigma) = \frac{3}{R}\frac{\partial}{\partial R}\left[R^{1/2}\frac{\partial}{\partial R}(R^{1/2}\Delta\mu)\right].$$

Eliminating $\Delta\Sigma$ we obtain the equation governing the growth of the perturbation

$$\frac{\partial}{\partial t}(\Delta\mu) = \frac{\partial\mu}{\partial\Sigma}\frac{3}{R}\frac{\partial}{\partial R}\left[R^{1/2}\frac{\partial}{\partial R}(R^{1/2}\Delta\mu)\right].$$

Not surprisingly, $\Delta\mu$ obeys a diffusion equation, but the interesting thing is that the diffusion coefficient is proportional to $\partial\mu/\partial\Sigma$ and in principle can be either positive or negative, since we have no information about v and hence about μ. If $\partial\mu/\partial\Sigma$ is positive we get the type of diffusive behaviour discussed in Section 5.2 and the perturbation decays on a viscous timescale. However if $\partial\mu/\partial\Sigma$ is negative, more material will be fed into those regions of the disc that are denser than their surroundings and material will be removed from those regions that are less dense, so that the disc will tend to break up into rings. This breakup of the disc on a timescale t_{visc} constitutes the *viscous instability*; more precisely stated: *steady disc flow is only possible provided* $\partial \mu / \partial \Sigma > 0$.

From (5.17) and (5.41) we note that $\mu = \nu\Sigma \propto \dot{M} \propto T^4(R)$, where $\dot{M}, T(R)$ are the local equilibrium values of the mass transfer rate and surface temperature, so that this requirement can be rewritten as $\partial \dot{M}/\partial \Sigma > 0$ or $\partial T(R)/\partial \Sigma > 0$. Indeed the first of these forms makes the stability requirement transparent: if, instead, a region of the disc had $\partial \dot{M}/\partial \Sigma < 0$, the local mass transfer rate would increase in response to a decrease in surface density, and vice versa. Equivalently, it is easy to show that heating dominates cooling to the right of the equilibrium curve in the T–Σ plane, and vice versa. Thus in regions with $\partial T/\partial \Sigma > 0$, a small increase or decrease in surface density will lead to a small rise or fall in temperature, and thus a return to the equilibrium curve on a local thermal timescale. If however $\partial T/\partial \Sigma < 0$, small changes in Σ result in larger changes in T (cf. Fig. 31).

Let us now consider dwarf nova outbursts. As we saw in Section 5.7 there is considerable observational evidence that accretion discs conforming roughly to the picture we outlined in Section 5.3 to 5.6 actually exist in cataclysmic variables and indeed provide most of their light in many cases. A subclass of these systems, the dwarf novae, undergo outbursts from time to time in which they brighten by several magnitudes. These are much smaller in amplitude than the outbursts of classical novae (hence the name), and recur at intervals of weeks to months. A typical outburst light curve is shown in Fig. 30.

It is by now generally agreed that dwarf nova outbursts are powered by a sudden increase in the accretion rate. The alternative mechanism of runaway thermonuclear burning of the accreted material on the white dwarf, which is believed to give rise to novae, cannot provide recurrence times of less than about 10^4 yr. The existence of dwarf nova outbursts therefore offers an ideal opportunity to test ideas about the time dependence of disc accretion, since it should be possible to explain the general shape of the outburst light curve Fig. 30. Two main types of model have been proposed:

(a) The mass transfer rate from the companion star is variable because the envelope of this star is dynamically unstable; thus, outburst corresponds to a sudden burst of mass transfer from this star.

(b) The viscosity in the disc varies: between outbursts (at 'quiescence') the viscosity is low, so that the viscous timescale is long and matter accumulates in the disc; at outburst the viscosity suddenly rises, the viscous timescale becomes short and the matter previously stored in the disc is rapidly deposited on to the white dwarf.

We examine the variable mass transfer model first. Models of the companion star instability suggest an increase in the mass transfer rate by factors $\sim 10^2$ on timescales of order 10^5 s, followed by a slower decay back to the quiescent state. To see what sort of changes in the disc structure result we must introduce this burst of

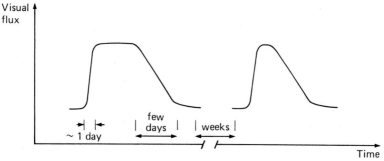

Figure 30. Visual light curve of a dwarf nova showing outbursts. Any variations on the orbital timescale ($\sim$ hours) due to e.g. eclipses are not shown: these are generally much smaller than the outburst amplitudes.

mass transfer into the time-dependent disc equations (5.68) as a source term in the diffusion equation 7. This can be done numerically and model light curves resembling Fig. 30 produced, provided α is suitably adjusted. In this model, the basic reason for the form of the light curve (rapid rise, flat top, slow decay) is the dependence (cf. 5.9) of the viscous timescale on the surface density gradient. Since, for the assumed parameters, the timescale of the mass transfer burst is always shorter than $t_{\rm visc}$, it is the latter which determines the timescale of changes in Σ and ultimately in luminosity. Thus, at the beginning of the outburst, the disc tries to reach a steady state corresponding to the new, high accretion rate $\dot{M}_{\rm high}$. This evolution proceeds on the viscous timescale $t_{\rm visc}$ (rise) $\sim l^2/\nu$, $l \ll R$ corresponding to the steep density gradients in the disc caused by the sudden influx of mass. One thus expects a rapid rise in luminosity, starting at long (visible) wavelengths and moving to the ultraviolet as the extra mass diffuses inwards on a timescale of order a day. This is in good qualitative agreement with observations. As the matter injected in the burst spreads out and gets accreted, the disc will decay back to its quiescent brightness, but on the longer timescale $t_{\rm visc}$ (decay) $\sim R^2/\nu$ given by the more even density distribution. Numerical modelling of this process shows that to get rise times of the observed order ($\sim$ 1 day), values of α in the range 0.1–1 are required. Such values of α are also needed to ensure that a sufficiently large mass of gas can be transported through the disc to power the outburst for the observed duration of $\lesssim$ weeks. This model has its problems: first, the basic cause of the instability in the companion star remains obscure. Further, in some dwarf novae, the disc has been observed to shrink in quiescence, rather than returning to a steady state. Also, the observed decline of the luminosity back to the quiescent level is much sharper than the slow exponential decay predicted by this model.

The basic idea underlying the disc instability model can be understood using the diffusion equation (5.6), and in particular applying our conclusion (p. 106) that steady disc flow is only possible if $\partial T/\partial \Sigma > 0$. Let us consider possible forms of the function $T(\Sigma)$ at a particular radius R of the disc. This will arise from solutions of the steady disc equations, such as (5.45) for example. In Fig. 31(a) we have $\partial T/\partial \Sigma > 0$ throughout. Suppose that the accretion rate $\dot{M}$ into the annulus at R is fixed at a particular value $\dot{M}_0$. Then the annulus will evolve on a viscous timescale to an

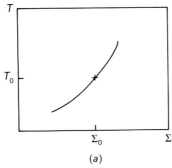 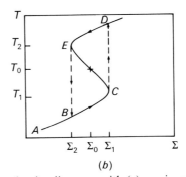

Figure 31. Effective temperature–surface density diagrams with (a) a unique, stable steady solution (T_0, Σ_0), and (b) a case where limit-cycle behaviour occurs, as the steady solution (T_0, Σ_0) lies on a region of the curve with $\partial T/\partial \Sigma < 0$.

equilibrium with $T(R) = T_0$, $\Sigma = \Sigma_0$, where T_0 is given by (5.41) with $\dot{M} = \dot{M}_0$, and Σ_0 is the corresponding solution of the steady disc equations. The annulus will remain in this state until $\dot{M}_0$ is varied, since the equilibrium state has $\partial T/\partial \Sigma > 0$ and is therefore stable. For the form of $T(\Sigma)$ shown in Fig. 31(b) there is also a unique equilibrium state T_0, Σ_0. However, for the assumed external parameters, this lies on the unstable $\partial T/\partial \Sigma < 0$ branch of the $T(\Sigma)$ curve: this means that the annulus can never settle into a steady configuration. If it starts, say, on the lower stable part of the curve (A–C), the low value of T (and hence $\dot{M}$) will cause it to evolve through states of increasing Σ in an attempt to raise $\dot{M}$ to $\dot{M}_0$. This continues until Σ reaches the value Σ_1 at point C: the only stable configuration for this value of Σ is the high-temperature state represented by point D, so any perturbation increasing Σ will cause the annulus to make a transition on a thermal timescale to this state. Here however the local value of $\dot{M}$ is larger than the externally-fixed value $\dot{M}_0$, so the annulus evolves towards lower T, Σ until it reaches Σ_2 (point E). Again T and $\dot{M}$ must adjust on a thermal timescale, this time to the lower branch (point B) so that the whole cycle B–C–D–E repeats. The annulus therefore displays limit-cycle behaviour, alternating long-lived (low viscosity) states of low $\dot{M}$ and short-lived (high viscosity) states of high $\dot{M}$, with rapid (thermal timescale) transitions between them.

Qualitatively then, the behaviour predicted by the $T(\Sigma)$ curve of Fig. 31(b) resembles the quiescence–outburst pattern of dwarf novae. Indeed, solutions for the vertical structure of steady discs with realistic opacity laws do tend to produce $T(\Sigma)$-curves with the required 'kinks', because the opacity of hydrogen increases dramatically when it becomes ionized near $T = 10^4$ K. However we must remember that this is so far a purely local instability at a fixed value of R. To turn this into a global instability capable of mimicking dwarf nova outbursts requires conditions in adjacent annuli to be arranged so that a local instability in one triggers the instability in the next, in a sort of grand domino effect. In fact outbursts of the observed duration and brightness can only be obtained by using the freedom to manipulate the effective value of the α parameter; in particular, if α is kept fixed, no large outburst results. The reason for this is that the times spent by a given disc annulus on either of the stable branches of the $\mu(\Sigma)$ curves are rather short: if the annulus jumps from one branch to the other, the disturbance does not propagate very far through the disc before the original ring jumps back again. This gives localized out-of-phase disturbances rather than a large outburst of the whole disc. The usual remedy is to increase the size of the local instability artificially by adopting a larger value of α (say 0.2) on the upper (hot, high $\dot{M}$) branch than on the lower (cooler, low $\dot{M}$) branch (say $\alpha = 0.05$). The nature of the outbursts depends on the external accretion rate $\dot{M}_e$ from the secondary star into the disc. For low $\dot{M}_e$ mass accumulates more rapidly near the centre of the disc, so that outbursts tend to start there, whereas for higher $\dot{M}_e$ mass accumulates more rapidly near the outer edge of the disc, and outbursts begin there instead. Given the freedom both in $\dot{M}_e$ and α it is hardly surprising that this type of model can be tuned to reproduce most of the features of observed outbursts, which therefore do not provide much of a test: it could be that the values of α used to fit a given outburst would be ruled out if we understood the disc viscosity properly. A more durable feature is that for $\dot{M}_e \gtrsim$ a few times 10^{16} g s^{-1} the whole of the disc is hotter than the hydrogen recombination temperature $\sim 10^4$ K (see 5.42); no outbursts can now occur, as the kinks in the $T(\Sigma)$ curves disappear. The disc-instability picture is thus in basic agreement with the observed fact that brighter cataclysmics do not seem to undergo outbursts. A major observational problem for this class of models is that the luminosity rise is predicted to happen almost simultaneously in the ultraviolet and the visible, whereas the observations show that the ultraviolet brightens about a day later than that in visible light. Also some outbursts, especially the superoutbursts of the SU UMa dwarf novae, would require very massive discs to form in order to power them. An intriguing possibility is that outbursts caused by disc instabilities might occasionally trigger a burst of mass transfer from the companion: in this event both types of mechanism proposed for dwarf nova outbursts would be involved in producing superoutbursts.

5.9 Tides, resonances and superhumps

We have seen that accretion discs act as efficient machines for transporting mass inwards and angular momentum outwards. What happens to the matter at the point where it encounters the central object is the subject of the next chapter; here we deal with the disposal of the angular momentum and the structure of the disc's outer edge. There must be a significant torque acting as a sink of angular momentum at this edge; in a steady state the secondary supplies the disc with angular momentum at a rate $\dot{M}(GMR_{\text{circ}})^{1/2}$ (cf Section 4.5), whereas the central star accretes angular momentum only at a rate $\dot{M}(GMR_*)^{1/2}$, which is in general substantially less. A good candidate for this torque presents itself quite straightforwardly: as the disc spreads to radii close to the Roche lobe of the primary, we can see from Fig. 11 that the orbits of test particles will be distorted more and more from circular by the influence of the secondary. Clearly the angular momentum is no longer a constant of the motion for these orbits as it is (approximately) for the near-Keplerian disc orbits further in; there can be exchange of angular momentum with the secondary star. This is not enough to provide the required torque however:

Tides, resonances and superhumps **107**

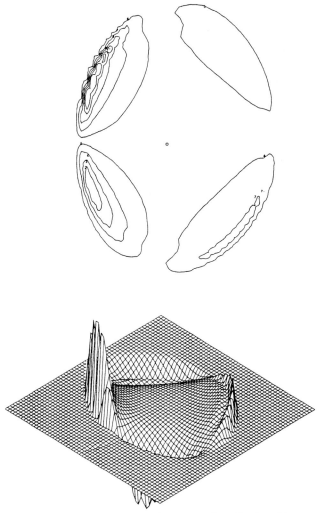

Figure 32. The tidal torques on an accretion disc in a binary system with mass ratio 0.5. The upper panel shows the torques as a contour plot, the primary being the small circle and the secondary lying off to the left along the approximate horizontal symmetry axis. The lower panel gives a surface plot showing the varying signs of the torque around the disc edge. The secondary is again to the left. (Figure by Dr R. Whitehurst.)

after all, for periodic orbits, the angular momentum must return to the same values it had on the previous orbit, so that there can be no net torque on a set of test particles populating all phases of these orbits.

The crucial element giving the torque is the viscosity, which makes all orbits slightly aperiodic, and in particular breaks the symmetry about the line of centres which otherwise causes the net torque to vanish. Another way of thinking of this is to regard the viscosity as slightly distorting the symmetry of the disc about the line of centres, so leading to a net gravitational torque on it. This is very similar to the

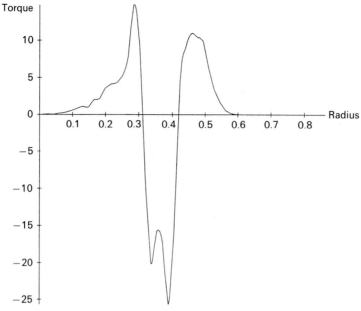

Figure 33. The average azimuthal torque derived from the results of Fig. 32. The disc is cut off at the tidal radius where the torque first changes sign. (Figure by Dr R. Whitehurst.)

torques in the Earth–Moon system cause by the tidal bulges on the two bodies, and is usually called the tidal torque. Thus the angular momentum balance can be expressed as

$$\dot{M}(GMR_{\rm circ})^{1/2} = \dot{M}(GMR_*)^{1/2} + G_{\rm tides} \qquad (5.71)$$

where $G_{\rm tides}$ is the tidal torque. This must hold over timescales short compared with that for orbital angular momentum loss, i.e. $J/\dot{J}$. Of course, by removing the angular momentum from orbits near the disc edge, $G_{\rm tides}$ prevents it spreading further and so defines its edge. By the reasoning above, this edge cannot be completely axisymmetric, or even symmetrical about the line of centres. Computation of $G_{\rm tides}$ is complex and must be performed numerically. Fig. 32 shows the result of a recent calculation. As can be seen, tidal effects are very strongly concentrated towards the edge of the disc; further, alternate quadrants of the disc gain and lose angular momentum from the secondary. The net effect of adding these torques to produce $G_{\rm tides}$ is shown in Fig. 33. As can be seen, $G_{\rm tides}$ is large and positive, and such a steep function of radius that it cuts off the disc at essentially the same average outer radius, independently of the value of R_*. This *tidal radius* $R_{\rm tides}$ depends on binary parameters, but is always found to be close to the estimate

$$R_{\rm tides} \cong 0.9 R_1 \qquad (5.72)$$

where R_1 is, as usual, the primary's Roche lobe radius.

The structure of the outer disc near the tidal radius governs the distinctive behaviour of a class of dwarf novae called the SU UMa systems. These undergo

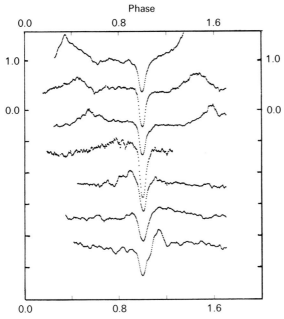

Figure 34. Superhumps observed during superoutbursts of the dwarf nova Z Cha. The figure shows orbital light curves, aligned by the deep central eclipse. The superhump is the large maximum seen near phase 0.35 in the upper curve and migrating to later phases, so that it appears near phase 0.1 in the lowest curve. This shows that the superhump has a period slightly longer than the orbital period. Data from B. Warner and D. O'Donoghue, with kind permission. The curves are assembled from different superoutbursts to give sufficient phase coverage.

outbursts in the way described in Section 5.7 above, but from time to time show rather longer events, called superoutbursts. The systems brighten much as in normal outbursts, but a photometric modulation of the optical light appears, known as superhumps. Fascinatingly, the superhumps do not have the orbital period, but one a few per cent longer (Fig. 34). The orbital periods P_{orb} of these objects are always found to be less than 3 h; because of the dearth of cataclysmics in the range 2–3 h this means that most have periods $\lesssim 2$ h. Conversely all dwarf novae in this period range are eventually found to undergo superoutbursts and reveal themselves as SU UMa systems.

Various models for superhumps were suggested in the years following their discovery, but the best clue as to their nature was provided by numerical simulations of discs as collections of interacting particles. These showed that for mass ratios $q \lesssim 0.25$–0.33 an injection of mass results in the disc adopting a rather eccentric shape, which precesses on a period slightly longer than P_{orb} (Fig. 35). This produces shear and dissipation, observable as a photometric modulation at the same period. These results are strongly reminiscent of the observed superhump phenomena; further, the fact that this behaviour occurs only for $q \lesssim 0.25$–0.33 would then explain why SU UMa systems all have $P_{orb} \lesssim 2$–3 h. For if the white dwarf masses

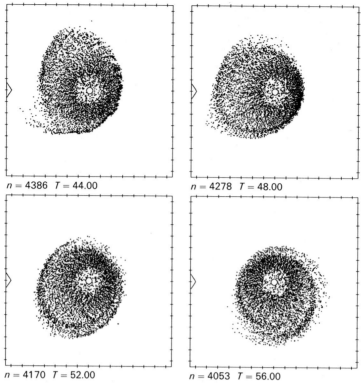

Figure 35. Numerical simulation of an accretion disc in an SU UMa system. The mass ratio is $q = 0.12$: the primary is the small circle, and part of the secondary's Roche lobe is visible at the left of each frame. n is the number of particles present in each frame and T the number of orbital cycles since the start of the simulation. The disc precesses on a period slightly longer than the orbit. (Figure by Dr R. Whitehurst).

are typically $\sim 0.6 M_\odot$ (which is plausible on other grounds), the secondaries are restricted to masses $\lesssim 0.2 M_\odot$. By (4.9) this requires $P_{\rm orb} \lesssim 2$ h. It remains to explain the behaviour seen in the simulations.

From the numerical simulations it is found that the particles in the precessing disc rim follow orbits with approximate periods very close to $P_{\rm orb}/3$; this strongly suggests that the superhumps are in some way caused by resonances between the binary orbit and the orbits followed by gas elements in the disc. Resonance occurs in a disc when the frequency of radial motion of a particle in the disc is commensurate with the angular frequency of the secondary star as seen by the particle. This condition ensures that the particle will always receive a 'kick' from the secondary at exactly the same phase of its radial motion, so allowing the cumulative effect of repeated kicks to build up and affect the motion significantly. We recall from the discussion of the tidal torque given above that the particle orbits will generally be somewhat aperiodic. This means that the orbits do not in general close, so that the particle is not at the same radial distance once it has completed an azimuthal motion through 2π about the white dwarf. The orbits can be closely

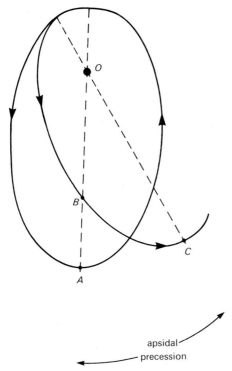

Figure 36. A precessing particle orbit about a mass at O. Starting from A, the particle completes a full revolution about O after a time $2\pi/\Omega$ (point B), where Ω is its mean angular frequency. However it reaches the same radial distance from O only at point C, at the later time $2\pi/(\Omega-\omega)$, where ω is the apsidal precession frequency (here shown prograde). The frequency of radial motion (the *epicyclic* frequency) is thus $\Omega-\omega$. Resonance with another orbiting body (e.g. the secondary star, if O is the primary in a binary system) occurs when this radial frequency is commensurate with the angular frequency $\Omega-\Omega_{\mathrm{orb}}$ of the secondary as seen by the particle (Ω_{orb} = secondary's angular frequency); this ensures that the same relative configuration of primary, secondary and test particle recurs periodically.

represented as ellipses whose axes precess at an *apsidal precession frequency* ω (Fig. 36) which depends on the particular orbit. If the mean frequency of angular motion in a given orbit is Ω (measured in a non-rotating frame) the azimuthal period is $2\pi/\Omega$, but the time for the particle to return to the same radial distance is somewhat longer, i.e. $2\pi/(\Omega-\omega)$. (Periodic orbits of course have $\omega = 0$ so that radial and azimuthal periods are the same.) Thus the frequency of radial motion, the *epicyclic* frequency, is $\Omega-\omega$. The particle sees the secondary star move with angular frequency $\Omega-\Omega_{\mathrm{orb}}$, where $P = 2\pi/\Omega_{\mathrm{orb}}$ is the orbital period of the binary. Thus resonance requires

$$k(\Omega-\omega) = j(\Omega-\Omega_{\mathrm{orb}}) \tag{5.73}$$

where j and k are positive integers. Cases with $k > 1$ arise if the symmetry of the binary orbit with respect to the particle orbit breaks down, so that successive kicks by the secondary in opposing directions do not completely cancel. An example of this would be a binary eccentricity e', which causes the secondary's distance from the white dwarf to vary by $\sim \pm e'a$, where a is the binary separation. In the case of a disc this symmetry-breaking is provided by the viscosity, which causes variations of order λ in particle orbits, where $\lambda \sim H$ is the viscous scale-length (see Sections 3.6 and 4.6). The apsidal precession frequency ω is in general much smaller than Ω and Ω_{orb}, so we see from (5.73) that resonances occur very near commensurabilities of the form $j:(j-k)$ between particle and binary frequencies. The effect of aperiodic motion ($\omega \neq 0$) is to shift the resonances slightly away from precise commensurabilities (similar effects can occur if we consider vertical motions of the disc particles). There is a large body of literature on resonances in the context of planetary satellites: perturbation expansions of the gravitational potential show that the disturbing effect of a (j, k) resonance builds up at a growth rate $\propto e^k$, where e is the eccentricity of the particle orbit. Since $e < 1$, the strongest resonances arise near the $j:j-1$ commensurability, followed by $j:j-2$ and so on. Near a given commensurability, those orbits with the largest values of e compatible with stability will resonate most strongly: these are generally aperiodic orbits with non-vanishing precession frequencies ω. Resonances thus act as filters picking out these extreme orbits provided the viscosity perturbs the disc orbits enough to populate them.

To get an idea of which resonances are possible in a disc we estimate the typical radii R_{jk} of resonant orbits near the $j:k$ commensurability. Even quite extreme aperiodic orbits do not differ greatly in size from Keplerian circles, so a good measure is provided by

$$R_{jk} = (GM/\Omega_{jk}^2)^{1/3} \tag{5.74}$$

where Ω_{jk} is the value of Ω given by the resonance condition (5.73). Using Kepler's law (4.1) with $P = 2\pi/\Omega$ gives

$$\frac{R_{jk}}{a} = \left(\frac{j-k}{j}\right)^{2/3} (1+q)^{-1/3}. \tag{5.75}$$

The possible resonances for a given mass ratio q are given by comparing this estimate with the tidal radius R_{tides} given by (5.72). Using (4.6a), with q replaced by q^{-1}, the result is shown in Fig. 37. Since we must have $R_{jk} < R_{\mathrm{tides}}$ it is immediately apparent that the strongest ($k = 1$) resonances can only occur for very small mass ratios: $j = 2, k = 1$ requires $q \lesssim 0.025$, with still smaller ratios required for larger j. These cannot occur in SU UMa systems, although some low-mass X-ray binaries have such extreme ratios. For a possible explanation of the superhumps in SU UMa systems we must turn to the next strongest ($k = 2$) resonances. The smallest radius is for $j = 3$, and is inside R_{tides} for $q \lesssim 0.33$. Larger values of j require mass ratios at least as extreme as for the 2:1 resonance. Thus the strongest resonance possible in the discs in SU UMa systems is that at $j = 3, k = 2$, and requires $q \lesssim 0.33$. From (5.73) we see that this will appear for particle orbits in the disc with periods close to $P/3$. This is precisely what is found in the simulations, and so is powerful circumstantial evidence that this resonance underlies the superhump phenomenon.

A full explanation of superhumps now requires an understanding of what

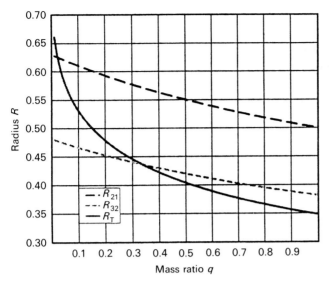

Figure 37. The resonant radii R_{21}, R_{32} and the tidal radius R_T. Resonant orbits can only exist for sufficiently small mass ratios q; the $j = 3$, $k = 2$ resonance (particle period $\frac{1}{3}$ of orbital) responsible for superhumps occurs only for $q \lesssim 0.33$.

happens near resonance. For *periodic* orbits this is well known: viewed in a reference frame corotating with the binary, resonances cause period-doubling of such orbits; i.e. the orbit *circulates* twice before closing on itself. Viewed in a non-rotating frame these orbits are seen to circulate three times before closing on themselves, thus following a three-petalled rosette. The period of this orbit is clearly three times that of the original periodic orbit, i.e. $6\pi/\Omega$. A similar result must hold for the precessing orbits ($\omega \neq 0$) which resonate most strongly. For $j = 3$, $k = 2$, equation (5.73) shows that $\Omega = 3\Omega_{\mathrm{orb}} - 2\omega$ for them, so that the circulating orbit resulting from the resonance has a radial period

$$P_{\mathrm{SH}} = 6\pi/\Omega \cong P_{\mathrm{orb}}(1 + 2\omega/3\Omega_{\mathrm{orb}}), \tag{5.76}$$

where we have used the fact that $\omega \ll \Omega_{\mathrm{orb}}$. The precession is prograde ($\omega > 0$), so the radial period of these resonant orbits is slightly longer than P_{orb}. We see that P_{SH} has just the properties of the observed superhump period. Of course this explanation is no more than a plausibility argument: in a fluid disc, it is not obvious which are the extreme orbits which resonate, nor whether the interaction of the various orbits will result in the radial period being observable. It does however strongly suggest that the results of the numerical simulations giving the superhump behaviour are believable.

An important question is why the superhumps only appear in superoutbursts, rather than in quiescence or normal outbursts. Clearly the resonance is only likely to be manifest if the relevant radius of the disc is well populated, which is much more likely at outburst or superoutbursts than quiescence. What seems to discriminate between superoutbursts and the shorter normal outbursts is the possibility that the former may represent episodes of enhanced mass transfer from the secondary star. In numerical simulations the perturbing effect of the gas stream from this star considerably shortens the growth time for the superhumps, allowing them to appear before the resonant region is drained of mass by accretion. Observations suggest that superoutbursts may actually be triggered by certain normal outbursts. It is thus possible that both outburst models described in Section 5.8 are relevant, in that disc instabilities may cause normal outbursts, which sometimes lead to variable mass transfer.

5.10 Discs around young stars

The process of star formation by the collapse of a protostellar cloud is a complicated phenomenon in which gravity, thermal pressure, rotation, magnetic fields, and microphysics all play important parts. It is quite possible that star formation is a 'bimodal' process with different emphasis on the above ingredients depending on the stellar mass (e.g. Shu, Adams & Lizano (1987)). The formation of stars with masses in excess of a few solar masses is likely to be a violent and efficient process in the sense that a significant fraction of the collapsing gas mass ends up in the newborn star in a dynamical timescale. Since rotation is present, it may well be that a disc of some sort forms while the protostar is dealing with its excess of angular momentum. Because young massive stars in giant molecular clouds are formed so fast, they are shrouded from view in their early life and by the time they become visible the disc is no longer there either because a proper disc never formed or because it has been destroyed by dynamical processes and/or winds. Fortunately the formation of stars with masses $\lesssim 2\,M_\odot$ is more accessible to observation and consequently better understood theoretically. We know that discs exist around low-mass young stars and we can explain a number of important observational properties of these systems using a simple model consisting of an accretion disc around a star of mass $\sim 1\,M_\odot$. The rest of this section is therefore confined to the study of low-mass young stars typified by those found in the Taurus star-forming region.

The evidence for discs around young stellar objects has become really compelling over the last few years (e.g. Beckwith *et al.* (1990)). The direct evidence comes from interferometric observations of ^{13}CO lines at mm wavelengths. These observations are capable of resolving spatially the disc in a couple of cases – HL Tau and DG Tau – where the disc is *seen* to extend to $\gtrsim 10^3$ AU. Furthermore, the observations demonstrate that the gas is in Keplerian rotation about a $\sim 1\,M_\odot$ star. These are in fact the only cases for which we have *direct* evidence of a circumstellar (accretion) disc in Keplerian rotation. The evidence for accretion taking place is indirect since the radial drift is unobservable. The indirect evidence in favour of discs is more extensive, being based on a large number of objects. The observed spatial distribution of polarization and the association of young stars with bipolar outflows and jets clearly indicates the presence of a preferred axis, presumably the axis of rotation of the disc. The broad spectral energy distribution and the correlation between the Doppler width of absorption features with wavelength can be understood by involving an accretion disc whose emission is due to viscous dissipation and – in some cases – reprocessing of the radiation from the central star. In the early phases of low-mass star formation, an opaque dusty cloud collapses forming in its shrouded interior an incipient star plus disc system. This infall lasts for about $\sim 10^5$ years and reveals, after some mass loss, a newly-formed T Tauri star (see Hartmann, Kenyon and Hartigan (1991)). We know relatively more about the formation of low-mass young stars because they are revealed to us in a relatively earlier stage of formation. In the same sense used above for describing massive star formation, the formation of low-mass stars is a gentle and inefficient process. Recent observational estimates suggest that after the young star has become visible, it accretes at least 5% to 10% of its mass from the circumstellar disc. The disc around the T Tauri star generally accretes on to the central star at a rate of $\sim 10^{-7} M_\odot/\text{yr}$ with typical variability by a factor of ~ 10 over months and years and with occasional outbursts where $\dot{M} \sim 10^{-4} M_\odot/\text{yr}$. These 'Fuor' eruptions – after the prototype FU Orionis – have a rise time typically ~ 1 yr, but in some cases the rise is gradual over a decade. The decay times are of the order of decades or centuries, and in the case of FU Ori itself, probably more than a millennium. Since the number of Fuor outbursts observed in the solar neighbourhood is approximately ten times larger than the estimated local star formation rate, one concludes that the average young star undergoes around ten Fuor eruptions in its lifetime. More modest outbursts of the type observed in EX Ori may occur frequently towards the end of the life of a classical T Tauri. The classical T Tauri phase ends when accretion stops and the system probably evolves into a weak-emission T Tauri star.

Currently the most promising model for the Fuor phenomenon is an accretion disc instability analogous to the limit cycle thought to occur in cataclysmic variables (cf. Section 5.8). Detailed hydrodynamic simulations of the limit cycle including the appropriate microphysics – in particular the formation and sublimation of grains – remain to be done. The outburst timescales are indicative of a viscosity parameter $\alpha \sim 10^{-3} - 10^{-2}$. The strengths of the disc model for Fuors could be summarized by saying that it is successful in providing a qualitative and quantitative account of the following observations: (a) the broad spectral distribution of the observed radiation, (b) the correlation of rotational broadening with wavelength, (c) the double-peaked absorption line profiles, and (d) the outburst energetics and timescales. Not everything is without problems: the temperature distribution during decline appears flatter than predicted by steady-state models; also the disc appears to be rotating faster at maximum light contrary to simple expectations. It is however quite likely that more realistic time-dependent models will resolve these difficulties.

5.11 Spiral shocks

When non-axisymmetric waves are excited in an accretion disc – for example by the tidal forces due to a companion star – the differential rotation rapidly shears the disturbances into spiral waves. The resultant spiral waves are in general trailing since the angular velocity increases inwards in most situations. Although it has been known for more than twenty years that such waves are capable of transporting angular momentum by gravitational coupling, it is only recently that the possibility of driving accretion by such waves has attracted considerable interest. This is due to two factors: the increasing power of modern vectorized and parallel computers that has allowed an expansion in the scope and spatial resolution of large scale numerical simulations of accretion flows in binary systems; and the recent emergence of a better physical understanding of the transport of angular momentum by such waves even in the absence of gravitational coupling.

It can be shown that *trailing* spiral waves carry negative angular momentum in the sense that a disc in which trailing spirals are present has a lower angular momentum than the unperturbed disc. As long as the waves remain linear, they do not interact with the fluid and the angular momentum of the fluid and the waves is conserved separately. However, these spiral waves steepen into shocks as they propagate inward and, as they dissipate, they communicate to the fluid within the angular velocity of the outer parts of the disc. The resultant transport of angular momentum outwards drives accretion *without* the requirement of *any* viscosity except of course within the shocks. The transport of angular momentum by more general non-axisymmetric disturbances – with and without the inclusion of gravity – finds applications in very diverse astrophysical problems such as star formation, discs around young stars, cosmogony, planetary rings, discs in accreting binaries, and active galactic nuclei (see also Section 7.6). We refer the reader interested in further details to recent review articles since a quantitative treatment of these problems is beyond the scope of this book. A summary of the physics involved in these and other angular momentum transport processes and references to recent work can be found in Morfill *et al.* (1991).

In the case of accretion discs in close binaries, the numerical simulations of mass transfer in a Roche potential have shown that it is possible to set up a steady two-armed shock pattern which leads to steady accretion. These shocks arise and are maintained as a result of the dominant $m = 2$ tidal perturbations at the edge of the disc. But as they propagate inwards, they reach a balance between amplification and dissipation, and yield an angular momentum transport rate which is independent of the strength of the wave excited at the edge. The efficiency of the process, when expressed in terms of an *effective* viscosity parameter α, can be compared to the values inferred from observations – in particular from the outburst timescales (Section 5.9). Unfortunately, the values obtained so far are of the order of $\alpha \sim 10^{-4}$–10^{-2}, too small for cataclysmic variables and probably also for low-mass X-ray binaries. It is possible, however, that future calculations with more detailed physics, including cooling, will yield higher values of α. In any case, the process may be relevant to young stars with companions or giant planets since low values of effective α are sufficient to drive the evolution of accretion discs around young stars.

6

Accretion on to a compact object

6.1 Introduction

So far, we have seen in some detail how an accretion disc provides an efficient machine for the extraction of up to half the available accretion energy, but it must be confessed that our understanding of how the remaining substantial fraction of the accretion luminosity is released near the central object is rather less advanced. We have already mentioned the existence of an inner boundary layer in the case where the disc extends down to the surface of the accreting star. For neutron stars, runaway nuclear burning may occur at the surface and give rise to the X-ray bursters. In many systems the accreting star possesses a magnetic field strong enough to disrupt the inner regions of the disc and channel the accretion flow on to the magnetic polecaps. In this event, the accretion flow more nearly resembles free-fall on to the stellar surface. A black hole, on the other hand, cannot possess an intrinsic magnetic field, so column accretion of this kind cannot occur. In this chapter we shall examine these modes of accretion in some detail.

6.2 Boundary layers

In discussing the inner boundary condition for steady thin accretion discs (Section 5.3) we postulated that the angular velocity $\Omega(R)$ in the disc remains very close to the Keplerian value $\Omega_K(R) = (GM/R)^{1/2}$ until the accreting matter enters a boundary layer of radial extent b just outside the surface $R = R_*$ of the accreting star. Within this boundary layer Ω must decrease from a value $\Omega(R_* + b) \cong \Omega_K(R_* + b)$ to the surface angular velocity $\Omega_* < \Omega_K(R_*)$. Thus we envisage $\Omega(R)$ as a function having the form shown in Fig. 38.

Our derivation of the inner boundary condition (5.16) on the disc assumed $b \ll R_*$. Let us now justify this assumption. We consider in the thin disc approximation the radial component of the Euler equation (2.3), as given by (5.25):

$$v_R \frac{\partial v_R}{\partial R} - \frac{v_\phi^2}{R} + \frac{1}{\rho}\frac{\partial P}{\partial R} + \frac{GM}{R^2} = 0. \tag{5.25}$$

In a thin Keplerian disc this equation is dominated by the centrifugal (v_ϕ^2/R) and gravity (GM/R^2) terms as we showed in Section 5.3. However, the boundary layer is by definition that region of the disc in which $v_\phi^2 < v_K^2 = GM/R$. Thus the gravity term in (5.25) must be balanced either by $v_R \partial v_R/\partial R \sim v_R^2/b$ or the pressure gradient

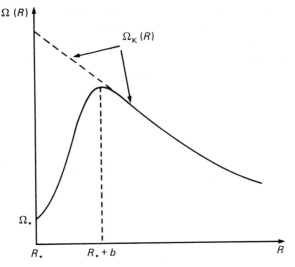

Figure 38. Distribution $\Omega(R)$ of angular velocity near the inner edge of an accretion disc around a star with surface angular velocity Ω_* less than the break-up value $\Omega_K(R_*)$. The region $R_* \leq R \leq R_* + b$, with $b \ll R_*$, is the boundary layer.

$\rho^{-1} \partial P / \partial R \sim c_s^2 / b$ where as usual we have taken $P = \rho c_s^2$ and set $\partial / \partial R \sim b^{-1}$ in the boundary layer. But we can infer that $c_s^2 > v_R^2$, since otherwise the inflow to the stellar surface would be supersonic and the 'news' of the presence of the stellar surface at $R = R_*$ could be communicated outwards to the disc. Hence in the boundary layer (5.25) is approximately

$$\frac{c_s^2}{b} \sim \frac{GM}{R_*^2}. \tag{6.1}$$

The boundary-layer size b is now given by noting that just outside $R_* + b$ the standard disc relation (5.23) implies a scale-height (assuming radiation pressure can be neglected)

$$H \sim c_s \left(\frac{R_*}{GM}\right)^{1/2} R_*. \tag{6.2}$$

Assuming that H and c_s just inside $R_* + b$ are similar to their values just outside, (6.2) and (6.1) can be combined to give

$$b \sim \frac{R_*^2}{GM} c_s^2 \sim \frac{H^2}{R_*}. \tag{6.3}$$

This justifies our assumption $b \ll R_*$, for

$$b \sim H^2 / R_* \ll H \ll R_*.$$

A schematic picture of the boundary layer geometry is given in Fig. 39. From the figure it is clear that radiation emitted by the boundary layer emerges through a region of radial extent $\sim H$ on the two faces of the disc. If the accretion rate, and therefore the density in this region, is high enough it will be optically thick and radiate roughly as a blackbody of area $\sim 2\pi R_* H \times 2$. We already know that the luminosity emitted by this area must be $\frac{1}{2}L_{\rm acc} = GM\dot{M}/2R_*$. Thus there is a characteristic boundary layer blackbody temperature $T_{\rm BL}$ given by

$$4\pi R_* H\sigma T_{\rm BL}^4 \sim \frac{GM\dot{M}}{2R_*}$$

where σ is the Stefan–Boltzmann constant. Comparison with the expression (5.42) for the characteristic disc blackbody temperature T_* shows that

$$T_{\rm BL} \sim \left(\frac{R_*}{H}\right)^{1/4} T_*. \tag{6.4}$$

Using (6.2) to eliminate H and remembering that $c_{\rm s}^2 \sim kT_*/\mu m_{\rm H}$ in this region, we get

$$T_{\rm BL} \sim T_*(T_{\rm s}/T_*)^{1/8} \tag{6.5}$$

where

$$T_{\rm s} = \tfrac{3}{8}\frac{GM\mu m_{\rm H}}{kR_*} \tag{6.6}$$

is the shock temperature (cf. (3.64)) we would have got for radial accretion on the star. $T_{\rm BL}$ exceeds T_*, but not by a large factor because of the very weak dependence of $T_{\rm BL}$ on $T_{\rm s}/T_*$ (equation (6.5)). For accretion on to a white dwarf, where the discussion given above has the greatest chance of being valid, we have

$$T_{\rm BL} \sim 1 \times 10^5 \dot{M}_{16}^{7/32} M_1^{11/32} R_9^{-25/32} \text{ K} \tag{6.7}$$

and $T_{\rm BL} \sim 3T_*$. Since the maximum temperature $T(R)$ in the disc is only $0.488T_*$ (Section 5.5), $T_{\rm BL}$ is significantly higher ($\sim 6\times$) than this maximum temperature. Equation (6.7) predicts that some cataclysmic variables ought to be soft X-ray sources as the tail of the blackbody at ~ 0.1 keV may be observable. This is indeed found to be the case, although most of the blackbody emission ($\lesssim 0.1$ keV) is hidden by interstellar absorption so that estimates of the total luminosity are very uncertain.

Because of the blackbody assumption made above we are able to avoid any discussion of the internal structure of the boundary layer. However, this is achieved at the expense of the assumption that the viscosity in the boundary layer can adjust itself suitably to make the structure self-consistent: for example $v_R < c_{\rm s}$ must be satisfied. Our ignorance of viscosity is crucial here, not least since the rate of shearing ($\propto d\Omega/dR$) in the boundary layer is very high. This problem becomes still more acute if the boundary layer is optically thin. Optically thin boundary layers are of especial interest for cataclysmic variables, since some of these are observed to be hard (~ 10 keV) as well as soft X-ray sources. We know (Section 5.7) that

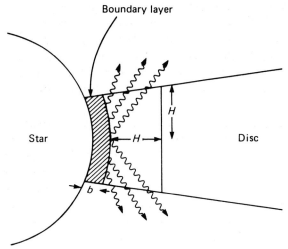

Figure 39. Schematic view (in a plane $\phi =$ constant) of an optically thick boundary layer (not to scale).

accretion discs roughly corresponding to the 'standard model' of Section 5.6 exist in these systems; in many systems there is no evidence such as periodic pulsing of the X-rays that the white dwarf has a magnetic field strong enough to disrupt the disc (see Section 6.3). Thus the disc presumably extends down to the white dwarf surface, with the boundary layer as the only plausible site for the X-ray production. As we have seen, the blackbody temperature $T_{\rm BL}$ for an optically thick boundary layer is too low to give hard X-ray emission, so optically thin boundary layers, with all their attendant complications and uncertainties, must be considered.

To date two types of X-ray production mechanism have been proposed. One uses the fact that the Kepler velocity $(GM/R_*)^{1/2}$ near the white dwarf surface is highly supersonic, and will, if the circulating gas can be strongly shocked, produce hard X-ray temperatures: from (3.58) we get

$$T = \tfrac{3}{16}\frac{\mu m_{\rm H}}{k}\frac{GM}{R_*} \cong 1.9 \times 10^8 M_1 R_9^{-1} \text{ K}.$$

The main difficulty here is in producing the strong shocks in the highly shearing boundary layer, rather than a series of weak shocks oblique to the flow. A two-stage process has been suggested, in which shocked blobs of gas expand into the path of circulating supersonic material, causing further strong shocks. The great complexity of the gas flow in the boundary layer has so far prevented a self-consistent treatment of this idea.

An alternative treatment shows that the optically thin boundary layer gas is likely to be thermally unstable to turbulent viscous heating (cf. Section 5.8), rapidly achieving hard X-ray temperatures ($\sim 10^8$ K). At the same time, this gas is likely to expand out of the boundary layer, and form a hard X-ray corona around most of the white dwarf. At high accretion rates $\dot{M}$ the instability will be largely suppressed, and the boundary layer luminosity mostly emitted as soft X-rays. The X-ray

Magnetized neutron stars and white dwarfs

emission from the dwarf nova SS Cygni shows behaviour which supports this picture. At quiescence (low $\dot{M}$) there is relatively bright, but very variable, hard X-ray emission. During optical outbursts (high $\dot{M}$) the hard X-rays are severely reduced, and the soft X-rays brighten. The hard X-ray spectra show no evidence for increased absorption, so that suppression is not the result of covering the hard X-ray region with cool material, as is appealed to in earlier models.

6.3 Accretion on to magnetized neutron stars and white dwarfs

The picture of the boundary layer accretion described above can only be relevant if the disc extends right down to the surface of the accreting star. Quite often this will not be the case, as white dwarfs and especially neutron stars often possess magnetic fields of an order ($\lesssim 10^7$ G and $\sim 10^{12}$ G respectively) strong enough to disrupt the disc flow, as shown schematically in Fig. 40. In general the interaction of the disc and magnetic field is exceedingly complex. Let us consider first the rather simpler case in which the stellar magnetic field disrupts an accretion flow which is quasi-spherical far from the star.

For a dipole-like magnetic field, the field strength B varies roughly as

$$B \sim \frac{\mu}{r^3} \tag{6.8}$$

at radial distance r from the star of radius R_*; here $\mu = B_* R_*^3$ is a constant (the magnetic moment) specified by the surface field strength B_* at $r = R_*$. Thus from (3.44) there is a magnetic pressure

$$P_{\text{mag}} = \left[\frac{4\pi}{\mu_0}\right]\frac{B^2}{8\pi} = \left[\frac{4\pi}{\mu_0}\right]\frac{\mu^2}{8\pi r^6} \tag{6.9}$$

increasing steeply as the matter approaches the stellar surface. This magnetic pressure will begin to control the matter flow and thus disrupt the spherically symmetric infall at a radius r_M where it first exceeds the ram and gas pressures of the matter (cf. equation (3.45)). For highly supersonic accretion, as expected on the basis of our discussion in Section 2.5, it is the ram pressure term ρv^2 which is important, with the velocity v close to the free-fall value $v_{\text{ff}} = (2GM/r)^{1/2}$ and $|\rho v|$ given in terms of the accretion rate $\dot{M}$ by (2.19):

$$|\rho v| = \frac{\dot{M}}{4\pi r^2}.$$

Thus setting $P_{\text{mag}}(r_M) = \rho v^2|_{r_M}$ we find

$$\left[\frac{4\pi}{\mu_0}\right]\frac{\mu^2}{8\pi r_M^6} = \frac{(2GM)^{1/2}\dot{M}}{4\pi r_M^{5/2}}$$

or

$$r_M = 5.1 \times 10^8 \dot{M}_{16}^{-2/7} M_1^{-1/7} \mu_{30}^{4/7} \text{ cm} \tag{6.10}$$

where μ_{30} is μ in units of 10^{30} G cm^3. We note that a neutron star with $B_* \simeq 10^{12}$ G,

Figure 40. Accretion disc around a magnetized neutron star or white dwarf.

$R_* \cong 10^6$ cm has $\mu_{30} \cong 1$, as does a white dwarf with $B_* \cong 10^4$ G, $R_* \cong 5 \times 10^8$ cm. Thus the order of magnitude of (6.10) suggests that observable effects are quite possible in systems where accretion takes place on to a magnetized neutron star or white dwarf. It is often convenient to replace $\dot{M}$ in (6.10) in terms of the accretion luminosity (1.3), which is more directly related to the observational quantities, especially for X-ray sources. Thus

$$r_M = \begin{Bmatrix} 5.5 \times 10^8 M_1^{1/7} R_9^{-2/7} L_{33}^{-2/7} \mu_{30}^{4/7} \text{ cm} \\ 2.9 \times 10^8 M_1^{1/7} R_6^{-2/7} L_{37}^{-2/7} \mu_{30}^{4/7} \text{ cm} \end{Bmatrix} \qquad (6.11)$$

with parametrizations appropriate for white dwarfs and neutron star accretors respectively ($L_{33} = L_{\text{acc}}/10^{33}$ erg s^{-1}, etc.). Of course the estimates (6.10) and (6.11) of r_M are rather crude. However, since P_{mag} is such a steep function of radius ($\sim r^{-6}$) we may hope they are acceptable at least in an order-of-magnitude sense. The main uncertainty arises from the plethora of instabilities which a plasma supported by a magnetic field is prey to. These may, for example, allow matter to slip through the field-lines before being channelled on to the stellar surface. The quantity r_M is known as the Alfvén radius. Within the Alfvén radius we expect that the matter will flow along field-lines (Section 3.7).

Let us return to the consideration of disc accretion. Here the condition for magnetic disruption at cylindrical radius $R = R_M$ must be that the torque exerted by the magnetic field on the disc at R_M should be of the order of the viscous torque $G(R_M)$. In a steady state (equation (5.12) et seq.) this is equal to the transport rate of specific angular momentum $\dot{M} R_M^2 \Omega(R_M)$. The main difficulty in this calculation is in trying to find an expression for the magnetic torque at R_M, for this involves the azimuthal component of $\mathbf{B}$ (B_ϕ), which in turn depends on how much the field-lines are distorted from a dipole-like configuration by the interaction with the disc and the effects of the instabilities referred to above. Several estimates of the resulting R_M have been made, of varying degrees of elaboration. Fortunately, because of the steep ($\sim R^{-6}$) dependence of magnetic stresses on radius, they give similar answers – typically slightly smaller than the spherical Alfvén radius r_M (equations (6.10) and (6.11)):

$$R_M \sim 0.5 r_M. \qquad (6.12)$$

Of course, the precise results must depend on the inclination of the dipole axis to the disc plane. In general, the accreting star and hence the magnetic field pattern rotate

Magnetized neutron stars and white dwarfs

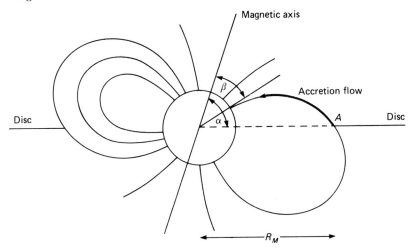

Figure 41. Accretion from a disc to the polecaps of a magnetized neutron star or white dwarf.

with an annular velocity Ω_* about an axis perpendicular to the disc, usually in the same sense as the disc angular velocity. Since inside $R = R_M$ the matter must flow along field-lines, if steady accretion is to occur we must have $\Omega_* < \Omega_K(R_M)$ where $\Omega_K(R_M) = (GM/R_M^3)^{1/2}$ is the Kepler angular velocity at R_M. Otherwise particles attached to field-lines at $R = R_M$ would spiral outwards to larger R, repelled by the 'centrifugal barrier' at R_M. A convenient way of stating this requirement is that the 'fastness parameter'

$$\omega_* = \Omega_*/\Omega_K(R_M) \tag{6.13}$$

should be < 1.

Combining (6.12) and (6.11), we see that for typical parameters ($M_1 \sim R_6 \sim L_{37} \sim \mu_{30} \sim 1$) a magnetized neutron star will have $r_M \sim R_M \sim 10^8$ cm, well outside the stellar radius $R_* \sim 10^6$ cm. Since L_{37} cannot exceed the Eddington limit value ~ 10 (equation (1.26)), and M_1, R_6 are always near unity, the radii r_M and R_M can only be less than $R_* \cong 10^6$ cm for rather low values of the energy magnetic field, $B_* \lesssim 10^9$ G. Thus we expect accretion on to neutron stars to be controlled by the magnetic field near the surface in many cases. Furthermore, R_M for a neutron star is always less than the circularization radius R_{circ} (e.g. equation (4.19)) for Roche lobe accretion in any plausible binary geometry. Since R_{circ} is a lower limit to the size of any accretion disc, we see that disc formation is not affected in this case. For wind accretion (cf. (4.38) *et seq.*), however, the question is much more difficult, and the whole subject of stellar wind accretion by magnetized neutron stars is one of considerable uncertainty.

Magnetically controlled accretion gives rise to a simple and immediately recognizable observational signature because the accretion flow is channelled on to only a small fraction of the stellar surface. To see how this works consider the situation depicted in Fig. 41, in which a disc is disrupted by a dipole-like field. We assume that at point A matter leaves the disc and follows the magnetic field-lines down to the polecaps of the accreting star. In polar coordinates (r, θ) with origin at the star's centre the field-lines have the approximate equations of dipole geometry $r = C \sin^2 \theta$ where C is a constant labelling all those field-lines emanating from a particular latitude on the star. At A we have $r = R_M$ and $\theta = \alpha$ so $C = R_M \sin^{-2} \alpha$. This field-line thus crosses the stellar surface $r = R_*$ at colatitude β given by

$$\sin^2 \beta = R_*/C = (R_*/R_M) \sin^2 \alpha.$$

If we neglect the possible effects of plasma instabilities, accretion cannot take place outside a polecap of half-angle β (see Fig. 41), since for larger colatitudes the relevant field-line crosses the disc plane inside $r = R_M$, which little matter can otherwise penetrate. The area of an accreting polecap is thus a fraction

$$f_{\text{disc}} \sim \frac{\pi R_*^2 \sin^2 \beta}{4\pi R_*^2} \cong \frac{R_* \sin^2 \alpha}{4 R_M}$$

of the total stellar surface; accretion may take place on the opposite polecap also. Thus in general only a fraction

$$f_{\text{disc}} \sim R_*/2R_M \tag{6.14}$$

of the stellar surface accretes. From (6.11) and (6.12) we find $f_{\text{disc}} \sim (10^{-1}\text{--}10^{-4}$ depending on L and μ. Like many of the estimates in this section, f_{disc} is subject to considerable uncertainty because of the complicated role of plasma instabilities. In particular there is evidence (from hard X-ray light curves) that f_{disc} as given by (6.14) may be an underestimate for the intermediate polars. Nonetheless it is clear that f_{disc} is certainly smaller than unity in most cases; we get a very similar estimate $f_{\text{sph}} \sim 0.5 f_{\text{disc}}$ for the case of quasi-spherical accretion simply by making the replacement $R_M \to r_M$. Because most of the accretion luminosity must be released close to the stellar surface and, as we have seen, only over a fraction of it, any rotation of the accreting star will produce a periodic modulation in the observed flux from such a system. The effect will be enhanced if the radiation from the accreting polecap is beamed (see Section 6.4). Pulsation periods P_{pulse} of this kind are observed in many X-ray binaries and intermediate polar systems, generally in X-rays, but sometimes optically, in the range

$$1 \text{ s} \lesssim P_{\text{pulse}} \lesssim 10^3 \text{ s}.$$

Because these periods are so short, it is relatively easy to obtain a sufficient data base to detect significant changes in P_{pulse}. These changes are of two main types. The first kind of change in P_{pulse} was noticed in the X-ray binary Cen X-3 very soon after its discovery by the Uhuru satellite in 1971. The 4.8 s pulse period is itself periodically modulated, by about one part in 1000, every 2.087 days. This modulation is easy to understand if we imagine the X-ray pulsations to act as a clock carried about by the neutron star. In its motion around the companion star, the clock frequency is periodically Doppler shifted just like the spectral lines of a spectroscopic binary. Hence one can construct a radial velocity curve from this 'pulse-timing analysis' and use it to determine the orbital period (2.087 days for Cen X-3) and a mass function in the standard way. In several X-ray binary systems, one can also measure radial velocity shifts in the spectral lines of the companion: these systems are then directly analogous to double-lined spectroscopic binaries and allow masses and inclinations to be determined. Whenever this can be done, the mass of the neutron star is always closely consistent with a value $\sim 1.4\,M_\odot$, satisfyingly close to that expected on evolutionary grounds. Furthermore, the non-compact star is usually found to be quite close to filling its Roche lobe. Because pulse-timing analysis is relatively 'clean' and free from most of the effects which can produce spurious eccentricities in radial velocity curves determined conventionally, results obtained in this way give reliable evidence that the neutron star orbits in some X-ray binaries are quite eccentric ($e \gtrsim 0.3$).

The second type of systematic change in pulse periods is more complicated. In many X-ray binary systems, P_{pulse} is observed to be decreasing steadily on a timescale of order 10^4 yr, sometimes with occasional short-lived increases in P_{pulse}. Presumably, the steady decrease in P_{pulse} (spinup) is due to the torques induced by the accretion process, while the occasional increases (spindowns) are caused either by fluctuations in the accretion torque or possibly by changes in the internal structure of the neutron star. With the observed pulse periods ($\gtrsim 1$ s) and likely values of R_M and r_M ($\sim 10^8$ cm, equations (6.11), (6.12) and (6.13)) the fastness parameter ω_* (6.13) is $\ll 1$. For any such 'slow rotators' which accrete from a disc, the straightforward accretion of angular momentum at R_M at the rate $\sim \dot{M}R_M^2 \Omega_{\text{K}}(R_M)$ discussed above is the dominant torque. For faster rotators ($\omega_* \sim 1$), other effects such as the interaction of the stellar magnetic field with parts of the disc at $R > R_M$, or the loss of angular momentum through the blowing-off of a 'wind' of particles along magnetic field-lines from the stellar surface can be important.

Thus in the case of a slow rotator accreting from a disc, if I is the moment of inertia of the star we have

$$I\dot{\Omega}_* = \dot{M}R_M^2 \Omega_{\text{K}}(R_M) = \dot{M}(GMR_M)^{1/2} \quad (6.15)$$

where we have assumed that Ω_* is in the same sense as the disc rotation (otherwise spindown occurs). Now using (6.11) and (6.12) and replacing $\dot{M}$ by RL_{acc}/GM as before, we get

$$\frac{-\dot{P}_{\text{pulse}}}{P_{\text{pulse}}} \cong 8 \times 10^{-5} M_1^{-3/7} R_6^{6/7} L_{37}^{6/7} \mu_{30}^{2/7} I_{45}^{-1} P_{\text{pulse}} \text{ yr}^{-1} \quad (6.16)$$

where I_{45} is the moment of inertia ($\sim MR_*^2$) of the accreting star in units of 10^{45} g cm^2. For a neutron star, $I_{45} \sim 1$, while a white dwarf has $I_{45} \sim 10^5$. Because of its much larger moment of inertia, a white dwarf is much harder to spin up, and (other things being equal) has a much smaller value of $-\dot{P}_{\text{pulse}}/P_{\text{pulse}}$. From (6.16), we see that $-\dot{P}_{\text{pulse}}/P_{\text{pulse}}$ depends mainly on the combination $P_{\text{pulse}} L^{6/7}$. Fig. 42 compares the measured spinup rates $-\dot{P}_{\text{pulse}}/P_{\text{pulse}}$ for ten pulsating X-ray sources with the theoretical predictions for neutron stars and white dwarfs. Generally, the agreement with (6.16) for the neutron star case is excellent, and all the measured points lie well within the likely theoretical scatter about the mean relation. By contrast, all of the $-\dot{P}_{\text{pulse}}/P_{\text{pulse}}$ values are far too large for the accreting objects to be white dwarfs (dashed line). This is strong confirmation of the identification of the

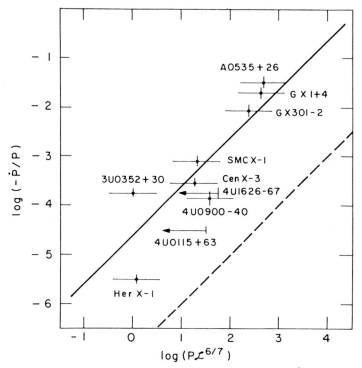

Figure 42. Observed values of the spinup rates $\dot P_{\rm pulse}/P_{\rm pulse}$ plotted versus $P_{\rm pulse} L^{6/7}$ for ten pulsating binary X-ray sources. The solid line is the best fit straight line with logarithmic slope 1, and is in good agreement with equation (6.16), the theoretical prediction for neutron stars. The dashed line is the expected relation for a 1 $M_\odot$ white dwarf. (Reproduced from Rappaport & Joss in *Accretion-Driven Stellar X-ray Sources*, eds. W. H. G. Lewin & E. P. J. van den Heuvel (1983), Cambridge University Press.)

accreting objects as magnetized neutron stars, and circumstantial evidence that the accretion occurs via a disc in some cases. (The caveat here is that we cannot calculate very accurately the spinup rate for wind accretion when there is no disc. Thus although the observed spinup rates are consistent with disc accretion they might also be consistent with some version of 'discless' accretion.)

Eventually, of course, any accreting star that is spun up for a sufficiently long time will enter the fast rotation regime $\omega_* \sim 1$. Here, as mentioned above, other torques balance and $P_{\rm pulse}$ attains an equilibrium value in which $\omega_* \sim 1$, remaining there until external conditions (e.g. the accretion rate) change. This could, for example, be the case for the white dwarfs in the intermediate polars we have just discussed. Another way of expressing the requirement $\omega_* < 1$ is to ask that the *corotation radius*

$$R_\Omega = \left(\frac{GMP_{\rm pulse}^2}{4\pi^2}\right)^{1/3} = 1.5 \times 10^8 P_{\rm pulse}^{2/3} M_1^{1/3} \text{ cm} \qquad (6.17)$$

where a particle attached to a field-line would rotate at the Keplerian rate, should

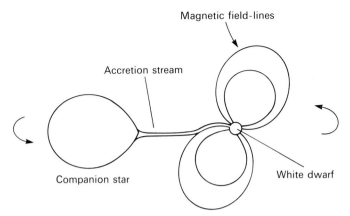

Figure 43. Schematic view of an AM Herculis system. The rotation of the strongly magnetic ($\gtrsim 10^7$ G) white dwarf is locked to that of the binary ($P \lesssim 4$ h). No accretion disc forms, matter impinging directly on the magnetosphere and following field-lines down to the white dwarf surface.

not be less than R_M. Thus at the equilibrium pulse period P_{eq} we have $R_\Omega \sim R_M$, so that from (6.11) and (6.12)

$$\left. \begin{array}{l} P_{eq} \sim 3 M_1^{-2/7} R_9^{-3/7} L_{33}^{-3/7} \mu_{30}^{6/7} \text{ s} \\ \sim 3 M_1^{-2/7} R_6^{-3/7} L_{37}^{-3/7} \mu_{30}^{6/7} \text{ s.} \end{array} \right\} \qquad (6.18)$$

Since $\mu_{30} \lesssim 1$ for neutron stars, it is clear that only a few of the X-ray binary accretors can be near their equilibrium periods: in some of these cases their interpretation as either fast or slow rotators seems possible.

In principle, one would like to understand the 'spin history' of all types of magnetized accreting stars and relate this to the observed pulse period distribution and spinup (spindown) rates. This is an ambitious undertaking, involving an understanding of all the various types of torque which can operate, as well as the evolution of the binary systems containing the objects, and observational selection effects on the P_{pulse}, $\dot{P}_{pulse}$ distributions. It is scarcely surprising therefore that no general consensus has been arrived at.

A particularly acute problem is posed by the large fraction of known pulsating X-ray sources with long pulse periods ($\gtrsim 10^2$ s) but which are spinning up on timescales (~ 50–100 yr) very short compared to their lifetimes ($\sim 10^3$–10^6 yr) as bright X-ray sources expected from evolutionary arguments. Most attempts to explain these facts begin in the same way. By an analogy with radio pulsars, one expects neutron stars in binaries to be formed following a supernova with a very short spin period of a fraction of a second. At such spin rates we have $R_\Omega \ll R_M$ (i.e. $\omega_* \gg 1$) and accretion is impossible because any matter approaching the neutron star (e.g. from the companion's stellar wind) will be thrown off by the centrifugal barrier at R_Ω. This process (the 'propeller mechanism') extracts rotational energy from the neutron star and spins it down. Although easy to describe, the coupling of matter and magnetic field involved in this process makes it hard to calculate its efficacy. At this point the theoretical 'histories' diverge, and we leave them.

Surface magnetic fields may be the cause of the quasi-periodic oscillations (QPOs) detected in the power spectra of low-mass X-ray binaries. These are low-amplitude (~ 1–10%) X-ray modulations, with typical frequencies of 5 to 60 Hz, which can vary dramatically with X-ray intensity and spectral hardness. As the name suggests, the peaks in the power spectra are rather broad, and these are clearly distinct from the strictly periodic modulations seen in the pulsing X-ray sources discussed above. The idea which has received the most attention is that QPOs represent the interaction of a rather weak ($\lesssim 10^{10}$ G) field on the central neutron star with transient patterns of orbiting blobs in the inner accretion disc. The QPO frequency is assumed to result from the beat between the neutron star spin and the Kepler frequency of the blobs. Coherent modulation at the spin period of the type seen in the pulsing sources is apparently undetectable, and considerable theoretical effort has gone into providing plausible reasons for this. The beat frequency model does provide at least qualitative explanations for some of the observed changes in the power spectra, and the idea of weak magnetic fields for the neutron stars in low-mass X-ray binaries is attractive for evolutionary reasons. A full discussion of this rapidly-developing subject would take us too far afield, and we refer the interested reader to the review by van der Klis (1989).

The case where the accreting magnetic star is a white dwarf (a magnetic cataclysmic variable) is particularly interesting, as the larger stellar radius means that μ can be much bigger than for neutron stars. In the most extreme case, a $0.5\ M_\odot$ white dwarf with $R_* = 10^9$ cm and a field $B_* = 3 \times 10^7$ G has $\mu = 3 \times 10^{34}$ G cm^3. Systems with these properties can be identified through the optical polarization produced by thermal cyclotron emission in the $\sim 10^7$ G magnetic field (see below) and are called AM Herculis systems, or polars. They are strong soft X-ray sources ($\lesssim 0.1$ keV), and this is generally how they are discovered. For such systems (6.10) gives $r_M \gtrsim 10^{11}$ cm, whereas from (4.2), with $q < 1$ we find a typical orbital separation $a \simeq 3.5 \times 10^{10} P_{\text{hr}}^{2/3}$ cm. The longest orbital period in any of the known AM Her systems is ~ 4 h, so $r_M > a$ for this class. It is hard to see how an accretion disc could form in this case, and indeed none of the usual disc indicators are present. It appears instead that the accretion stream couples directly to the white dwarf magnetic field at some point, the flow being along field-lines down to the surface thereafter (we consider the resulting accretion process in detail in the next section). Further, the white dwarf is observed to spin synchronously with the binary orbit, so that it is phase-locked to the secondary star (Fig. 43). Since it constantly accretes angular momentum from the secondary, this implies the existence of a counterbalancing torque enforcing synchronism. The most likely candidate mechanism for this is the interaction of the white dwarf's dipole field with that of the secondary, which for quite plausible surface fields ~ 100 G has a similar moment. The present limits on the relative rotation rates of the white dwarf and the orbit do not rule out the possibility that the white dwarf dipoles slowly liberate around the equilibrium position with periods ~ 30 yr.

For somewhat weaker magnetic fields or longer orbital periods the result $r_M > a$ will disappear. One would thus expect the white dwarfs to have spin periods P_{pulse} shorter than P_{orb}. Systems of this type are found through their hard X-ray emission (few keV), which shows the pulse period, and are called intermediate polars. The optical emission shows both this period and also the orbital sideband period $(1/P_{\text{pulse}} - 1/P_{\text{orb}})^{-1}$, indicating that some of the X-ray emission heats an object which is fixed in the binary frame. The accretion process in such non-synchronous systems might be expected to be intrinsically more messy than in the AM Her systems, where the field is so dominant that it imposes order. Indeed, it is at present unclear whether most of the accretion flow takes place via disc, rather than crashing directly into the rotating magnetosphere of the white dwarf. The simplest condition for a disc to form is that the initial gas stream's minimum distance of approach $\sim 0.5 R_{\text{circ}}$ to the white dwarf should exceed the size of the obstacle presented by the magnetosphere (cf. Section 4.5). The latter is not easy to work out, as it depends on whether instabilities allow matter to cross field-lines or not, but estimates suggest a value $\sim 0.37 r_M$. Using (4.17) and (6.10) one then finds that the simplest disc formation condition fails if $\mu \gtrsim 10^{33}$ G cm^3, which includes most intermediate polars. Independent evidence which may support this tentative conclusion comes from the fact that several of these systems seem to arrange their spin periods very precisely to be about 10% of P_{orb}. This is clearly very different from the equilibrium disc relation (6.18), which implies $P_{\text{pulse}} \sim \dot M^{-3/7}$, as both evolutionary theory and deductions from observations suggest that $\dot M$ increases roughly as P_{orb}. The observed tight relation $P_{\text{pulse}} \cong 0.1 P_{\text{orb}}$ follows naturally if the white dwarf accretes exactly the orbital specific angular momentum $(GMR_{\text{circ}})^{1/2}$ from the companion. For in equilibrium this will be equal to the Keplerian specific angular momentum $(GMR_\Omega)^{1/2}$ at the corotation radius, giving the equilibrium relation as simply $R_\Omega = R_{\text{circ}}$. From (6.17) and (4.17, 4.1) this gives exactly the observed relation $P_{\text{pulse}} \cong 0.1 P_{\text{orb}}$. It is clear that the requirement of accreting exactly the orbital specific angular momentum is met if there is no disc; if a disc is present the question is more complicated, and requires an understanding of how the white dwarf couples back some of its angular momentum to the disc. Some systems do not obey this relation, and probably accrete via discs in a similar way to the pulsing X-ray binaries described earlier. They may have somewhat weaker magnetic fields, particularly if the pulse period is short ($\leqslant 100$ s), except for the evolved system GK Per, where a disc must be present because of the unusually long orbital period (2 days).

The orbital evolution of magnetic cataclysmic variables is a further fascinating area: we have seen that the AM Herculis systems are distinguishable by the fact that r_M formally exceeds the binary separation a. Yet it is clear that this inequality depends on the orbital period; not only does a increase as $P_{\text{orb}}^{2/3}$, but the average accretion rate $\dot M$ is likely to be higher at longer periods (see Section 4.4), which from (6.10) tends to decrease r_M also. Thus it is easy to show that for field strengths typical of AM Her systems, we do not expect $r_M > a$ for $P_{\text{orb}} \gtrsim 4$ h, and indeed no AM Her systems have seen at such periods. But since P_{orb} decreases in the course of a short-period binary's evolution (Section 4.4), the observed AM Her systems must have descended from some other type of magnetic cataclysmics, presumably non-synchronous. The fact that the intermediate polars almost all have orbital periods > 3 h, and most > 4 h, makes it extremely tempting to identify them as the progenitors of the AM Herculis systems. However this obviously requires them to have the same magnetic field strengths, making their lack of polarization at visible wavelengths very puzzling. The question is still not finally settled, but a currently favoured explanation is that some selection effect, possibly the inhibition of X-ray emission, prevents us discovering the non-synchronous progenitors of the AM Her systems at all. Alternatively, some of the intermediate polars may have similar magnetic fields to the AM Her systems, but something, perhaps dilution by unpolarized light, prevents us detecting polarization in them.

As well as the dynamical effects of the stellar magnetic field on the accretion flow (cf. Section 3.7), a further process involving magnetic fields that is important, because it gives an independent idea of their likely strengths, is the emission of cyclotron radiation by electrons gyrating around the field-lines, often as a result of their thermal motions. For a non-relativistic electron, the Larmor frequency (3.46) implies that most of the radiation is emitted as a spectral line centred at the fundamental cyclotron frequency

$$\left.\begin{aligned}\nu_{\text{cyc}} &= \frac{eB}{2\pi m_e c}[c] = 2.8 \times 10^{13} B_7 \text{ Hz} \\ &= 2.8 \times 10^{18} B_{12} \text{ Hz}\end{aligned}\right\} \quad (6.19)$$

where $B_7 = B/10^7$ G, etc. If the electron's speed of gyration is v, the strengths of emission in the harmonics $2\nu_{\text{cyc}}, 3\nu_{\text{cyc}}, \ldots$ are reduced by factors $\sim (v^2/c^2), (v^2/c^2)^2, \ldots$ below that of the fundamental. A fact that is very important for observational purposes is that radiation emitted at angles close to the field-lines is circularly polarized, while emission orthogonal to the field-lines is linearly polarized in the plane of the electron's orbit. The defining observational property of the AM Her stars (giving rise to the alternative designation 'polar') is the presence of quite strong polarization of the optical light, up to 30% in some cases, which varies smoothly and periodically as the accreting white dwarf rotates. The relative phasing of linear and circular polarization curves is broadly consistent with that expected for thermal cyclotron radiation produced in a small region (cf. (6.14)) near the magnetic polecap of the white dwarf, provided that the angles of the magnetic axis to the rotation axis, and of the latter to the line of sight, are chosen appropriately for any individual system.

This simple picture assumes that the emitting gas is optically thin to its own cyclotron radiation. This is very likely to be invalid for thermal cyclotron emission at the fundamental, although not at higher harmonics. One can also show this by using the standard Larmor formula for the emission, estimating the line width of the fundamental from thermal Doppler broadening, and using Kirchhoff's law to derive an absorption coefficient κ_ν. Because the emission is proportional to the magnetic energy density $(B^2/8\pi)[4\pi/\mu_0]$, so is κ_ν. If the polarized optical light from the AM Her stars was emitted at the fundamental, field strengths $\sim 10^8$ G (equation (6.19)) would be required. The resulting values of κ_ν show that even for very small electron column densities the fundamental has a large optical depth. Because the emission at higher harmonics is less, so is the absorption and hence the

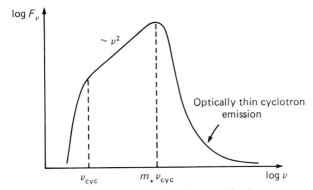

Figure 44. Spectrum of a cyclotron 'line'.

optical depth. Nevertheless, for plausible accretion column parameters (see Section 6.4) one expects the first few harmonics to be self-absorbed up to some 'last' harmonic m_*. The actual value of m_* is rather geometry dependent and can differ for different viewing angles (see below), although it is generally in the range $1 \lesssim m_* \lesssim 10$.

Since the emitting electron gas is thermal, the emergent intensity in the frequency range $v_{\rm cyc}$ to $m_* v_{\rm cyc}$ is roughly blackbody. For the AM Her stars $hv_{\rm cyc}$ (~ 1 eV) is much less than likely values of kT (~ 10 keV) in the emitting region, so the Planck function can be approximated by the Rayleigh–Jeans form, and the expected emission is shown schematically in Fig. 44. Hence the cyclotron line gets broadened to a $\sim v^2$ continuum between $v_{\rm cyc}$ and $m_* v_{\rm cyc}$. Above $m_* v_{\rm cyc}$ the intensity drops sharply as the optical depth becomes very small. Thus the total emission of cyclotron radiation is roughly

$$L_{\rm cyc} = \pi A \int_{v_{\rm cyc}}^{m_* v_{\rm cyc}} \frac{2kTv^2}{c^2} dv$$

$$\cong 2\pi A k T m_*^3 v_{\rm cyc}^3 / 3c^2$$

for $m_* \gtrsim 2$, where A is the emitting surface area. Note that this is very sensitive to m_*. As the blackbody part of the cyclotron emission is unpolarized, most of the polarized emission comes from the most intense optically thin region ($\tau_v \sim 1$), i.e. at $m_* v_{\rm cyc}$. Hence some idea of m_* is needed to derive the field strength from polarization measurements. For typical values of m_*, one finds fields of a few $\times 10^7$ G for the AM Her stars – in good agreement with estimates from Zeeman splitting of spectral lines. Because the total emission $L_{\rm cyc}$ as well as the spectral shape (Fig. 44) so is sensitive to m_*, considerable effort has gone into trying to estimate it. This of course entails solving the transfer problem for the (polarized) cyclotron radiation, and at present only solutions corresponding to rather idealized emitting regions are available. These show that $L_{\rm cyc}$ is in general small compared to other losses from the emission region near the white dwarf (see Section 6.4). The radiation patterns of the linearly and circularly polarized components are in qualitative agreement with the observed polarization curves.

A very different example of cyclotron radiation is provided by the X-ray

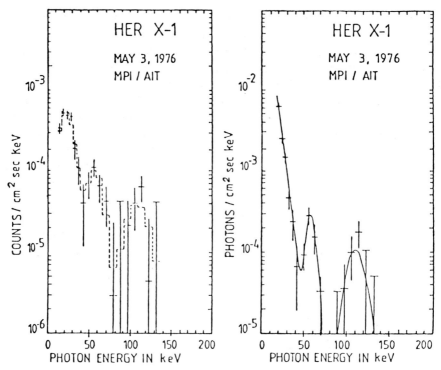

Figure 45. Hard X-ray spectrum of Her X-1 obtained during the pulse phase of the 1.24 s pulsation. The left-hand figure show a raw count-rate spectrum. The right-hand figure show a spectrum deconvolved from the instrumental response, assuming cyclotron lines in emission at 58 and 110 keV. (Reproduced from Kirk & Trümper in *Accretion-Driven Stellar X-ray Sources*, eds. W. H. G. Lewin & E. P. J. van den Heuvel (1983), Cambridge University Press.)

spectrum of the pulsating X-ray source Her X-1 (Fig. 45(*a*)). This shows, in addition to a steeply falling continuum, a narrow feature in the 40–60 keV region. Because the slope of the continuum is large, alternative interpretations of the feature as an emission or absorption line are possible (Figs. 45(*b*), (*c*)). For the emission line interpretation the line centre is at about 60 keV, while from the absorption line hypothesis one gets a line energy of ~ 40 keV. Since no abundant element has atomic bound states with an energy separation of anything approaching this magnitude, the spectral line cannot be due to an atomic bound–bound transition. The only plausible alternative is that the line is caused by the emission or absorption of radiation by the cyclotron process with a fundamental corresponding to the line centre energy. From (6.19), we have

$$h\nu_{\text{cyc}} \cong 12 B_{12} \text{ keV} \tag{6.20}$$

so that the magnetic field must be in the range $(3–5 \times 10^{12}$ G, which, satisfyingly, is just of the order expected for a neutron star. Tentative confirmation of this idea is provided by the possible detection of a further spectral feature at ~ 110 keV in Her X-1, which can be interpreted as the first harmonic.

The basic difference between the cyclotron process here and that in the AM Her stars is that for Her X-1 the line energy, ~ 50 keV, is a significant fraction of the electron rest-mass energy, $m_e c^2 \sim 500$ keV. This means that the interaction of the electrons and the radiation must be treated quantum-mechanically. When the Schrödinger equation describing the behaviour of an electron in a strong magnetic field is solved the electron energy is no longer allowed to take a continuum of values, as in classical mechanics, but is quantized in the so-called *Landau levels*. The line energy (6.20) reappears as the energy gap between the two lowest ($n = 1$, $n = 2$) Landau levels, while the higher Landau levels are the energies ($n = 3, 4, \ldots$). Cross-sections for emission and absorption of photons of various frequencies and polarizations can be worked out in an analogous fashion to the calculation of atomic bound–bound transition rates. Once this is done, one has a similar radiative transfer problem as in the case of the AM Her stars described above. Similarly, here, the overall aim is to predict the radiation patterns of the emergent flux as a function of photon energy. Since the measurement of polarization in the hard X-ray region is technically difficult, most effort has gone into trying to predict the X-ray pulse profiles in and near the cyclotron resonance in order to compare with observation. This problem is even harder than for the AM Her stars and, similarly, has only been solved for idealized emitting regions. For example, not surprisingly, there is no clear decision yet possible as to whether the absorption or emission line interpretation of the cyclotron hypothesis is more likely.

6.4 Accretion columns: the white dwarf case

We have seen above that many white dwarfs and probably most neutron stars possess magnetic fields strong enough to disrupt the accretion flow and to channel it directly on to a small fraction ((6.14) *et seq.*) of the stellar surface near the magnetic poles. The resulting configuration is often known as an *accretion column*. Because the accretion flow is roughly radial, column accretion is fairly similar to spherically symmetrical accretion, although there are important differences, particularly concerning the escape of radiation from the column. Note also that the accretion column could be a 'hollow' rather than a 'filled' cylinder, depending on how the matter attaches itself to the field-lines. Since it is essentially one-dimensional in character, column accretion is intrinsically simpler to study than the boundary-layer accretion processes we considered in Section 6.2.

Let us consider first the accretion column problem for white dwarfs since, although complicated enough, it is somewhat simpler than the neutron star case, which will be considered later. The results will be relevant to the AM Her and intermediate polar systems (see Section 6.3).

In order to set about constructing models of the accretion column, we need to know if the magnetic fields ($\lesssim$ a few $\times 10^7$ G) in these systems produce any other effects besides the channelling of the accretion flow on to the small fraction f (see e.g. (6.14)) of the stellar surface. This we do by estimating various quantities characterizing the conditions there. For accreting magnetized white dwarfs, temperatures T of order 10^8 K and densities N_e of order 10^{16} cm^{-3} are not untypical (see below). Therefore the gas pressure $N_e kT \sim 10^8$ dyne cm^{-2} $\ll$ the magnetic pressure $[4\pi/\mu_0] B^2/8\pi \sim 4 \times 10^{12} B_7^2$ dyne cm^{-2}, with $B = 10^7 B_7$ G. We find from (3.52) Larmor radii

$$\left.\begin{array}{l} r_{\text{Le}} \sim 2 \times 10^{-5} B_7^{-1} \text{ cm,} \\ r_{\text{Li}} \sim 10^{-3} B_7^{-1} \text{ cm,} \end{array}\right\} \quad (6.21)$$

for electrons and ions respectively. Equations (3.8) and (3.25) (with $\ln \Lambda = 19$, equation (3.24)) give

$$\left.\begin{array}{l} \lambda_{\text{Deb}} \sim 7 \times 10^{-4} \text{ cm} \\ \text{and} \\ \lambda_{\text{d}} \sim 3.6 \times 10^4 \text{ cm} \end{array}\right\} \quad (6.22)$$

for the Debye length and electron mean free path. The estimates (6.21), and (6.22) imply

$$\left.\begin{array}{l} r_{\text{Le}} \ll \lambda_{\text{d}} \\ \text{and} \\ r_{\text{Li}} \gtrsim \lambda_{\text{Deb}}. \end{array}\right\} \quad (6.23)$$

The first of these relations implies that electron transport processes, such as thermal conduction, will be strongly suppressed in directions orthogonal to the field-lines: we need only consider such processes taking place along the field-lines, so that the column is one-dimensional in this sense also. The second equation of (6.23) means that most of the simple plasma kinetic theory we described in Chapter 3 is valid, since it implies that the trajectory of an ion during a 'collision' is approximately a straight line. However, we shall see later that important effects can arise when this is no longer true. We have seen already in Section 6.3 that thermal cyclotron radiation is not usually an important energy loss from the accretion column (although it is very important in a 'diagnostic' sense). Thus for typical conditions the effects of the magnetic field on the gross structure of a white dwarf accretion column are confined to channelling the accretion on to the polecap and causing electron transport processes to take place only along field-line.

From our discussion of spherical accretion (Section 2.5) we expect the accreting matter to be highly supersonic and essentially in free-fall above the polecaps. Since, in order to accrete, the infalling material must be decelerated to subsonic velocities, we expect some sort of strong shock to occur in the accretion stream. In the shock the kinetic energy of infall $\frac{1}{2}v_{\text{ff}}^2 = GM/R$ per unit mass must be randomized and turned into thermal energy. Because of their much greater mass ($m_p \gg m_e$) the ions bring with them almost all of this kinetic energy supply. Hence it is of vital importance to ask where the ions will be stopped by the outer layers of the accreting white dwarf. If this stopping happens at large optical depths (i.e. below the 'photosphere') as occurred for an unmagnetized neutron star in our simple example of Section 3.5, the resulting emission will be largely blackbody; if not, the strong shock must occur 'out in the open', with the possibility of much higher radiation temperatures (cf. Chapter 1). Even in this case, a substantial fraction of the accretion luminosity will be either radiated or transported towards the white dwarf, absorbed and re-emitted, presumably in roughly blackbody form. This re-emitted luminosity will, for typical accretion rates $\dot M \gtrsim 10^{16}$ g s^{-1}, be of order 10^{33} erg s^{-1} (cf. (1.4a)), and thus greatly exceed any intrinsic luminosity (typically $\ll 10^{31}$ erg s^{-1}) that the white dwarf may have as a result of left-over thermal energy in its degenerate core. Hence the 'photosphere' referred to above must be a small fraction ($\sim f$) of the white dwarf surface at the base of the column, with a temperature of the order

$$T_b = \left(\frac{L_{acc}}{4\pi R^2 f \sigma}\right)^{1/4} = 1 \times 10^5 \dot M_{16}^{1/4} f_{-2}^{-1/4} M_1^{1/4} R_9^{-3/4} \text{ K} \tag{6.24}$$

(with $f_{-2} = 10^2 f$).

The accretion flow will be brought to a halt in a region of the white dwarf envelope where the gas pressure (cf. (2.2))

$$P_b = \frac{\rho_b k T_b}{\mu m_H} \tag{6.25}$$

is of the same order as the ram pressure

$$P_{ram} = \rho v^2 \tag{6.26}$$

of the infalling matter. Here ρ_b is the density at temperatures T_b in the envelope, while ρ is the density in the steam just above the region where stopping occurs, i.e. where the gas velocity $v = (2GM/R)^{1/2}$. Since the inflow is roughly radial, the continuity equation (cf. (2.19)) implies

$$4\pi R^2 f(-\rho v) = \dot M \tag{6.27}$$

so that from (6.26)

$$P_{ram} = 4 \times 10^7 \dot M_{16} f_{-2}^{-1} M_1^{1/2} R_9^{-5/2} \text{ dyne cm}^{-2}, \tag{6.28}$$

$$-v = v_{ff} = 5.2 \times 10^8 M_1^{1/2} R_9^{-1/2} \text{cm s}^{-1}. \tag{6.29}$$

Now from (6.25) we can find the density ρ_b in the region where stopping occurs:

$$\rho_b = \frac{\mu m_H}{k T_b} P_{ram} = 3 \times 10^{-6} \dot M_{16}^{3/4} f_{-2}^{-3/4} M_1^{1/4} R_9^{-7/4} \text{ g cm}^{-3}. \tag{6.30}$$

Our procedure now is to use the estimates (6.24) and (6.30) of T_b to estimate the opacity of the stellar material in the 'stopping' region. This in turn will tell us the mean free path λ_{ph} of an 'average' photon, i.e. with frequency $\nu \sim kT_b/h$. If λ_{ph} is short compared to the lengthscale λ_s over which ions can be stopped by the stellar material, a self-consistent solution will exist in which most of the accretion luminosity is radiated as blackbody radiation at $\sim T_b$. If $\lambda_{ph} \gg \lambda_s$, on the other hand, the accreting matter must be shocked and decelerated above the 'photosphere'.

For T_b, ρ_b of the orders indicated by (6.24) and (6.30), standard opacity tables show that the Rosseland mean is dominated by Kramers' opacity:

$$\kappa_R \cong 5 \times 10^{24} \rho_b T_b^{-7/2} \text{ cm}^2 \text{ g}^{-1}. \tag{6.31}$$

Note that the constant here is larger than in (5.44), which is appropriate for the rather lower densities and temperatures encountered in disc models. Combining (6.31), (6.24) and (6.30) we find

$$\lambda_{\rm ph} = \frac{1}{\kappa_{\rm R}\rho_{\rm b}} \cong 7\times 10^3 \dot{M}_{16}^{1/8} f_{-2}^{-1/8} M_1^{3/8} R_9^{-7/8} \text{ cm.} \qquad (6.32)$$

From (3.34), on the other hand, we find

$$\lambda_{\rm s} \cong 3\dot{M}_{16}^{-3/4} f_{-2}^{3/4} M_1^{7/4} R_9^{-1/4} \text{ cm} \qquad (6.33)$$

for the lengthscale over which the infalling ions are stopped. (This is the correct lengthscale for energy exchange also, as the estimate (6.24) implies an electron thermal speed of order 10^8 cm s^{-1}, comparable to the infall velocity (6.29) of the ions: these are therefore not suprathermal.) The estimates (6.32) and (6.33) imply that the protons are stopped above the photosphere unless the accretion rate per unit area is rather low, i.e. unless

$$\dot{M}_{16} f_{-2}^{-1} \lesssim 1.4\times 10^{-4} M_1^{11/7} R_9^{6/7}. \qquad (6.34)$$

From (1.4a) this implies an accretion luminosity

$$L_{\rm acc} \lesssim 2\times 10^{29} f_{-2} M_1^{18/7} R_9^{-1/7} \text{ erg s}^{-1}$$

which is much lower than the bolometric luminosities of observed cataclysmic variables. We must conclude therefore that the infalling matter goes through a strong shock before reaching the photosphere in all cases of interest to observation.

At the shock the accretion stream is not, of course, brought completely to a halt: from (3.54) we see that if the shock is adiabatic the gas just below it has a velocity

$$-v_2 = \frac{v_{\rm ff}}{4} = 1.3\times 10^8 M_1^{1/2} R_9^{-1/2} \text{ cm s}^{-1}. \qquad (6.35)$$

However, because the shocked gas is now hot, with characteristic (shock) temperature

$$T_{\rm s} = 3.7\times 10^8 M_1 R_9^{-1} \text{ K} \qquad (6.36)$$

(using (3.64), with $T_{\rm s} = T_2$), the velocity v_2 is *sub*sonic, and the gas can be decelerated by pressure forces and settle on to the white dwarf. It is the structure of this hot, settling region (Fig. 46) which now concerns us, for this determines the characteristics of the radiation we see.

Ideally, we might try to discover the structure of this region by 'switching on' an accretion stream and following the development of the shock and associated cooling mechanisms by using the gas dynamics equations (Chapter 2), until a steady state corresponding to the accretion rate $\dot{M}$ is reached. This is a formidable undertaking, for several reasons. First, the sheer complexity of even the one-dimensional equations in the time-dependent case, especially when allowance is made for all the possible cooling mechanisms, makes such a calculation difficult. Second, it is hard to set up boundary conditions at the 'base' of the column which

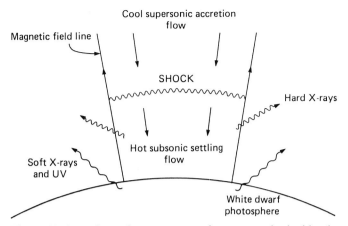

Figure 46. Accretion column geometry for a magnetized white dwarf.

are general enough not to select one particular type of steady solution. Finally, it is by no means clear that the time-dependent flow will reach an authentic steady state; in reality the flow may be continuously disturbed by instabilities. Hence the approach generally adopted has been to look for steady accretion column solutions of various types and subsequently, in some cases, to study their stability.

In a steady state, the accretion energy released below the shock must be removed from the post-shock column at the same rate ($L_{\rm acc}$) at which it is deposited. This removal must be effected by a combination of cooling mechanisms such as radiation and particle transport processes and it determines the height of the shock above the white dwarf surface. If the cooling were inadequate (i.e. less than $L_{\rm acc}$) the post-shock gas would simply expand adiabatically and raise the shock: if, on the other hand, cooling became too effective the shock would move in towards the surface, reducing the volume of the cooling region. Thus in order to construct an accretion column model we must specify how the shock-heated material cools. Inserting this into the energy equation of the post-shock gas, the dynamical equations (Chapter 2) then, in principle, tell us the run of density, temperature, velocity, etc. in the post-shock gas; by matching this solution to the (heated) white dwarf photosphere we determine the position of the shock relative to the star. Finally, we must check that the solution so constructed is self-consistent, i.e. that the cooling mechanism specified at the outset is indeed the dominant one for the column structure found. Note that this procedure leaves open the possibility of distinct self-consistent solutions for the same white dwarf parameters and accretion rate: for example, radiation-dominated solutions in which particle transport processes are negligible, and vice versa. This is the price we must pay for seeking steady solutions rather than investigating the full time-dependent problem. Some insight into which, if any, steady solution is actually realized in a given solution against small perturbations. But we should note that the system might also choose one of the solutions for 'historical' reasons, in the sense that the precise manner in which the accretion process starts up might favour one or another type of steady solution.

138 *Accretion on to a compact object*

Let us estimate the physical conditions in the post-shock gas in order to decide what cooling processes are potentially important. From the estimate (6.35) of the gas velocity, and the continuity equation (6.27) we find a post-shock density

$$\rho_2 = 6.2 \times 10^{-10} \dot{M}_{16} f_{-2}^{-1} M_1^{-1/2} R_9^{-3/2} \text{ g cm}^{-2} \quad (6.37)$$

corresponding to an electron density

$$N_2 = 5.9 \times 10^{14} \dot{M}_{16} f_{-2}^{-1} M_1^{-1/2} R_9^{-3/2} \text{ cm}^{-3} \quad (6.38)$$

assuming a fully ionized gas with cosmic abundances. At the temperatures $\gtrsim 10^8$ K (see (6.36)) characteristic of the post-shock gas, the only realistic opacity source is electron scattering. The shock height D will be found to be $\lesssim R$ in all realistic cases (see below) so curvature of the white dwarf surface can be neglected. The cylindrical radius of the column is therefore approximately

$$d = 2f^{1/2}R = 2 \times 10^8 f_{-2}^{1/2} R_9 \text{ cm}. \quad (6.39)$$

Thus the 'horizontal' optical depth is

$$\tau_{\text{es}}(d) \sim N_e \sigma_T d = 8 \times 10^{-2} \dot{M}_{16} f_{-2}^{-1} M_1^{-1/2} R_9^{-1/2} \quad (6.40)$$

so that radiative cooling will be optically thin unless the accretion rate per unit area is very high. On the other hand, the radial optical depth through the column encountered by photons emitted from the post-shock gas can be large for a high value of $\dot{M}/f$: thus all this radiation will leave through the sides of the column after at most a few scatterings. This is a basic difference from spherical accretion geometries, where all the radiation must make its escape through the full radial extent of the accretion flow (Fig. 47).

Because T_s (equation (6.36)) is so high, radiative cooling of the post-shock gas will be dominated by free-free (bremsstrahlung) emission. The volume emissivity is

$$4\pi j_{\text{br}} \cong 2 \times 10^{-27} N_e^2 T_e^{1/2} \text{ erg s}^{-1} \text{ cm}^{-3} \quad (6.41)$$

for a gas of cosmic abundances with electron temperature T_e and electron density N_e. By our considerations of the last paragraph, this cooling is optically thin, so that an element of shocked gas will be cooled by this process in a characteristic time

$$t_{\text{rad}} \sim \frac{3 N_e k T_e}{4\pi j_{\text{br}}} \cong \frac{2 \times 10^{11} T_e^{1/2}}{N_e} \text{ s}. \quad (6.42)$$

Replacing N_e by N_2 (equation (6.38)) and T_e by T_s (equation (6.36)) we find

$$t_{\text{rad}} \sim 7 \dot{M}_{16}^{-1} f_{-2} M_1 R_9 \text{ s}. \quad (6.43)$$

During this time, the gas element will be settling on to the white dwarf from the shock. Since it starts with velocity $-v_2 = v_{\text{ff}}/4$ (equation (6.35)) and decelerates, it would travel a distance

$$D_{\text{rad}} \sim -v_2 t_{\text{rad}} \cong 9 \times 10^8 \dot{M}_{16}^{-1} f_{-2} M_1^{3/2} R_9^{1/2} \text{ cm} \quad (6.44)$$

before cooling to a temperature of the order of the photospheric value T_b (equation 6.24)) if its cooling were due to radiation alone. Thus (6.44) gives an estimate of the shock height D when radiation alone cools the post-shock gas. In general, there may

Accretion columns: the white dwarf case

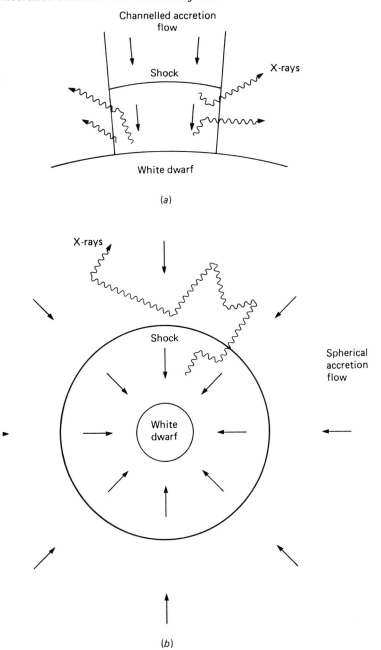

Figure 47. Escape of X-rays from (*a*) an accretion column and (*b*) spherical accretion. In (*a*) the X-ray photons are scattered at most once or twice before escaping, while in (*b*) they must traverse the entire accretion flow.

be other, possibly more efficient, cooling processes operating which would reduce the cooling time and hence reduce D: therefore (6.44) is an upper limit to the shock height.

For most AM Her and intermediate polar systems, observations suggest $\dot{M}_{16} f_{-2}^{-1} \gtrsim 1$, so that (6.44) implies $D \lesssim R$, as asserted above.

The estimates of the last paragraph show that the post-shock region of the accretion column is hot, optically thin and lies close to the white dwarf surface. This last property means that, however the accretion luminosity is removed from the column, a substantial fraction of it must be re-emitted from the white dwarf surface with characteristic blackbody temperature T_b (equation (6.4)). This suggests a possible candidate for a further cooling mechanism: the radiation field characterized by T_b will consist largely of 'soft' photons with energies ($\sim kT_b$) much smaller than those of the hot infalling electrons and ions. The scattering of these soft photons by the energetic electrons involves a transfer of energy from electrons to photons which need not be negligible. The photons will be 'upscattered' to higher energies, while the electron gas will be cooled. The analogous process does not cool the ions directly because of their very small scattering cross-section. It is important to distinguish the occurrence of this process from the existence of the scattering optical depth in the column: the latter depends on the probability with which each photon encounters an electron, whereas here we are concerned with the likelihood of each electron encountering a photon.

The theory of Compton scattering shows that a non-relativistic electron of energy E loses energy at an average rate

$$-\dot{E} \cong \tfrac{8}{3} \sigma_T U_{\rm rad} \frac{E}{m_e c}$$

in a radiation field of energy density $U_{\rm rad}$ (see the Appendix). Thus we can define a Compton cooling time

$$t_{\rm Compt} = \frac{E}{-\dot{E}} = \frac{3 m_e c}{8 \sigma_T U_{\rm rad}}.$$

This cooling process will be more important than direct free–free emission in the post-shock gas only if $t_{\rm Compt} < t_{\rm rad}$: using (6.42) we can write

$$\frac{\text{free–free}}{\text{Compton}} = \frac{t_{\rm Compt}}{t_{\rm rad}} = \frac{7.5 \times 10^{-5} N_e}{U_{\rm rad} T_e^{1/2}}. \tag{6.45}$$

Equation (6.45) shows that Compton cooling is at its most effective when the post-shock gas is hottest (large T_e) and least compressed (low N_e). We know that the gas cools below the shock, and, since it decelerates, the continuity equation (cf. (6.27)) tells us that it becomes more compressed (see below). Hence the immediate post-shock conditions are the most favourable to Compton cooling. To estimate the ratio (6.45), we need some idea of the radiation energy density $U_{\rm rad}$. If the soft radiation arises solely from the accretion process, we can write

$$U_{\rm rad} \cong \frac{f_{\rm soft} L_{\rm acc}}{4 \pi R^2 fc} \tag{6.46}$$

where $f_{\rm soft}$ is the fraction ($\leqslant 1$) of the accretion luminosity which is re-radiated from the white dwarf photosphere, since there will be little dilution of this radiation by scattering and geometrical effects. Now using (6.36) and (6.38) to estimate the immediate post-shock conditions, (6.46) and (6.45) give

$$\frac{\text{free–free}}{\text{Compton}} = 6.5 f_{\text{soft}}^{-1} M_1^{-2} R_9^2 \left(\frac{T_e}{T_s}\right)^{-1/2} \left(\frac{N_e}{N_2}\right). \tag{6.47}$$

This estimate shows that Compton cooling will only be significant, even near the shock, if the white dwarf is very massive ($\sim 1.4\,M_\odot$ and hence very small in radius, $R_9 \sim 0.2$). Of course, the fraction f_{soft} is itself determined by the column structure. If, for example, the accretion column cools only by radiation we would expect on geometrical grounds $f_{\text{soft}} \sim 0.5$. (Actually, as discussed below, $f_{\text{soft}} \sim 0.3$ is more likely because of reflection of hard photons off the white dwarf surface.) Hence for a $1\,M_\odot$ white dwarf ($R_9 = 0.5$) we find

$$\frac{\text{free–free}}{\text{Compton}} \sim 3\text{–}5$$

even at the shock. Compton cooling would be more important if the white dwarf had a high intrinsic luminosity, so that formally $f_{\text{soft}} \gg 1$ in (6.47). This would be the case if, for example, the accreted material underwent steady nuclear burning on the white dwarf surface. As we shall see shortly, this is rather unlikely.

The final radiative cooling process we must consider is the thermal cyclotron emission discussed in Section 6.3. As mentioned there, estimates of the effectiveness of this process are difficult, but generally imply fairly small luminosities L_{cyc} compared to L_{acc}, unless the accretion rate $\dot{M}$ is rather low, or the magnetic field strength B is rather high, or both. When cyclotron emission does dominate over free–free we can show that the electron temperature T_e will be somewhat lower than the ion temperature T_i. This happens because the cyclotron process cools the electrons faster than collisions with the hot post-shock ions can heat them unless there is a significant temperature difference $T_i - T_e$. The rate at which the ions heat the electrons in an effort to bring about equipartition is

$$\Gamma_{\text{eq}} \cong 2 \times 10^{-17} N_e^2 T_e^{-3/2} (T_i - T_e) \text{ erg s}^{-1} \text{ cm}^{-3}. \tag{6.48}$$

Under normal circumstances, the energy-loss rate of the electrons can be balanced by Γ_{eq} with only a very small temperature difference $|T_i - T_e| \ll T_e$. For example, if the electrons lose energy by free–free emission, (6.48) and (6.41) show that the electron–ion energy balance

$$\Gamma_{\text{eq}} = 4\pi j_{\text{br}}$$

implies

$$T_i - T_e \sim 10^{-10} T_e^2.$$

This reflects the fact that the radiation emitted in free–free Coulomb collisions is such a small fraction of the energy exchange between the particles (see Chapter 3). Thus a gas cooling solely by free–free will keep $T_i \cong T_e$, draining energy out of electrons and ions at the same rate. If however the electron loss rate is very large,

(6.48) shows that a significant difference $T_i - T_e$ is required to balance the losses: thus the electrons will equilibrate at a lower temperature than the ions. Thermal cyclotron radiation can be powerful enough to do this if the accretion rate is low; Compton cooling would have a similar effect if it were dominant. Hence dominant cyclotron cooling can reduce T_e significantly below T_s, and ultimately suppress hard X-ray emission altogether ($T_e \lesssim 10^7$ K) if it is strong enough. It is possible that at least one known AM Her system is of this type: it is intrinsically faint, and a weak X-ray source, while showing every sign of strong cyclotron emission.

We can conclude that, for the 'normal' conditions of a not too massive white dwarf with $\dot{M}$ not too low and B not too high, free–free emission of hard X-rays is the only important radiative cooling process.

At long last we are in a position to find an explicit column solution. We have seen that the column structure is fixed once we specify the cooling processes. As a particularly simple example, let us assume that the cooling is purely radiative; as we have seen, in most cases of interest the emission is dominated by thermal bremsstrahlung emission (equation (6.41)). We know that electron and ion temperatures will remain very close to a common value T in this case, so we can use the gas dynamics equations of Chapter 2. Furthermore, we expect the shock height D to be of the order of D_{rad} (equation 6.44) in this case. Since normally $D_{\text{rad}} < R$, we can adopt a one-dimensional geometry, with vertical coordinate z, rather than a quasi-spherical one; moreover the gravitational force density $g = GM/R^2$ is effectively constant over the region of interest. With these simplifications, the equations of continuity, momentum and energy for a steady accretion flow become

$$\left.\begin{aligned}
\rho v &= \text{constant}, \\
\rho v \frac{dv}{dz} &+ \frac{d}{dz}\left(\frac{\rho k T}{\mu m_{\text{H}}}\right) + g\rho = 0, \\
\frac{d}{dz}\left[\rho v\left(\frac{3kT}{\mu m_{\text{H}}} + \frac{v^2}{2} + gz\right) + \frac{\rho v k T}{\mu m_{\text{H}}}\right] &= -a\rho^2 T^{1/2}.
\end{aligned}\right\} \quad (6.49)$$

Here we have made use of the perfect gas law $P = \rho k T/\mu m_{\text{H}}$ and written $4\pi j_{\text{br}} = a\rho^2 T^{1/2}$ (cf. (6.41)). Further, we have neglected entirely the effects of thermal condition, in line with our expectation that transport processes will be negligible: we shall check this at the end of the calculation. Using $\rho v = \text{constant}$ in the third of equations (6.49), we can subtract the momentum equation and obtain a simplified energy equation:

$$\tfrac{3}{2}v\frac{dT}{dz} + T\frac{dv}{dz} = -\frac{a\mu m_{\text{H}}}{k}\rho T^{1/2}. \quad (6.50)$$

The second of equations (6.49) can be written in the integrated form

$$P + \rho v^2 = P_{\text{ram}} + g\int_0^D \rho\, dz,$$

where $z = 0$ is the base of the column ($v \sim 0$) and D is the shock height. Of course, we do not yet know how ρ depends on z, and so we cannot perform the integration

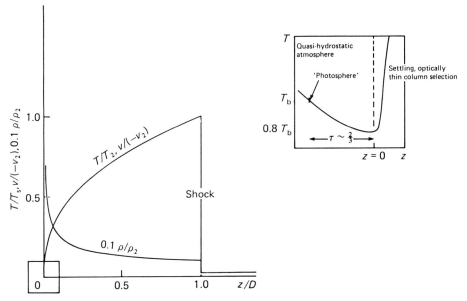

Figure 48. Simple radiative accretion column as discussed in the text. Inset: the base of the column near $z = 0$, showing how the column solution matches to a quasi-hydrostatic 'atmosphere' solution having effective temperature T_b at the point where $T \cong 0.8 T_b$.

on the right hand side which represents the effect of the weight of cooling gas in the column. However, the estimate (6.44) of D suggests that this term will be much smaller than the ram pressure at the shock $P_{\rm ram} = \rho_1 v_1^2$. Immediately behind the shock, the jump conditions $v_2 = \frac{1}{4}v_1$ and $\rho_2 = 4\rho_1$ show that the gas pressure P is three-quarters of the ram pressure $P_{\rm ram}$. If we can neglect the integral, P reaches the value $P_{\rm ram}$ at $z = 0$. Thus P varies only by a factor $\frac{4}{3}$ in the region of interest. This suggests that we adopt the simple approximation

$$P = {\rm constant} = P_{\rm ram}$$

to gain some idea of the post-shock structure. (In fact, the equations can be integrated numerically without this simplification, and give qualitatively similar results.) With $P = $ constant, we have from the perfect gas law $\rho T = $ constant, which when combined with the continuity requirement $\rho v = $ constant shows that

$$v/v_2 = T/T_s = \rho_2/\rho. \tag{6.51}$$

Using this in (6.50), gives

$$\tfrac{5}{2} T^{3/2} \frac{dT}{dz} = \frac{\mu m_{\rm H} a}{k} T_s^2 \frac{\rho_2}{(-v_2)}$$

so that

$$T^{5/2} = \frac{\mu m_{\rm H} a}{k} \frac{\rho_2 T_s^2}{(-v_2)} z + {\rm constant}.$$

Near the base of the column ($z = 0$) we expect $v \sim 0$, so $T \sim 0$ and the constant in this equation must be very small. Strictly speaking, several of our assumptions break down when T becomes of the order of T_b, since re-heating by the photospheric flux will balance the cooling and eventually cause T to reach a minimum value just below T_b, before becoming optically thick and matching on to the white dwarf atmosphere; however it can be shown that this is not serious for the gross features of the solution. Thus we write

$$\left. \begin{array}{c} \dfrac{T}{T_s} = \left(\dfrac{z}{D}\right)^{2/5} \\ \\ D = kT_s^{1/2}(-v_2)/(\mu m_H a\rho_2). \end{array} \right\} \quad (6.52)$$

where

Using (6.44) and the definition of a, it is easy to see that

$$D \sim \tfrac{1}{3} D_{\text{rad}}.$$

The structure we have found is shown graphically in Fig. 48.

Our next task is to check for self-consistency the assumptions made in deriving this solution. First, we neglected the weight term $g\int_0^D \rho\,dz$ in the integral of the momentum equation by comparison with the ram pressure term P_{ram}. From our solution (6.51) and (6.52), we have

$$g\int_0^D \rho\,dz/P_{\text{ram}} = \tfrac{5}{3}(GM/R^2)\rho_2 D/(4\rho_2 v_{\text{ff}}^2) = \tfrac{5}{12}(D/R).$$

This is clearly justified when, as usual, $\dot{M}/f$ is sufficiently large that $D < R$. Second, we neglected the term ρv^2 by comparison with $P \cong P_{\text{ram}}$. We find

$$\rho v^2 / P_{\text{ram}} \sim \rho v_2^2 (z/D)^{2/5}/(4\rho_2 v_2^2) = \tfrac{1}{4}(z/D)^{2/5} < 1.$$

Finally, (2.6) shows that the conductive flux into the white dwarf is

$$|q| = 4 \times 10^{-7} z^{2/5} D^{-7/5} \text{ erg s}^{-1} \text{ cm}^{-2}.$$

This vanishes at $z = 0$, so there is no conduction into the white dwarf; moreover, the neglect of the divergence of this term in the energy equation is justified, since

$$|dq/dz|/(\rho v\, dv/dz) \sim 10^{-7}/(\rho_2 v_2 D) \sim 10^{-12}$$

using (6.44), (6.37) and (6.35). Hence the simple radiative column structure represented by (6.51) and (6.52) is a quite reasonable approximation. Numerical solutions of the full set (6.49), some taking account of the corrections due to spherical geometry, have been obtained: from our work above, it is no surprise that these differ only slightly from our simple analytic solution.

Let us see what such radiative column solutions predict for observations of AM Her and intermediate polar systems. We can show, from (6.51) and (6.52), that the shocked gas radiates almost all its luminosity L_{acc} as hard X-rays with characteristic photon energies $kT_s \sim 60$ keV (for $M_1 = 1$, $R_9 = 0.5$). This result arises since the gas is subsonic and thus has in the main only internal energy to draw on to provide the

Accretion columns: the white dwarf case

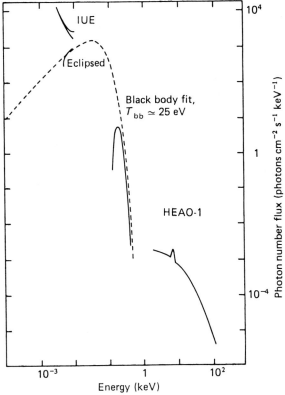

Figure 49. UV and X-ray continuum spectrum of AM Herculis, showing the large (and uncertain) soft X-ray excess.

luminosity. Thus the energy removed from a given element of gas is provided essentially by the temperature drop $T_s - T$ at each point of the column. Half of the total available luminosity ($L_{\rm acc}$) is thus radiated by gas at temperatures $T \gtrsim 0.5 T_s$, and three-quarters at $T \gtrsim 0.25 T_s$, etc. Hence almost all of $L_{\rm acc}$ initially appears as hard X-rays. Since $D < R$, one-half of these are emitted towards the white dwarf surface and are intercepted. Of this luminosity $\frac{1}{2} L_{\rm acc}$, a fraction $a_{\rm x}$ (typically ~ 0.3) is reflected upwards, while the remaining $\frac{1}{2}(1 - a_{\rm x}) L_{\rm acc}$ is absorbed to re-emerge as the 'blackbody' component at $\sim T_{\rm b}$. Hence for radiative solutions

$$f_{\rm soft} \sim \tfrac{1}{2}(1 - a_{\rm x}) \sim 0.35$$

or

$$\frac{L_{\rm X}}{L_{\rm soft}} \sim \frac{\tfrac{1}{2}(1 + a_{\rm x})}{\tfrac{1}{2}(1 - a_{\rm x})} \sim 2. \tag{6.53}$$

The AM Her and intermediate polar stars are indeed found to be hard X-ray and soft X-ray-extreme ultraviolet sources in general. However, the $L_{\rm X}/L_{\rm soft}$ ratio is always found to be much smaller than the estimate (6.53). Fig. 49 shows observations of AM Her itself. Because of the effects of interstellar absorption the emission between ~ 10 and 100 eV is unobservable; but even a very conservative interpolation between the ultraviolet and soft X-rays implies

$$\frac{L_X}{L_{soft}} \sim 0.1. \tag{6.54}$$

A blackbody interpolation would give $L_X/L_{soft} < 10^{-2}$. Thus instead of the hard X-rays being $\sim \frac{2}{3}$ of the total luminosity of the system, as predicted by (6.53), observation implies that they are no more than 10%.

The very large discrepancy between (6.53) and (6.54) is a serious problem for radiative column models: it is easy to convince oneself, for example, that the column geometry we have discussed does not allow the presence of large volumes of cool gas close to the accretion column which would 'degrade' the hard X-rays. Such a model would probably conflict with the evidence of the soft X-ray lightcurves, which show that the soft X-ray source indeed occupies an area of the suggested size ($\sim 4\pi R^2 f$) close to the white dwarf. Thus early attempts to salvage radiative column models centred on the possibility of steady nuclear burning of the accreted matter on the white dwarf surface. This is at first sight an attractive idea, since (1.1) and (1.2) show that the nuclear burning luminosity could be ~ 50 times the accretion luminosity L_{acc}; thus if the burning region were optically thick this would provide a natural explanation for the ratio (6.4). The main difficulty with this idea is that *steady* burning, rather than an explosive 'runaway' event (such as may power classical novae), does not seem to be possible at the luminosities required by observation. Detailed calculations show that steady burning is impossible unless either

$$L_{burn} \lesssim 2 \times 10^{32} f \text{ erg s}^{-1}$$

or

$$L_{burn} \gtrsim 4 \times 10^{37} f \text{ erg s}^{-1}.$$

Observations require $0.1 \gtrsim f \gtrsim 10^{-2}-10^{-3}$ and $L_{soft} \sim 10^{33}-10^{34}$ for most AM Her systems; hence L_{soft} falls squarely in the gap between the two possible stable burning regimes.

Since there was nothing inconsistent or unreasonable in the physical assumptions behind the radiative column solutions, it seems sensible to ask if they are unstable. Quite complex stability calculations have been performed. First, these show that there is a 'bouncing' instability in which the shock height D oscillates between a small value and $\sim D_{rad}$ with characteristic timescale $\sim t_{rad}$. The origin of this instability is the nature of the bremsstrahlung cooling rate (6.41). If a gas element is slightly compressed, raising N_e, it tends to cool more rapidly (cf. (6.42)); this in turn reduces its temperature T and indeed its pressure $N_e kT$. Thus the element tends to get compressed still further since it is at a lower pressure than its surroundings, leading to a rise in N_e and still more rapid cooling, and so on. At the same time the gas element is settling on to the white dwarf on the timescale t_{rad}, so this simple analysis is not enough to show that the instability actually grows. But full global investigations of time-dependent radiative columns do show this. Thus after a time $\sim t_{\rm rad}$, the shock structure collapses on to the white dwarf, and the incoming accretion stream has to 'start again', trying to set up a steady radiative column structure because this is the only way we are allowing it to cool. This occurs on a timescale $\sim t_{\rm rad}$, when the whole cycle starts again. Whatever the details involved, it is clear that the assumption that the primary cooling process is the emission of bremsstrahlung X-rays means that (6.53) holds in a time-averaged sense, so that there is still a conflict with observation.

One way of escaping the prediction (6.53) would be to ensure that the primary hard X-ray emission is all absorbed by cool surrounding material; a sufficiently narrow, dense column would act like a point source, powering the photosphere of the accretion region from below. As all the photons must escape through large amounts of cool ($\sim T_{\rm b}$) material the hard X-ray emission would all be degraded, allowing very large observed ratios of soft to hard emission. A quantitative check of this idea gives an interesting result. First, to ensure that $\gtrsim 10$ keV X-rays are all absorbed and degraded requires a column density $\gtrsim 10^{25}$ cm^{-2}; below ~ 10 keV the photons are absorbed photoelectrically by bound–free absorption in the K-shells of heavy elements, while higher energy photons are degraded by repeated Thomson scattering ($\tau_{\rm es} \gtrsim 7$). This result tells us that the accretion flow must penetrate a few scale heights into the (heated) white dwarf atmosphere before being stopped. We can see this simply as follows. The pressure and density in this atmosphere must obey the hydrostatic condition

$$\frac{dP}{dz} = -g\rho \tag{6.55}$$

with $g = GM/R^2$ the surface gravity (M and R = white dwarf mass and radius), since the atmosphere has negligible mass or vertical extent. Defining the scale height H by $dP/dz \sim P/H$ and using $P \sim \rho c_S^2$, $c_S \sim (kT_{\rm b}/\mu m_{\rm H})^{1/2}$ shows that

$$H \sim \frac{kT_{\rm b}}{\mu m_H g} \sim 10^5 T_5 \text{ cm}, \tag{6.56}$$

for typical parameters, where $T_5 = T_{\rm b}/10^5 \sim 1-3$. If the temperature does not vary rapidly with z, (6.55) shows that ρ varies as $e^{-z/H}$. The definition of the photosphere is that the optical depth measured from infinity is of order unity, i.e.

$$\int_{z_{\rm ph}}^{\infty} \kappa_R \rho \, dz \sim 1$$

i.e.

$$\kappa_R \rho H \sim 1, \tag{6.57}$$

where κ_R is the Rosseland mean opacity. For the conditions in the accretion region this is close to the electron scattering value (0.34), so (6.57) shows that the electron scattering depth is of order unity over one scale height. The requirement $\tau_{es} \gtrsim 7$ for the effective disappearance of the hard X-rays thus amounts to requiring the accreting material to penetrate ~ 7 scale heights below the surrounding photosphere before being stopped. This is not enough of course: we need also that

$$d < D \lesssim 7H$$

so that the column is both narrow and short enough to prevent hard X-ray photons avoiding the absorbing matter. Using (6.39) and (6.56) these conditions yield the constraints $T_5 \lesssim 5$, $\dot{M} \lesssim 10^{15}$ g s^{-1}. This shows that such 'buried column' solutions are only available for rather small accretion rates. However, there is a simple way out of this restriction: adding together a number of such buried columns. Provided these are separated by distances $\gtrsim 7H$ on the white dwarf surface we can regard them as independent. They amount to a total accretion flow with a higher $\dot{M}$ than allowed above, but still with a very strong excess of soft X-rays. The only condition left to check is that the ram pressure in each of the columns is enough to penetrate to the required depth in the atmosphere. From (6.57) the photospheric density is $\rho_{ph} \sim 3 \times 10^{-5}$ g cm^{-3}, giving a photospheric pressure $P_{ph} \sim 3 \times 10^8$ dyne cm^{-2}. The very small values of $f \lesssim 10^{-7}$ for each of the individual buried columns imply from (6.28) ram pressures considerably exceeding this. Thus photospheric penetration will indeed occur, but be limited to a few scale heights because of the exponential increase of atmospheric pressure with depth.

The analysis of the last paragraph shows that it is possible to construct column solutions which have soft excesses, as required by observation, although the splitting of the accretion flow into many separate buried columns seems contrived. A more realistic version of this idea is to suppose that instabilities in the accretion flow above the white dwarf surface break it up into well-separated blobs. A favoured site for this is the region near r_M where the flow threads the magnetic field. Blobs which follow field-lines will be compressed sideways as $d \propto (R/r_M)^{3/2}$, but stretched lengthways as $l \propto (r_M/R)^{1/2}$ because the leading edge accelerates faster than the trailing edge in free fall. Provided that a given blob is not too short to begin with, it is likely to be much longer than the free–free cooling length D when it reaches the surface. Provided also that the blob's mass exceeds the mass of atmosphere it has to sweep up before stopping a few scale heights below the photosphere, it is legitimate to treat it as a buried column of the type discussed above. A full analysis on these lines shows that accretion regions giving soft excesses can be constructed without placing extreme requirements on the instabilities producing the blobs. In particular the instabilities do not need to compress the blobs above the densities already expected in the accretion flow, but merely have to create gaps between them of width $> 7H$ at the white dwarf surface. Material of lower densities may be present in these gaps and elsewhere, giving the observed hard X-rays in the usual way. Figure 50 show a schematic view.

At present progress in understanding blob solutions is limited by our lack of knowledge of the instabilities which divide the flow. Thus we cannot predict the number or density spectrum of the blobs in any given case. The blob picture is however very useful in understanding some phenomena in a qualitative way. In

Accretion columns: the white dwarf case

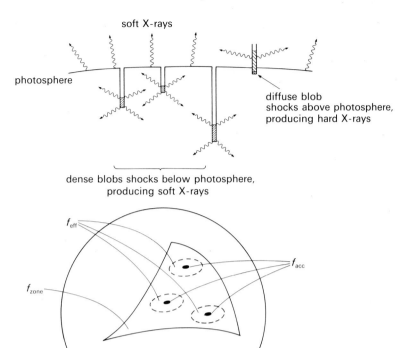

Figure 50. Highly inhomogeneous accretion on to a magnetic white dwarf. Upper figure: blobs of varying density and length hit the surface. Sufficiently dense blobs penetrate the photosphere and radiate almost all their energy as soft X-rays. Lower figure: schematic picture of the three fractional areas f_{acc}, f_{eff} and f_{zone}. The fractional area over which blobs accrete, f_{acc}, is always much smaller than that over which the soft X-rays emerge (f_{eff}). The geometrical region over which blobs land is characterized by f_{zone}; in normal accretion states of AM Herculis systems $f_{eff} \lesssim f_{zone}$, while in anomalous states $f_{eff} \ll f_{zone}$.

particular, it is clear that we must now distinguish between three surface fractional areas which were all previously denoted by f for the case of a single accretion column. If we confine ourselves to considering dense blobs penetrating the photosphere these are: f_{acc}, the instantaneous fractional area over which blobs accrete, f_{eff}, the effective fractional area radiating most of the accretion luminosity as soft X-rays, and f_{zone}, the total zone fractional area over which the blobs are scattered (Fig. 50). In all cases there is a strict hierarchy between these quantities:

$$f_{acc} \ll f_{eff} < \text{ or } \ll f_{zone} < 1.$$

Clearly f_{eff} is the important quantity for the soft X-ray spectrum, and f_{zone} that for the soft X-ray light curves. The two possible inequalities between f_{eff} and f_{zone} give rise to very different types of soft X-ray light curves, especially if the instantaneous

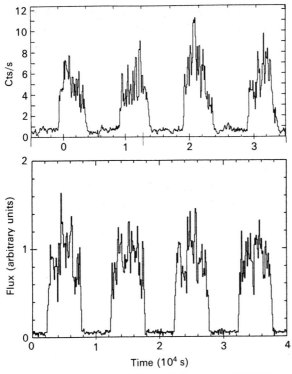

Figure 51. Top: soft X-ray light curve of AM Herculis in an anomalous state (J. Heise, A. C. Brinkman, E. Gronenschild, M. Watson, A. R. King, L. Stella & K. Kieboom (1985) *Astron. Astrophys.* **148**, L14). Bottom: simulated soft X-ray light curve with random accretion of blobs dense enough to penetrate the photosphere. (J. M. Hameury & A. R. King, (1988) *Mon. Not. R. astr. Soc.* **235**, 433). On average blobs are accreting on the main polecap at any given instant. A second diametrically opposed pole accretes 5% of the flow on to the main pole.

photosphere for each region heated by an individual blob bulges or splashes above the surrounding surface, as seems very likely. If the accretion zone is only sparsely populated by accreting blobs, i.e. $f_{\text{eff}} \ll f_{\text{zone}}$, we see each individual splash as a rather luminous event. The light curve is thus very noisy. Even more strikingly, we see the sides of all the splashes, and thus most of the emitted soft X-ray flux, as soon as the accretion zone rotates into view at all, giving an effective square-wave light curve. All of this is very like what is observed during the 'anomalous' X-ray states of some of the AM Her systems. From time to time (intervals of at least months) some of the AM Her systems are observed to change their soft X-ray light curves quite markedly from the usual quasi-sinusoidal shapes, remaining in these states for months. Often these changes involve large phase shifts of the brightest parts of the light curve, suggesting that accretion may switch between the two magnetic poles to some extent. Fig. 51 gives a comparison between a simulation involving about 15 accreting blobs and observations of an anomalous state of AM Her itself. If the number of blobs is increased so that f_{eff} begins to approach f_{zone}, the individual splashes begin to shadow each other and the accretion region behaves like a roughened surface. The predicted soft X-ray light curves are quasi-sinusoidal, just as observed during 'normal' X-ray epochs. It is easy to show that anomalous-type square-wave light curves are predicted if the number N of instantaneously accreting blobs satisfies

$$N \lesssim f_{\text{zone}}/f_{\text{eff}}, \tag{6.58}$$

the light curves changing to quasi-sinusoidal if the inequality is reversed.

6.5 Accretion column structure for neutron stars

Our discussion of column structure has so far been concerned with white dwarf accretion; an equally pressing problem is the analogous process for neutron stars. In the white dwarf case, the basic physical processes are comparatively better understood whereas at the very extreme physical conditions of a neutron star surface, with free-fall velocities $\sim c/2$, magnetic fields $B \sim 10^{12}$ G, and sometimes luminosities $\sim L_{\text{Edd}}$, great complications are introduced into the microscopic and macroscopic physics. Since, as we have seen, our understanding of white dwarf accretion is far from complete, it is hardly surprising that it is still less so for neutron stars. Indeed it is probably fair to say that most of the difficulties and uncertainties of the white dwarf case remain while new ones are added.

Thus, for example, it is unclear in both cases whether the accretion column is a 'hollow' or 'filled' cylinder. Additionally, however, it is far from clear that a shock of any kind occurs, even when the accretion flow is smoothly homogeneous. The reason for this is that the deflection mean free path λ_d (equation (3.19)) in the accretion flow is always very large compared to any other length in the problem, especially the neutron star radius $R \sim 10$ km. To see this, we use the continuity equation (6.27) with $-v = v_{\text{ff}} \cong c/2$ (equation (6.29)) to find the number density

$$N \cong 10^{16} \dot{M}_{16} f^{-1} \text{ cm}^{-3}.$$

From (3.19), with $m \cong m_p$ and $v \cong c/2$, we find

$$\lambda_d \cong 5 \times 10^{11} \dot{M}_{16}^{-1} f \text{ cm}$$

with $\ln \Lambda = 20$. A limit on $\dot{M} f^{-1}$ comes from the consideration that accretion on to a fraction of the neutron star surface cannot yield an accretion luminosity L_{acc} greatly in excess of the Eddington value $f L_{\text{Edd}}$ (equation (1.2)) appropriate to that fractional area. Using (1.2b) and (1.4b) we find

$$\dot{M}_{16} f^{-1} \lesssim 10^2$$

so giving

$$\lambda_d \gtrsim 5 \times 10^9 \text{ cm} \gg R_* \cong 10^6 \text{ cm}$$

for protons with the energy of free-fall. Thus a collisional ion shock of the type we have considered up to now is impossible in the neutron star case. Unfortunately, this does not mean that the possibility of *some kind* of shock, defined as a region where fluid quantities change almost discontinuously, can be entirely disregarded: the existence of shocks having a thickness much smaller than the deflection mean free path λ_d has been known from observation for some time. Such shocks are called *collisionless*: the most notable example is the bow shock where the solar wind impinges on the Earth's magnetosphere. Satellite observations show that the shock thickness is $\lesssim 10^3$ km, while λ_d for the measured physical conditions is of the order 10^8 km.

The possibility greatly complicates the discussion of neutron star accretion. The physics of collisionless shocks is ill-understood even in the case where there is no magnetic field, let alone in the $\sim 10^{12}$ G fields typical of neutron star polecaps. The main problem is that some process other than viscosity, which operates over scales $\sim \lambda_d$, must be found to dissipate the ordered energy and randomize it. Our ignorance of how such shocks work means that we do not know the conditions under which they form: hence both possibilities (shock or no shock) have to be considered.

A further set of complications is introduced by the fact that for fairly modest effective accretion rates $\dot{M}_{16} f^{-1} \lesssim 10^2$ (see above) the radiation pressure due to the accretion luminosity will become dynamically important: indeed the observed X-ray luminosities of many pulsing neutron star sources is of the order of the Eddington value $\sim 10^{38}$ erg s^{-1}. To all of these problems must be added finally the fact that the accretion takes place in a highly magnetized ($\sim 10^{12}$ G) environment, so that every microscopic physical process must be calculated accordingly. We have mentioned already (in Section 6.3) some of the problems which arise in discussing the transfer of radiation in such a medium; another important effect is on the Coulomb encounters of charged particles, making the calculation of the stopping length much more complex than the simple treatment of Chapter 3.

As one would expect given this formidable list of difficulties, no fully self-consistent column solution has yet been given which incorporates all of these effects. In particular, as we mentioned in Section 6.3, detailed treatments of the radiative transfer problem have been given only for rather simple *ad hoc* geometries, while models of the column structure always simplify the transfer problem. The ultimate goal, of course, is to combine all these effects in order to predict, for example, the X-ray pulse profile seen by an observer at some angle to the rotation axis as a function of the X-ray energy. So far, progress has been made in constructing self-consistent column models in three cases: we shall describe these briefly.

If the accretion rate is less than about 10^{17} g s^{-1}, (1.4b) shows that the accretion luminosity L_{acc} will be $< 10^{37}$ erg s^{-1} and thus *subcritical* (i.e. less than the Eddington value (1.2b) $\sim 10^{38}$ erg s^{-1}). Hence for such lower-luminosity objects we may expect that complications due to radiation pressure may be safely neglected. This subcritical accretion regime divides into two cases depending on whether or not a collisionless shock forms. If no shock forms the infalling matter bombards the accretion column, compressing it by its ram pressure and heating it by giving up its infall energy via multiple Coulomb collisions. Since, in the absence of a shock, electrons and ions have the same (free-fall) velocity $\sim c/2$, both of these effects are dominated by the ions because of their much greater masses. Hence the fundamental heating process is the loss of energy by infalling protons. In Section 3.5 we considered this process for an unmagnetized plasma and showed that the stopping power $\rho\lambda_s$ is inversely proportional to the square of the energy of the incident proton (3.35). For an infall speed $\sim c/2$, we find

$$y_0 = \rho\lambda_s \sim 50 \text{ g cm}^{-2}. \tag{6.59}$$

Further, in Section 3.5 we showed that such a proton would probably reach significant electron scattering optical depths $\tau \cong y_0 \sigma_T/m_p$ before releasing its energy. The very strong magnetic fields ($\gtrsim 10^{12}$ G) inferred for the polecaps of accreting neutron stars mean that the derivation of the stopping 'length' y_0 given in Section 3.5 is not strictly valid; in particular, the inequality $r_{Li} \gtrsim \lambda_{Deb}$ (equation (6.23)) is now reversed. For 'typical' parameters $N_e = 10^{23}$ cm^{-3}, $B = 5 \times 10^{12}$ G, we find from (3.47) and (3.8)

$$r_{Li} = 1.5 \times 10^{-9} \text{ cm} < \lambda_{Deb} = 1.7 \times 10^{-7} \text{ cm}.$$

This means that the trajectory of the ion during a scattering event will not be a straight line, as implicitly assumed for the unmagnetized case. The result of allowing for this effect in a detailed treatment using the Fokker–Planck equation is to reduce y_0 to ~ 30–40 g cm^{-2}. With y_0 known as a function of energy the heating rate Γ_{Coul} due to Coulomb collisions can be taken as roughly uniform within the stopping region defined by

$$y(z) = \int_z^\infty \rho(z') \, dz' \leqslant y_0. \tag{6.60}$$

For a column of cross-sectional area A the total heating rate per unit area must be L_{acc}/A: hence

$$\left. \begin{array}{l} \Gamma_{Coul} \cong \dfrac{L_{acc}\rho}{Ay_0} \text{ erg s}^{-1} \text{ cm}^{-3} \quad \text{for } y \leqslant y_0 \\[2mm] \Gamma_{Coul} = 0 \quad \text{for } y > y_0. \end{array} \right\} \tag{6.61}$$

For a steady column solution, Γ_{Coul} must be balanced by the total cooling rate by radiation from the heated electrons. Two processes are at work here; thermal bremsstrahlung and Compton cooling (cf. the discussion preceding (6.45)) – the latter being important because the electron temperature will be found to increase outwards. Both processes are affected by the strong magnetic field (in technical language the appropriate cross-sections have resonances at $\nu = \nu_{cyc}$, equation (6.19)), so that cyclotron radiation is automatically included. The Compton cooling rate involves a knowledge of the radiation energy density U_{rad} (cf. the equations preceding (6.45)) which in turn entails a simultaneous treatment of the radiative transfer problem. For example, the simplest approximation is to assume local thermodynamic equilibrium, so that

$$U_{rad} = (4\sigma/c) T_{rad}^4.$$

Under such conditions the radiative transfer equation takes the diffusion form

$$F = -\frac{c}{3}\frac{dU_{\text{rad}}}{d\tau} \quad (6.62)$$

where F is the total flux of radiation and τ the optical depth (see the Appendix). For the highly ionized column the only opacity is electron scattering, so

$$d\tau = (\sigma_{\text{T}}/m_{\text{p}})\,dy$$

while F is given by the total heating rate of material below the point under consideration:

$$F(y) = \int_{y_0}^{y} \Gamma_{\text{Coul}}\,dz = \frac{L_{\text{acc}}}{Ay_0}(y-y_0) \quad (6.63)$$

using (6.61). Hence U_{rad} can be found from (6.62) and (6.63):

$$\frac{-cm_{\text{p}}}{3\sigma_{\text{T}}}\frac{dU_{\text{rad}}}{dy} = \frac{L_{\text{acc}}}{Ay_0}(y-y_0)$$

so that U_{rad} is quadratic in y, with an appropriate boundary condition at $y = 0$. Thus the Compton cooling rate and hence the energy balance equation can be expressed in terms of the radiation temperature T_{rad}, the electron temperature T_{e}, y and ρ.

More elaborate treatments of the radiative transfer problem can also be used to obtain U_{rad}, and hence the energy equation. Finally, the column must be in hydrostatic equilibrium under the combined effects of gas pressure (in the outer, non-degenerate layer), the weight of the column material, and the ram pressure of the infalling matter: thus

$$P = \frac{\rho k T_{\text{e}}}{\mu m_{\text{H}}} = \left(\frac{GM}{R_*^2} + \frac{\rho_0 v^2}{y_0}\right)y, \quad 0 < y \leqslant y_0$$

$$= \frac{GM}{R_*^2}y + \rho_0 v^2, \quad y > y_0 \quad (6.64)$$

where $v = (2GM/R_*)^{1/2}$ is the infall velocity, and ρ_0 the density of the infalling matter. Numerical solution of the energy and pressure balance equations, together with whatever approximate transfer equation is adopted, now yields the density and temperature structure of the column, together with its radiation field. Fig. 52 shows the results of a recent calculation of this type, taking magnetic effects approximately into account. For a typical atmosphere, the total stopping 'length' $\sim$ 30–60 g cm^{-2} corresponds to a rather thin plane parallel section of the polecap, with vertical extent $\Delta z \ll R_*$.

If we assume that a subcritical accretion flow on to a neutron star undergoes a collisionless shock, the resulting column structure bears a qualitative resemblance

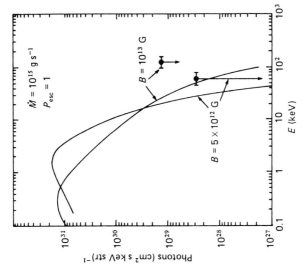

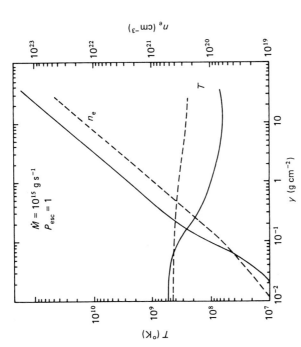

Figure 52. The structure and spectrum of neutron star atmospheres heated by an infalling stream of matter via Coulomb collisions. Left-hand figure: temperature T and electron density n_e versus column density $y = \int_z^\infty \rho(z') dz'$ for an accretion rate $\dot{M} = 10^{15}$ g s^{-1} and magnetic fields $B = 5 \times 10^{12}$ gauss (solid curves) and 10^{13} gauss (dashed curves). Right-hand figure: emergent photon number spectrum $(F_\nu/h\nu)$ for the same parameters as the left-hand figure. The cyclotron line is indicated by the arrow; the crossbar measures the number of photons in the line, not its profile. (Reproduced from P. Meszaros, A. K. Harding, J. G. Kirk & D. J. Galloway (1983), *Astrophys. J. Lett.*, **266**, L33.)

to the radiative columns discussed above for white dwarf accretion, especially if $\dot{M}$ is taken low enough ($\lesssim 10^{16}$ g s^{-1}) that the column is optically thin in the transverse direction to Thompson scattering. The biggest difference is that the lack of collisional coupling between ions and electrons in the shock means that the ion and electron temperatures jump independently; hence an equation like (3.64) must be written for each species, i.e.

$$\left. \begin{array}{c} T_{\rm i} = \frac{3}{16} \frac{\mu m_{\rm H}}{k} v_{\rm ff}^2 \cong 3.7 \times 10^{11} M_1 R_6^{-1} \text{ K} \\ \\ \text{and} \\ \\ T_{\rm e} = \frac{5}{16} \frac{m_{\rm e}}{k} v_{\rm ff}^2 \cong 5.5 \times 10^{8} M_1 R_6^{-1} \text{ K} \end{array} \right\} \quad (6.65)$$

(cf. (6.36)). The factor $\frac{5}{16}$ rather than $\frac{3}{16}$ in $T_{\rm e}$ arises because the electrons have only one degree of freedom (motion along **B**) rather than three.

The enormous disparity ($\sim m_{\rm p}/m_{\rm e}$) in $T_{\rm i}$ and $T_{\rm e}$ means that the ions will heat the electrons at a rate similar to $\Gamma_{\rm eq}$, (6.48) (modified by magnetic effects), in an effort to bring about equipartition. This effect is opposed by radiative cooling of the electrons via bremsstrahlung and cyclotron emission. In fact, the latter dominates the emission, producing at each height in the column a broadened 'line' around the local cyclotron frequency. The shock height D is determined by the time required to equalize ion and electron temperatures, since the ions have all the energy which must ultimately be radiated by the electrons. Thus

$$D \sim (v_{\rm ff}/4) \, t_{\rm eq} \sim \lambda_{\rm eq}. \quad (6.66)$$

Since from (6.19)

$$v_{\rm cyc} \propto B \sim B_*(r/R_*)^{-3}$$

where B_* is the surface field, there is a significant variation of $v_{\rm cyc}$ over the column height D if $\dot{M} \lesssim 10^{15} f_{-3}$ (where $f_{-3} = f/10^3$) and a continuous spectrum which drops sharply at frequencies above the cyclotron frequency $v_{\rm cyc} B_*$ results. For $\dot{M} \gtrsim 10^{15} f_{-3}$ g s^{-1}, by contrast, D is sufficiently small that $v_{\rm cyc}$ does not vary greatly over the emission region and the emergent X-ray spectrum is a broadened 'line' centred on $v_{\rm cyc}(B_*)$: see Fig. 53. Because, especially at low accretion rates, the shocked column protrudes a large distance $D \sim R_*$ from the stellar surface, radiation is emitted over a much larger solid angle than in the Coulomb-heated case described earlier: this is sometimes referred to as a 'fan' as opposed to 'pencil' beam of radiation.

The two models we have described above both neglect radiation pressure: thus they cannot be applied to the more luminous X-ray sources such as Sco X-1, Her X-1 or Cen X-3. For such sources one can make instead the approximation that radiation pressure dominates gas pressure, so that the infalling gas is stopped by photons acting on the electrons through scattering, while the protons are decelerated because they are closely tied to the electrons by Coulomb forces. In the

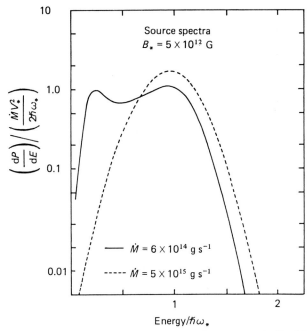

Figure 53. Emergent spectra from an optically thin neutron star accretion column heated by a collisionless shock, with a magnetic field $B = 5 \times 10^{12}$ gauss and accretion rates $\dot{M} = 6 \times 10^{14}$ g s^{-1} and 5×10^{15} g s^{-1}. The photon energy is measured in units of the energy $\hbar\omega \, (= h\nu_{\text{cyc}})$ of cyclotron photons at the surface of the neutron star, and the area under each curve has been normalized to unity. (Reproduced from S. Langer & S. Rappaport (1982), *Astrophys. J.*, **257**, 733.)

simplest treatment of this approximation, all magnetic effects, other than the one-dimensional channelling of the accretion on to the polecaps, are neglected, so that in particular the electron scattering cross-section, and hence the effects of radiation pressure, are treated as isotropic. Thus, if the radiation field at a point in the accretion column has energy density U_{rad}, the corresponding pressure exerted is given by the standard result $P_{\text{rad}} = \frac{1}{3} U_{\text{rad}}$. In the transverse directions this pressure is easily resisted by the magnetic pressure $[4\pi/\mu_0] B^2/8\pi$; moreover, the same force confines the infall velocity **v** to be in negative z-direction, where z is as usual the coordinate measured along the magnetic field, assumed normal to the stellar surface. Thus in close analogy with the first two of equations (6.49) we can write the steady-state equations of continuity and momentum for the infalling gas as

$$\left. \begin{array}{l} \rho v = \text{constant}, \\ \rho v \dfrac{\partial v}{\partial z} + \dfrac{1}{3}\dfrac{\partial U_{\text{rad}}}{\partial z} + g\rho = 0. \end{array} \right\} \quad (6.67)$$

As in the white dwarf case, the gravity term $g\rho = GM\rho/R_*^2$ in the momentum equation may easily be shown to be negligible, so that we can integrate, obtaining

$$U_{\text{rad}} = 3\rho|v|(v_{\text{ff}} - v). \quad (6.68)$$

158   *Accretion on to a compact object*

Here we have used the facts that $\rho v =$ constant, and that the infall velocity must have approximately the free-fall value $v_{\rm ff} = (2GM/R_*)^{1/2}$ sufficiently far above the stellar surface that $U_{\rm rad} \cong 0$. As before, the constant ρv is given by the accretion rate $\dot{M}$ and column cross-sectional area A:

$$\dot{M} = -A\rho v.$$

Note that although v is in the $-z$-direction, both v and $U_{\rm rad}$ will in general depend on the 'transverse' coordinates in the column (see below).

To make further progress we need an energy equation relating $U_{\rm rad}$ and v. As in the case of the Coulomb-heated column (equations (6.62) and (6.63)) this must take the form of an equation relating $U_{\rm rad}$ to the radiative flux, plus an equation specifying how the flux is modified by the energy production due to accretion. Since $U_{\rm rad}$ can depend on the transverse coordinates in the column, in particular the cylindrical radial coordinate R, the radiative transfer equation in the diffusion form analogous to (6.62) would be

$$\mathbf{F} = -\frac{c}{3}\frac{m_{\rm p}}{\sigma_{\rm T}\rho}\nabla U_{\rm rad}.$$

However, the radiative flux vector $\mathbf{F}$ will also be modified by the tendency of the (mildly relativistic) infalling gas to scatter radiation preferentially downwards, implying an extra term on the right hand side:

$$\mathbf{F} = -\frac{c}{3}\frac{m_{\rm p}}{\sigma_{\rm T}\rho}\nabla U_{\rm rad} + \mathbf{v}U_{\rm rad}. \tag{6.69}$$

(By using (6.69) to eliminate $\nabla U_{\rm rad}$ from the second equation of (6.67) it will be seen that the term in $\mathbf{F}$ represents the net photon momentum flux before scattering and the term in $U_{\rm rad}\mathbf{v}$ that after scattering.) We must still specify how $\mathbf{F}$ is changed by the energy release due to accretion. The simplest assumption is that the production of radiation is 'prompt', i.e. any diminution of the kinetic energy of the infalling matter is instantaneously converted to radiation. In this case, we have

$$\nabla \cdot \mathbf{F} = -\rho \mathbf{v} \cdot \nabla(v^2/2) = \rho v \frac{\partial}{\partial z}\left(\frac{v^2}{2}\right). \tag{6.70}$$

Now let us write $Q = v^2/v_{\rm ff}^2$. Then (6.68) can be written in either of two forms:

$$U_{\rm rad} = 3|\rho v|v_{\rm ff}(1-Q^{1/2}) = 3\rho v_{\rm ff}^2(Q^{1/2}-Q).$$

We substitute the first form into the $\nabla U_{\rm rad}$ term in (6.69), and the second form into the $\mathbf{v}U_{\rm rad}$ term. Using the fact that $\rho v =$ constant we obtain

$$\mathbf{F} = \frac{cv_{\rm ff}^2 m_{\rm p}}{2\sigma_{\rm T}}\nabla Q + 3\rho v v_{\rm ff}(Q^{1/2}-Q).$$

Inserting this into (6.70) produces finally

$$\nabla^2 Q = \frac{\sigma_T \rho |v|}{cm_p} \frac{\partial}{\partial z}(6Q^{1/2} - 5Q) \tag{6.71}$$

where, of course,

$$\nabla^2 \equiv \frac{\partial^2}{\partial R^2} + \frac{1}{R}\frac{\partial}{\partial R} + \frac{\partial^2}{\partial z^2}.$$

Given suitable boundary conditions, (6.71) can be solved numerically to find the infall velocity v as a function of height and cylindrical radius R. From (6.67) we can find ρ, using $\rho = -\dot{M}/Av$, and from (6.68) get U_{rad}. Finally, (6.69) gives the radiation field $\mathbf{F}$. The boundary conditions for (6.71) must involve specifying $Q = v^2/v_{ff}^2$ at the stellar surface ($z = 0$), far from the star ($z = \infty$), and on the cylindrical boundary of the column ($R = R_c$). Clearly, $Q \to 1$ as $z \to \infty$, and Q becomes small as $z \to 0$. We can approximate this last condition by setting $Q = 0$ at $z = 0$; although this makes $\rho \to \infty$ as $z \to 0$, most of the accretion energy has already been released at this point. At $R = R_c$, Q should really be given by some condition on $\mathbf{F}$ appropriate to the optical depth structure of the column. However, if the column is optically thick in the transverse (R) direction (i.e. $(\sigma_T/m_p) R_c \rho > 1$), U_{rad} will be much smaller at the boundary $R = R_c$ than in the interior: we may therefore set $U_{rad} = 0$ ($Q = 1$) there. This is closely analogous to setting the temperature equal to zero at the surface of a stellar model; as in stellar structure calculations, we expect this approximation to have little effect except near the surface $R = R_c$.

In Fig. 54 the contours of constant Q, and hence constant velocity, are shown for a particular case. It will be seen that the velocity decreases rather sharply (from $0.95v$ to $0.32v$) across the comparatively narrow region between contours a and c; within contour c the flow is almost at rest. Since $\rho v = $ constant, ρ increases by a factor ~ 3 across this region. This is the kind of behaviour found in shock waves (cf. Section 3.8), although here it is the pressure of locally produced radiation rather than viscous forces which causes the velocity and density transitions. Thus the term 'radiative shock' for the region between a and c seems quite appropriate. The near-stagnant flow within contour c means that there is a 'mound' of slowly settling high-density material at the base of the column. From (6.68), the radiation density there is $U_{rad} = 3\rho|v|v_{ff}$. For a typical case, $R_c = 10^5$ cm ($= 0.1R_*$), $\dot{M} = 10^{17}$ g s^{-1} ($L_{acc} = 0.1L_{Edd}$) and $v = c/2$, we find $\rho|v| = 3 \times 10^6$ g cm^{-2} s^{-1} and hence $U_{rad} = 1.4 \times 10^{17}$ erg cm^{-3}. If owing to the large optical depths ($\tau(R_c) \cong (\sigma_T/m_p)\rho R_c = 7.5$ even at $z = \infty$) this radiation is thermalized, the equivalent blackbody temperature is

$$T_b = \left(\frac{c}{4\sigma}U_{rad}\right)^{1/4} = 6.6 \times 10^7 \text{ K}$$

in line with our expectations (cf. Chapter 1). At sufficiently high densities, the matter at the base of the column, which must have a temperature $\sim T_b$, will exert

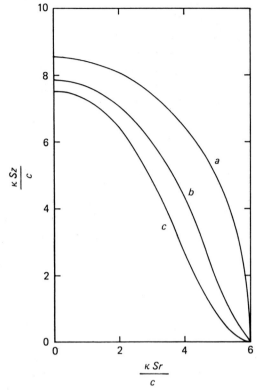

Figure 54. The base of a neutron star accretion column in which deceleration of the infalling matter by radiation pressure is dominant. The figure shows the contours of constant velocity v measured in terms of $Q = v^2/v_{\rm ff}^2$, where $v_{\rm ff}$ is the free-fall velocity for the cases (a) $Q = 0.9$, (b) $Q = 0.5$, (c) $Q = 0.1$. The horizontal and vertical scales are measured in terms of the variables $\kappa Sr/c$, $\kappa Sz/c$, where $\kappa = \sigma_{\rm T}/M_{\rm p}$, $S = \rho|v|$ and r, z are the usual cylindrical coordinates (see text). (Reprinted by permission from K. Davidson (1973), *Nature Phys. Sci.*, **245**, 1. Copyright © 1973 Macmillan Journals Limited.)

a gas pressure greater than the radiation pressure, and our approximation which takes into account radiation pressure *only* will break down. However, this requires a density $\rho_{\rm b}$ such that

$$\frac{\rho_{\rm b} k T_{\rm b}}{\mu m_{\rm H}} \sim \frac{U_{\rm rad}}{3}$$

i.e.

$$\rho_{\rm b} \sim 5 \text{ g cm}^{-3}.$$

At such densities, $|v| \sim 4 \times 10^{-5} v_{\rm ff}$, so virtually all of the kinetic energy $\frac{1}{2}\rho v_{\rm ff}^2 = \frac{1}{2}\rho|v|v_{\rm ff}$ of infall has been released. Hence the approximation of negligible gas pressure is justified throughout.

The kind of approach we have outlined here can be refined by introducing magnetic effects into the cross-sections and by a more detailed treatment of the radiative transfer problem. Qualitatively, the results are quite similar, with, of course, a better description of the spectrum and pulse pattern of the X-ray emission. Given a model, we may check when radiation pressure effects become important and find the limiting luminosity, at which the radiative pressure balances gravity. The approach of the limiting luminosity is signified by the tendency of the radiative shock to move away to distances $z \gtrsim R_*$ as we raise $\dot{M}$ towards the limiting accretion rate. In general, the limiting luminosity is of order $(A/4\pi R_*^2)L_{\text{Edd}}$, with radiation pressure effects first important for $L \gtrsim 7 \times 10^{36}$ erg s^{-1}, although these results depend somewhat on the precise geometry of the column (hollow vs. filled, etc.).

Despite the considerable complexity of present accretion column models for neutron stars, it is probably fair to say that they still neglect too many important physical effects for detailed comparison with X-ray observations to be worthwhile. In particular, the difficult problems of column geometry and radiation transfer in the presence of strong magnetic fields will continue to be active areas of research.

6.6 X-ray bursters

About 30 X-ray sources are now known to exhibit 'bursting' behaviour, in which their X-ray luminosities from time to time brighten by large factors ($\sim 10\times$), with rise times ~ 1–10 s (Fig. 55). Because these events involve X-ray emission, are very luminous and occur on short timescales, there is immediately a *prima facie* suspicion that compact objects and possibly accretion are involved. We shall see that there are strong grounds for identifying the site of the burst phenomenon as accreting neutron stars. To do so we need to describe some of the observed properties of bursts in more detail.

X-ray bursts divide naturally into two distinct categories, Types I and II. Type I bursts recur at intervals of hours to days, or possibly longer. Between bursts, the sources are usually detectable through their persistent X-ray emission. The fundamentally important fact about Type I bursts is that the persistent luminosity outweighs the luminosity in bursts, averaged over long time intervals, by factors ~ 100. Thus, although the luminosity at the peak of a Type I burst is L_X(burst) $\sim 10 L_X$(persistent), the sources only spend about one part in $\sim 10^3$ of their time in a burst state, resulting in

$$\frac{L_X(\text{Type I bursts, averaged})}{L_X(\text{persistent})} \sim \frac{10^{-3} L_X(\text{burst})}{L_X(\text{persistent})} \sim 10^{-2}. \tag{6.72}$$

Type I bursts have been detected from all burst sources while only one source (MXB 1730-335) is so far known to show Type II bursts and these only at certain epochs. During these epochs, the Type II bursts recur so frequently that they account for almost the total luminosity of the source; this behaviour gives the source its other designation as the Rapid Burster, and these frequently recurring bursts are designated Type II bursts.

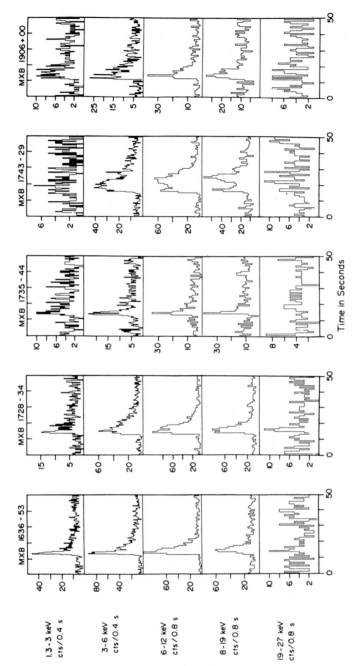

Figure 55. Profiles of Type I X-ray bursts from five different sources. Note that the gradual decay (burst 'tail') persists longer at low energies than at high, indicating that the burst spectrum is characterized by steadily decreasing blackbody temperatures. (Reproduced from W. H. G. Lewin & P. C. Joss in *Accretion-Driven Stellar X-ray Sources*, eds. W. H. G. Lewin & E. P. J. van den Heuvel (1983), Cambridge University Press.)

X-ray bursters

The relation (6.72) receives an elegant explanation if we make the hypothesis that Type I bursts result from runaway thermonuclear burning of matter, of which the steady accretion by the compact object gives rise to the persistent emission. For by (1.1) and (1.2), we have

$$\frac{L_X(\text{Type I bursts, averaged})}{L_X(\text{persistent})} \sim \frac{\Delta E_{\text{nuc}}}{\Delta E_{\text{acc}}}$$

$$\lesssim \frac{0.007 R_* c^2}{GM} \qquad (6.73)$$

since ΔE_{nuc} cannot exceed the value $0.007c^2$ erg g^{-1} for hydrogen burning. Clearly, the only candidate for the compact object is a neutron star: a white dwarf has $\Delta E_{\text{acc}} < \Delta E_{\text{nuc}}$ which would imply $L_X(\text{Type I bursts, averaged}) > L_X(\text{persistent})$, while a black hole has no hard surface on which the thermonuclear reactions can occur. For a neutron star with $M \sim 1\, M_\odot$, $R_* \sim 10$ km, (6.73) implies

$$\frac{L_X(\text{Type I bursts, averaged})}{L_X(\text{persistent})} \lesssim 0.04 \qquad (6.74)$$

which is clearly compatible with the rough estimate (6.72), especially if burning occurs with less than the maximum efficiency. By contrast, Type II bursts cannot be due to runaway nuclear burning since, by the same argument which leads to (6.74), they would have to be accompanied by strong persistent X-ray emission (the Rapid Burster emits Type I bursts also, so it presumably involves a neutron star).

The reasoning of the last paragraph does not of itself demonstrate that burst sources involve neutron stars; it merely shows that the ratio of burst to persistent luminosity could be explained naturally if this were so, provided that thermonuclear 'flashes' of the required duration can indeed be generated on the stellar surface. Observation, however, provides further persuasive arguments in favour of the neutron star hypothesis. The X-ray spectra of bursts are found to be roughly blackbody in form. Thus the received X-ray flux F_ν has the form

$$F_\nu = \frac{A}{4\pi D^2} \pi B_\nu(T) \qquad (6.75)$$

where D is the distance to the source and A its radiating area, assuming isotropic emission. During a burst, the spectrum is observed to 'soften' in a way consistent with a temperature T steadily decreasing from $\sim 3 \times 10^7$ K to about half that value while the area A remains roughly constant. To get an estimate of A we need some idea of the distance D. Now burst sources are not distributed uniformly over the X-ray sky: most of them lie on lines of sight which pass close to the galactic centre. They are presumably therefore a stellar population loosely arranged about the galactic centre, and must lie at distances D of order 10 kpc from us. Setting $D = 10$ kpc in (6.75) and assuming the radiating area to be a spherical surface of radius R, the spectral fits imply $R \sim 10$ km for all sources. Hence the dimensions of the burst emission regions are characteristic of neutron star surfaces. One may push

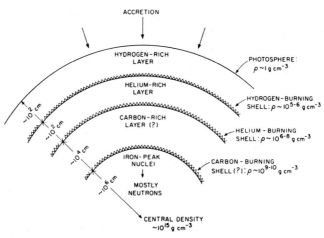

Figure 56. Schematic sketch of the surface layers of a non-magnetic accreting neutron star. (Reproduced from W. H. G. Lewin & P. C. Joss in *Accretion-Driven Stellar X-ray Sources*, eds. W. H. G. Lewin & E. P. J. van den Heuvel (1983), Cambridge University Press.)

this analysis further, for we might expect the burst process to be self-regulating in that the peak luminosity of a burst could not greatly exceed the Eddington value (1.2b) before radiation pressure would disrupt whatever conditions are required for the burst to occur. This idea turns the peak luminosity of a burst into a standard candle. It is found that all ten burst sources for which spectral fits are possible have very similar ratios of radiating area A to total peak flux $F_{max} = A\sigma T^4_{max}/4\pi D^2$. If the average peak luminosities $4\pi D^2 F_{max} = A\sigma T^4_{max}$ are assumed to take the value

$$L_{Edd} = 1.8 \times 10^{38} \text{ erg s}^{-1}$$

appropriate to a 1.4 $M_\odot$ object (assuming its surface is hydrogen rich) the observed values of T_{max} imply an average source radius 7 km, with only a 20% scatter about this mean.

The arguments given above make it difficult to doubt the neutron star hypothesis, and provide a tight set of constraints that any model of Type I bursts must satisfy. Models in which the bursts result from thermonuclear runaway burning at the surface of a neutron star are able to account for the gross observed properties. It is found that the outer, hydrogen burning layer (Fig. 56) of an accreting neutron star is stable against runaway burning, merely adding a small ($\sim$ a few %) contribution to the persistent X-ray luminosity, which is, of course, provided mainly by accretion. Further in, a helium burning shell *can* be unstable in certain ranges of accretion rates and neutron star core temperatures (Fig. 56) giving rise to 'flashes' of the requisite energy and frequency. The rise and decay times of the bursts also seem to be explicable in this type of model, despite its severe simplifications, such as spherical symmetry and the neglect of magnetic fields. Of course, there are many detailed observational properties, most notably the sometimes unexpectedly rapid recurrence of Type I bursts, the origin of which is not yet fully understood.

However, one can be optimistic that refinements of the present highly idealized calculations, including, for example, the deviations from spherical symmetry brought about by surface magnetic fields, will gradually account for these. (Since in no case is the persistent X-ray emission observed to be pulsed, these surface fields are presumably rather weaker ($\lesssim 10^{11}$ G) than in the pulsing X-ray sources.)

The order of magnitude of accretion rates required to explain both the thermonuclear flashes (Fig. 56) and the persistent X-ray emission clearly requires that the neutron stars emitting the bursts must be members of close binary systems. It seems likely that the companion stars must be rather low-mass, late type main-sequence stars. Most of the optical identifications for X-ray burst sources do not reveal the presence of normal stellar absorption lines, but in a few cases lines characteristic of G to K-type main-sequence spectra superimposed on a reddish continuum can be detected when the X-ray flux is very weak. When the persistent X-rays are bright the optical spectra are very blue, sometimes with emission lines. All of this fits quite well with the idea of a low-mass companion: at typical distances $D \sim 10$ kpc, such companions would be very faint. Moreover, when the X-ray flux is high, they will be outshone by large factors by reprocessed optical emission resulting from the absorption of X-rays by the accretion disc. Simultaneous optical and X-ray bursts from some sources have been observed, supporting this idea. It seems very likely, in fact, that burst sources are members of the class of galactic bulge sources discussed in Chapter 5. Although the faintness of the optical sources makes the usual proofs of binary nature (e.g. radial-velocity variations) difficult, and we have seen in Chapter 5 that there may be selection effects against observing X-ray eclipses, the consistency of the picture we have sketched above makes the binary hypothesis reasonably secure. This is probably also true of the bright X-ray sources located in globular clusters in our galaxy, which have very similar properties to the low-mass X-ray binaries, although direct evidence for this will be very difficult to find because most of the clusters are centrally condensed.

The one large gap in what will be recognized as our otherwise fairly convincing picture of the burst sources concerns the Type II bursts of the Rapid Burster. As we noted above, these cannot be due to thermonuclear flashes, and must be attributed to some form of unsteady accretion. Any model of this process must explain a number of well-documented properties of Type II bursts. First, Type II bursts have rise times ~ 1 s and can last from a few seconds up to several minutes. Again the X-ray spectra are reasonably well fitted by blackbody spectra; for Type II bursts, however, the assigned temperature remains constant at $\sim 1.8 \times 10^7$ K (compared to $\sim 3 \times 10^7$ K at the peak of Type I bursts from the same object), rather than decreasing as in Type I bursts. The scale of the emitting surface decreases as the burst decays, from ~ 16 km down to $\lesssim 10$ km for an assumed distance of 10 kpc. For moderately energetic Type II bursts, the total energy involved, assuming this is emitted isotropically, is $\gtrsim 4 \times 10^{38}$ erg: this total burst energy E is approximately proportional to the waiting time Δt to the next burst. Thus a strong burst is followed by a longer than usual wait until the next burst.

There have been many attempts to provide accretion instability mechanisms to account for these properties; so far, none has gained wide acceptance. For example, a viscous instability model, like a minutely scaled-down version of the viscous-variation mechanisms for dwarf nova outbursts discussed in Section 5.8, has some attractive features: the timescale $t_{\text{visc}} \sim$ 1–10 s (5.63) is of the right order to explain the burst profiles if the instability is assumed to operate very close to the neutron star surface ($R_{10} \sim 1\text{--}2 \times 10^{-4}$ in (5.63)) and to be characterized by $\alpha \sim 1$. Moreover, from (5.41) and (5.42), the blackbody temperature is of just the observed order. It would remain effectively constant as the burst of enhanced accretion passed through the inner disc region, while the effective radius would, as observed, decrease. Finally, one might hope to explain the E–Δt relation in terms of the time needed to fill up the inner disc region and re-establish the conditions for the viscous transition. However, such a model raises as many questions as it answers, for the effects of radiation pressure, the nature of the viscosity and the reason for its 'unstable' behaviour, as well as the structure of any boundary layer, are all highly uncertain and clearly important. Other types of instability, such as thermal and especially magnetospheric ones, have also been suggested, with similar lack of conspicuous success. Clearly, whatever process does provide the Type II bursts must require rather special conditions to function at all, for we observe ~ 30 burst sources, but Type II bursts have been convincingly identified from only one. Moreover, the Rapid Burster itself emits Type II bursts only when conditions seem to be just right, a state of affairs which recurs about every 6 months and lasts of the order of a few weeks.

6.7 Black holes

The possibility of observing black holes of stellar mass via the radiation emitted by infalling matter was, along with the rise of X-ray astronomy, one of the main reasons for the upsurge in interest in accretion over the last twenty-five years. Many of the early treatments of the subject concentrated on this possibility largely to the exclusion of neutron star, and especially white dwarf, accretion. Yet so far in this book we have made comparatively little mention of black hole accretion, and discussion of it in the stellar-mass case now occupies a rather less prominent position in the current astrophysical literature than in earlier years. The reason for this is threefold: (i) it was realized that accreting neutron stars and black holes of comparable masses do not differ very significantly in many of their observable properties, (ii) the enormous expansion of satellite astronomy and its repercussions on ground-based observation disclosed a multitude of objects in which accretion was clearly of great importance and relatively easy to study, but which did not contain black holes, and (iii) only one widely accepted candidate for a system containing a stellar-mass black hole was known until relatively recently. Nonetheless, there is no doubt that the existence of black holes is compatible with all the laws of physics, and calculations leading to their formation as the endpoint of the evolution of high-mass stars have been performed.

The proper description of the immediate vicinity of a black hole involves the use of general relativity, Einstein's theory of gravitation. (Indeed, strictly speaking this is also true of a neutron star, although the corrections are fairly small.) For the

Black holes

purposes of studying accretion, however, a few general-relativistic results may be extracted and 'standard' (i.e. non-general-relativistic) physics may be otherwise applied as usual. From general relativity, we find that the boundary (or event horizon) of a black hole of mass M is at a circular radius (the Schwarzschild radius)

$$R_{\text{Schw}} = \frac{2GM}{c^2} = 3M_1 \text{ km} \tag{6.76}$$

unless the hole is rotating at a very high rate, analogous to break-up velocity for a normal star. There is no hard surface at R_{Schw}; the event horizon marks off a region from which no physical effects can propagate into the rest of the Universe, apart from the gravitational pull of the mass inside R_{Schw} and the rather academic possibility of electrostatic effects due to a net electric charge on the hole. Thus any matter crossing into $r < R_{\text{Schw}}$ can only influence the exterior through its mass: in particular, no radiation can escape from within R_{Schw}. At distances $r \gg R_{\text{Schw}}$ the gravitational field of the hole is indistinguishable from that of a 'normal' star of the same mass M. At distances of a few R_{Schw}, the effects of general relativity become important, for example modifying the binding energy of circular orbits from the Newtonian value. At $3R_{\text{Schw}}$ the modified 'potential' describing circular orbits has a maximum and the circular orbits do not exist within this radius (see Section 9.2). Small perturbations immediately cause matter in this orbit to fall into the hole.

These results are directly applicable to the case of an accretion disc around a black hole. At distances $R \gg R_{\text{Schw}}$, under the circumstances discussed in Section 5.3, the disc will be Keplerian. For $R \sim$ a few R_{Schw} the azimuthal velocity v_ϕ, and hence the binding energy, deviate from the Keplerian values by terms of order R_{Schw}/R, until at $R = 3R_{\text{Schw}}$ the matter is removed from orbit by instabilities as fast as it arrives. It reaches $R = R_{\text{Schw}}$ in the free-fall time $\sim (R_{\text{Schw}}^3/GM)^{1/2} \sim 10^{-4}$ s, and therefore radiates very little before disappearing behind the event horizon. Thus the maximum energy per unit mass which can be extracted by this sort of accretion is the specific binding energy of the innermost stable circular orbit. For a low angular momentum hole this is $0.057c^2$ erg g^{-1}, using $R = 3R_{\text{Schw}} = 6GM/c^2$ we see that this is rather less than the $GM/2R \cong 0.083c^2$ which would result if a Newtonian orbit were possible at $R = 3R_{\text{Schw}}$, since the 'potential' is less than the Newtonian value (see Section 7.7, Fig. 58). If the hole has been spun up by accretion of angular momentum, the efficiency is increased up to a theoretical maximum of $0.42c^2$ erg g^{-1}, although values approaching this are only attained within a few parts in 10^{-3} of the 'break-up' angular momentum $J_{\text{max}} = GM^2/c$. (It can be shown that accretion of matter with angular momentum always increases M fast enough to keep $J < J_{\text{max}}$.) Note also that the efficiency $0.057c^2$ erg g^{-1} for disc accretion on to a non-rotating hole is actually rather *less* than the *total* accretion yield $\cong 0.19c^2$ erg g^{-1} from a $1 M_\odot$ neutron star with radius 10 km $\cong 3R_{\text{Schw}}$. This is because a neutron star has a hard surface, so that all of the potential energy at $R \cong 3R_{\text{Schw}}$ is released as radiation rather than just the binding energy of a circular orbit. (The quoted yield includes the general relativistic correction to the Newtonian specific potential energy (GM/R) erg g^{-1}.)

These considerations represent essentially all that distinguishes the discussion of accretion on to stellar-mass black holes from accretion on to unmagnetized neutron stars. In fact, the black hole case is simpler to treat, as one has no boundary layer or torques to worry about, and the boundary condition is that the viscous torque $G(R)$ (equation (4.23)) vanishes at $R = 3R_{\text{Schw}}$ (equation (5.17) with $R_* = 3R_{\text{Schw}}$). As one can show that spherical infall is likely to produce rather low accretion efficiencies (Section 7.8), this largely reduces stellar-mass black hole accretion to a subcase of neutron star accretion.

Despite this, there is widespread agreement that three known binary systems contain black holes. The basic line of argument is as follows.

(i) The objects are (at times) bright X-ray sources (10^{37}, 10^{38} erg s^{-1}), implying the presence of a compact object; a neutron star, black hole, or possibly a white dwarf.

(ii) A white dwarf cannot have a mass exceeding the Chandrasekhar limit ($\simeq 1.4\ M_\odot$); for neutron stars too there is a maximum possible mass which is less certain but probably less than $3\ M_\odot$.

(iii) Spectroscopic evidence implies that the compact components in these systems have masses exceeding $3\ M_\odot$ and therefore must be black holes.

Clearly step (iii) is the crucial one; in all three cases the systems are observed to be single-lined spectroscopic binaries. The Doppler effect on the absorption lines from the non-compact companion stars shows that their radial velocities vary on the orbital periods. The resulting mass limit is simplest in the case of A0620-00, which is a low-mass X-ray binary (a soft X-ray transient in fact), with a period $P = 7.7$ h. The companion star must have a circular orbit about the centre of mass, so the radial velocity amplitude $K_2 = 457$ km s^{-1} implies

$$\frac{2\pi}{P} a_2 \sin i = K$$

where a_2 is the distance from the centre of the companion to the centre of mass and i is the inclination of the binary plane to the plane of the sky. Using $a_2 = M_1 a/(M_1+M_2)$ and Kepler's law (4.1) shows that the combination

$$f_2 = \frac{M_1^3 \sin^3 i}{(M_1+M_2)^2} M_\odot = \frac{K_2^3 P}{2\pi G},$$

the 'mass function', is an observable quantity with the dimensions of a mass. Clearly the value of the mass function f_2 is itself an absolute lower limit to the mass of star 1, the compact object ($M_1 = f_2$ only if $M_2 = 0$ and $\sin i = 1$). For A0620-00 this limit is about $3.2\ M_\odot$, very close to the likely theoretical upper limit on the mass of a neutron star. With realistic estimates for M_2 and i however the spectroscopic mass limit exceeds $7\ M_\odot$. The two other black hole candidates (Cyg X-1 and LMC X-3) are early-type systems in which the companion may be at least as massive as the compact object. The mass functions are rather smaller ($0.25\ M_\odot$ and $2.3\ M_\odot$ respectively), and indirect arguments using the lack of X-ray eclipses are used to show that in these systems too the compact object mass probably exceeds the neutron star mass limit.

Table 1. *Compact accreting binary systems*

Companion star \ Compact star	White dwarf	Neutron star	Black hole
Early type, massive	None known	Massive X-ray binaries (e.g. pulsing X-ray sources)	Cygnus X-1 LMC X-3
Late-type, low-mass	Cataclysmic variables (e.g. dwarf novae)	Low-mass X-ray binaries (e.g. galactic bulge sources)	A0620-00

6.8 Accreting binary systems with compact components

In this chapter and the previous one we have discussed a number of different types of binary system in which a compact object accretes from a 'normal' stellar companion. According to how the mass transfer is effected (Roche lobe overflow, stellar wind, etc.) and properties such as the magnetic field of the compact star, a large variety of possible modes of accretion arise. Let us try to bring order to what is perhaps a confusing diversity by classifying the systems according to the compact and 'normal' components (Table 1). We can go further than this, and distinguish among the systems on the basis of the magnetic field of the compact component. Moreover, in many cases it is possible to decide if the mass transfer is by Roche lobe overflow or stellar wind, and whether an accretion disc or accretion columns (or both) are likely to form. This is summarized in Table 2.

The distribution of systems in these tables is far from uniform. This may be partly due to selection effects, but is more likely to be the result of binary evolution clearly favouring some combinations over others. Naively, one might expect white dwarfs, which are the evolutionary endpoint for low-mass stars, to be found in low-mass, late-type systems; and neutron stars and black holes, born in supernova explosions of massive stars, to be found in massive, early-type systems. While the first expectation seems to be fulfilled, neutron stars, in particular, can be found with a wide range of companion stars. Clearly, the complicating effects of mass transfer (the direction of which may be reversed in some epochs compared to others), mass loss, and angular momentum loss are at work here.

Table 2. *Modes of accretion in compact binary systems*

Companion star	Compact object					
	White dwarf		Neutron star			Black hole
	Unmagnetized	Magnetized	Unmagnetized	Magnetized		
Early-type, massive { O, B supergiant: wind, Roche lobe (?), disc (?)			Unpulsed massive X-ray binaries	Rapidly pulsing massive X-ray binaries [C]		Cygnus X-1 LMC X-3
Be star: wind, eccentric (?), no disc (?)			Unpulsed X-ray binaries	Slowly pulsing X-ray binaries, [C] only		
Late-type, low-mass ($\leq 2M_\odot$) { Roche lobe, disc (except AM Hers and some IPs?)	Classical novae Dwarf novae Nova-like variables	Intermediate polars [C] disc? AM Hers ([C] only, no disc)	Galactic bulge sources, bursters, globular cluster sources (?)	Her X-1 + two others [C], pulsed		A0620-00

[C] denotes the presence of accretion columns.

7
Active galactic nuclei

7.1 Observations

Accretion on to stellar mass objects occurs in a wide variety of systems and yields a wide variety of observational behaviour. While there may be many arguments over detailed models, the broad basis of these differences is largely understood. *Active galactic nuclei* also come in many observed forms. From an observational viewpoint, they can be defined as apparently stellar sources but with non-thermal spectra, and, in cases where they can be determined, significant redshifts. Beyond this, we find a wide variety of properties, which we shall classify in more detail below. But in these cases it is not at all clear how these differences arise, or, indeed, whether one is even dealing with variants of a single basic model. We shall argue that the sources are all manifestations of accretion on to supermassive black holes (of order $10^8 \, M_\odot$), but the argument is by no means certain. Furthermore, for stellar-mass objects, at least in some cases, we have a complete picture of the system even if some of the details are missing. In no case do we have anything comparable for active nuclei, and with present knowledge, both observational and theoretical, any attempt to produce a complete model would be more likely to produce complete fiction. That is not to say that there are no aspects of active nuclei that appear to be fairly well understood, but those that are do not include the nature of the basic energy source. Thus we have to try to extract from the available data what clues we can to the nature of the central engine. This we shall do in this chapter and the next, admittedly not with conspicuous ultimate success.

We consider first the observed properties of active nuclei in as far as they are required for subsequent discussion, giving first of all (subsections (*a*)–(*c*)) a survey of observed properties and secondly (subsection (*d*)) an outline of the classification of observed sources.

(a) Continuum emission

Where the galactic environment of an active nucleus cannot be discerned, or where it was not discerned in the original identification, the object is referred to as a quasar. About 10% of optically selected quasars show strong radio emission and hence appear in surveys of strong radio sources. In the earlier literature, especially, this distinction is emphasized by referring to radio quasars as quasi-stellar sources and non-radio ones as quasi-stellar objects (QSOs). In fact, in neither

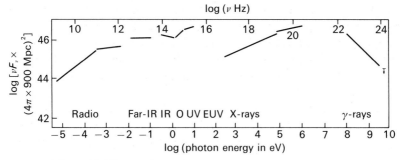

Figure 57. The continuum spectrum of the quasar 3C 273 plotted as $\log \nu F_\nu$ against $\log \nu$ (based on M. H. Ulrich (1981), *Space Science Review*, **28**, 89 and B. Wilkes (unpublished)).

of these cases does the radio emission dominate the energy output (although it may do so in the nuclei of radio *galaxies*). In some cases the greatest luminosity is emitted in the infrared, while in others X-rays may be more important, and in the case of the quasar 3C 273 the peak power occurs at low-energy γ-rays. In Fig. 57 the spectrum of 3C 273, which was the first quasar found, is plotted as $\log \nu F_\nu$ against $\log \nu$. Since the flux in the waveband $[\nu_1, \nu_2]$ is

$$F_{12} = \int_{\nu_1}^{\nu_2} F_\nu \, d\nu = \int_{\nu_1}^{\nu_2} \nu F_\nu \, d(\ln \nu),$$

equal areas under the graph correspond to equal energies. Note that optical quasars were first found in surveys of excessively blue objects relative to a blackbody spectrum, although in absolute terms quasars are 'red' objects. More recent surveys depend on the identification of spectral lines in objective prism plates.

(b)   *Line spectra*

The spectra of active nuclei usually show strong emission lines which appear to be of two distinct types. In quasars and some galactic nuclei (Seyfert type 1 galaxies, broad line radio galaxies) very broad lines are present with widths in the range $0.05 \lesssim \delta\nu/\nu \lesssim 0.1$ at the base of the line, the so-called full width at zero intensity (FW0I). In addition, many objects have narrower emission lines with $\delta\nu/\nu \sim 0.002$ at half the maximum intensity (FWHM). Some narrow line quasars have only narrow lines as do Seyfert type 2 galaxies and narrow emission line galaxies (NELGs) while in the quasar 3C 273 the narrow lines are very weak. Nearby Seyferts are observed to possess also an extended emission line region (EELR) of narrow lines on kpc scales which may represent the interstellar medium in the host galaxy illuminated by a cone of radiation from the active nucleus. Since the part of the quasar spectrum that falls in the fixed optical band at which ground observations can be made depends on the quasar redshift, the prominent emission lines are redshift dependent. In low-redshift objects the Balmer lines of hydrogen dominate, while at higher redshifts ultraviolet lines of C, Mg and the Lα hydrogen line are shifted into the optical.

Observations 173

Objects in which emission lines are entirely absent, or only very weak, but which manifest the other quasar properties are usually called BL Lac objects or Lacertids, since the variable 'star' BL Lac was the first of this type to be found. In these cases, emission line redshifts cannot be determined and they must be distinguished from galactic objects by, for example, absorption line redshifts or similarity to other Lacertids with respect to polarization and variability.

Many quasars (and Lacertids) show absorption lines in their spectra. In some cases, these can be attributed to absorption at zero redshift in the halo of our galaxy. In others, there is little doubt that the broad troughs and P-Cygni profiles can be attributed to absorbing clouds associated with the quasars.

About 10% of quasars belong to this class of broad absorption line quasars (referred to as BALs or BALQs in the literature). But the majority of the absorption lines are narrow and are generally believed to arise from intervening clouds at intermediate redshifts. In many quasars these lines fall into several sets or systems with different redshifts, suggesting as many intervening clouds.

(c)   Polarization and variability

Some quasars and Lacertids exhibit substantial variability over short timescales (days to years) in various wavebands. Extreme examples are the quasar OJ 169 which varied in X-rays by a factor 3 over 100 minutes and the Lacertid AO 0235+164 which varied in the radio and optical by a similar factor over a month. Sometimes variability in different wavebands is correlated, sometimes not. In general, a high percentage polarization of the optical and radio emission is associated with high variability. This is the case in almost all Lacertids, most of which show high polarization. Quasars which exhibit these properties are called optically violent variables (OVVs). All OVVs are highly polarized but rare exceptions to the converse do exist (e.g. PHL 957). As a class, quasars associated with double radio sources are found to be less variable than those lacking extended radio structure. Generally speaking, the amplitude of variability is larger and the timescale shorter in a given object at higher radiation frequencies.

The highly polarized quasars have lower (or zero) polarization in their emission lines than in the continuum. Quasars with lower polarization ($\lesssim 3\%$) have the same polarization in the lines as in the continuum at neighbouring frequencies, and in these cases the rising polarization to the blue region of the spectrum points to scattering by dust as the polarizing source. In highly polarized quasars, the polarization appears to be associated with the emission mechanism.

(d)   Summary of active galactic nuclei
Seyfert galaxies

With hindsight we know that the first active nuclei were discovered as bright nuclei in short exposure plates of spiral galaxies by Carl Seyfert in 1943. These are characterized by a non-stellar continuum and high ionization emission lines. The class is divided into Type 1 Seyferts and Type 2 Seyferts although intermediate types have been defined; the Type 1 Seyferts and intermediate types have emission lines with broad bases which often exhibit variability on timescales of tens of days, similar to that of the UV continuum. All types have a narrower emission line component but appear to be distinguished by characteristically higher [OIII]/Hβ intensity ratios in the Type 2 Seyferts. In these also the broad component appears to be missing, but spectropolarimetry of some nearby bright Type 2 Seyferts has shown a broad base to the permitted lines in polarized light. Type 2 Seyferts appear to be stronger radio sources but only the brightest have so far been detected in X-rays. In Type 1 Seyferts, on the other hand, X-rays may dominate the energy output, with a characteristic (or 'canonical') spectrum $F_\nu \propto \nu^{-0.7}$ in the 2–10 keV band; rapid variability is observed in X-rays with factor 2 changes over timescales of hours. In low luminosity sources the variability appears chaotic with no characteristic timescales. In about half of the Seyfert Type 1 nuclei where its detection would be possible there is an excess of soft X-rays above the extrapolated $\nu^{-0.7}$ power law.

Quasars

Low luminosity quasars are indistinguishable from bright Type 1 Seyferts, so there seems to a continuum of similar objects separated into two classes by observational limitations (the failure to resolve low surface brightness spiral galaxies around bright quasars). A subset of quasars, about 10%, are strong radio sources, ('radio loud' quasars) which, like radio galaxies, may be compact or exhibit the classical double lobe structure. There is some evidence that the radio loud quasars occur in elliptical galaxies whereas radio quiet ones are in spirals. In a number of radio loud quasars the radio emitting plasma shows apparent superluminal velocities. Variability, where observed, tends to be on timescales of years.

BL Lacs (Lacertids)

These show weak lines, if any, but possess something like the characteristic continua of quasars, but, where the distance can be determined from weak absorption lines in an underlying galaxy, at lower luminosity. There are subclasses of radio loud Lacertids and radio quiet ones, the latter discovered through their X-ray emission. The X-ray spectrum is probably softer in the 2–10 keV range than that of Seyferts and quasars but the soft excess in the EUV – soft X-ray range below 0.5 keV is not seen. High polarization is a characteristic of the radio and optical emission. Variability is observed at all wavelengths and timescales down to a few minutes but at low amplitude. Lacertids are often associated with apparently superluminal jets. Optically violent variables (OVVs) are somewhat similar to Lacertids except for their variability and the presence of stronger emission lines. Highly polarized quasars (HPQs) are also similar, except that they are not necessarily variable.

Obviously these classes overlap. The terms active nucleus or active galactic nucleus (AGN) are used to refer to these objects as a whole, although one sometimes distinguishes between quasars and active galaxies, and occasionally the term quasar itself is used as all-embracing.

The distances of active galaxies

A form of activity usually (but not always) distinguished from that above occurs in galaxies undergoing bursts of star formation ('starburst' galaxies). This is the interpretation of a number of galaxies with large infrared luminosity which were revealed by the IRAS (infrared astronomy satellite) sky survey and are therefore sometimes referred to as IRAS galaxies. (The infrared emission arises from the reprocessing of starlight by dust, although the IRAS galaxies might also include cases where the dust reprocesses emission from a quasar nucleus.) Again, there is probably some overlap between starburst galaxies and active nuclei. Indeed, the two phenomena may be related. But we shall exclude from our discussion the notion that starbursts are responsible for all activity.

We now turn to the evidence that active nuclei are at the distances deduced from the redshifts of their line spectra according to standard cosmology, hence that they are intrinsically luminous objects of small size. We then give several arguments which led to estimates of the mass associated with an active nucleus of $10^8 M_\odot$. From amongst possible models we establish a preference for accretion on to a massive black hole as the energy source. In Chapter 8 we present the arguments, not at present conclusive ones, that the accretion occurs through a disc. The final chapters explore various models for the central engine.

7.2 The distances of active galaxies

The cosmological interpretation of the redshifts of active nuclei is now almost universally accepted. Doubters by and large do not accept the standard cosmological picture either; but this degree of doubt is not at present profitable. Here we give several arguments in support of the exclusively cosmological origin of the redshifts on the basis of standard cosmology.

In the standard general relativistic cosmological model the redshift $z = (\lambda_o - \lambda_e)/\lambda_e$ of a spectral line emitted at a rest wavelength λ_e and observed at λ_o is related to the *luminosity distance* d_L of the source by

$$d_L = cH_0^{-1}[z + \tfrac{1}{2}(1-q_0)z^2 + \ldots] \tag{7.1}$$

where H_0, the Hubble constant, measures the present rate of expansion of the Universe and q_0, the deceleration parameter, measures the rate at which the expansion is now slowing down. The first term in the power series in z is the non-relativistic Doppler shift if we interpret z in terms of a recession velocity, $z = v/c$, and the subsequent terms represent relativistic corrections, important for $z \gtrsim 1$. The precise meaning of the distance being measured here is that it gives the diminution in total flux for a source luminosity L radiating isotropically:

$$\int F_\nu \, d\nu = L/4\pi d_L^2. \tag{7.2}$$

For the quasar 3C 273 this gives $L \gtrsim 6 \times 10^{46}(H_0/100 \text{ km s}^{-1} \text{ Mpc}^{-1})$ erg s^{-1}, some 10^2 times larger than that from a giant galaxy and emitted from a volume probably at least 10^6 times smaller. This and similar energy requirements for other quasars initially led some astronomers to question the cosmological interpretation of the redshift, since the power requirements could be considerably reduced if quasars were assumed to be relatively nearby objects. On the other hand, energies of a similar order were familiar to radio astronomers in both radio galaxies and QSSs (radio quasars), the radio galaxies being at known distances. In addition, apparently violent events were recognized in the centres of otherwise normal galaxies associated with luminosities of order 10^{45} erg s^{-1} and large total energy content. In retrospect, therefore, the power requirements of quasars are large, but unexceptional, and do not provide strong grounds for doubting the distances inferred from their redshifts. In fact, we can advance several arguments in support of this interpretation.

(a)   Association with clusters

A number of quasars have been found to be associated with clusters of galaxies at the same redshift and for these the redshift must therefore be entirely cosmological since chance associations are too unlikely. Alternatively, the probability of finding a galaxy within 2 Mpc of a quasar with the same redshift to within 500 km s^{-1}, to allow for peculiar motion relative to the Hubble expansion, is far higher than random. Since 2 Mpc is the radius of a small group of galaxies, this suggests that quasars are often associated with groups of galaxies. In particular, 3C 273 appears to be a member of a small group.

Attempts have been made to support a non-cosmological interpretation of redshifts with the claim that associations of galaxies and quasars at *different* redshifts are found with probabilities far higher than chance coincidence. Quasars might then be objects ejected from relatively nearby galaxies with at least part of the redshift attributable to Doppler motion. Early attempts suffered from *a posteriori* statistics, associating improbabilities with events only after they had been found. More recent investigations also appear to be less than convincing. In any case, such an argument cannot overcome the energy problem for those objects in which the redshift is indisputably cosmological, so, if valid, it would require two populations of rather different types of object with the same observed properties, and this is generally held to be unlikely.

(b)   The continuity argument

The range of luminosities of quasars and Seyfert 1 galaxies overlap with the result that, at least with regard to their optical spectra, a high-luminosity Seyfert removed to a greater distance would be indistinguishable from a low-luminosity quasar. There also appears to be a continuity of X-ray spectra ranging from quasars at one extreme to, in this case, possibly Seyfert 2 and narrow emission line galaxies (NELGs). Indeed, several quasars show optical nebulosity around the central nucleus and in some cases, including 3C 273, this nebulosity has typical stellar absorption lines, at the same redshift as the emission line system, which can be attributed to a surrounding galaxy. The nearer Lacertids too exhibit associated absorption features, identified in these cases with giant elliptical galaxies.

(c) Absorption lines

Absorption lines at redshifts z_{abs} greater than the emission line redshift z_{em} in the same object presumably arise from absorbing material along the line of sight falling into the active nucleus. In many cases, narrow metal lines are observed with $z_{em} - z_{abs} < 0.001$ which are also presumably formed in material associated with the active nucleus. At larger $z_{em} - z_{abs}$, there are sometimes multiple systems of redshifts widely separated compared with the velocity widths of the lines. Early statistical analyses gave rise to apparently preferred redshift differences in these multiple systems and led to the idea of 'line-locking'. According to this, emission line photons from one cloud could be absorbed in a *different* atomic transition in another cloud having a suitable relative velocity. The momentum transfer would tend to accelerate the clouds together, with small velocity dispersion.

However, it is possible to estimate the distances of these absorption clouds from the active nucleus using measurements of the relative strengths of the observed absorption lines. The balance between the various ionization species in these clouds is obtained by equating the rate at which ionizing photons depopulate an ionization state of an atomic species and the rate at which recombinations with free electrons repopulate it. For these purposes, ions can be considered to be in their ground state since the waiting time for a radiative transition to take the ion from an excited state to the ground state is generally much less than that for a further ionization. Therefore the abundances N of r and $r+1$ times ionized states of an atom X are given by

$$\int_{v_{ion}}^{\infty} a(v)(F_v/hv) N(X^r) \, dv \sim \alpha N_e N(X^{r+1}),$$

where α (cm^{-3} s^{-1}) is the recombination rate of electrons, number density N_e, and the ion X^{r+1}, and $a(v)$ (cm^2) is the cross-section for the absorption of ionizing photons of frequency $v \geq v_{ion}$; $F_v \, dv$ is the energy flux of radiation between frequencies v and $v + dv$, so $(F_v/hv) \, dv$ is the number flux. Thus, at a distance D from a source of ionizing photons we have, approximately,

$$\frac{n(X^{r+1})}{n(X^r)} \sim \left(\frac{a(v_{ion})}{\alpha h v_{ion}}\right) \frac{L_{ion}}{4\pi D^2 N_e},$$

where L_{ion} is the luminosity of the source above the threshold frequency v_{ion} required for ionization of X^r. Consequently, apart from atomic constants, the relative populations of ionization states are controlled by the *ionization parameter*

$$\xi = L_{ion}/4\pi D^2 N_e, \tag{7.3}$$

or, equivalently, $U = \xi/hv_{ion}$ (see also Section 8.6).

The value of ξ or U for the absorption line clouds can be obtained from observations of the relative strengths of X^{r+1} and X^r lines (by a curve of growth analysis). Furthermore, an upper limit on the electron density in these clouds is set by the absence of the fine structure C II* λ1335.7 line when strong C II λ1334.5 is seen. This is because a free electron density $N_e \gtrsim 1$ cm^{-3} would be sufficient to populate the fine structure level significantly by collisional excitation. Thus

$$D \gtrsim (L_{\text{ion}}/4\pi\xi)^{1/2}$$

and estimates of L_{ion} by direct observation of the continuum emission yield distances in the range 0.5–3 Mpc.

Assuming spherical symmetry, the energy required to produce the observed mass outflow, with a velocity sufficient to produce the measured redshift differences and a column density sufficient to produce the absorption lines at these distances then turns out to exceed 3×10^{59} erg, considerably greater than the energy estimated, for example, for the compact radio sources. Absorption in intervening material not associated with the active nuclei, and for which the redshift differences are cosmological, is therefore the preferred explanation for the majority of the absorption lines. Material in the haloes of intervening galaxies or in clouds in the intergalactic medium could be responsible.

In high-redshift quasars, with $z_{\text{em}} \gtrsim 2$, the Lα $\lambda 1216$ line is shifted into the optical where it can be seen from the ground. A large number of absorption lines longward of the Lyman limit are then seen which are attributed to Lα absorption in clouds at various redshifts along the line of sight, the so-called 'Lyman α forest'. This interpretation is confirmed by the high incidence of pairs of lines at the wavelength ratio of 1.2 which are interpreted as Lα, Lβ ($\lambda 1025$) pairs. These lines are most easily understood as arising in clouds in the intergalactic medium, since the observed lack of clustering in redshift suggests that in this case the absorbing material is not associated with galaxies.

We conclude that at least those active nuclei with absorption lines lie behind clouds of absorbing material at cosmological distances.

(d)   *Gravitational lenses*

When a light ray passes within a distance R of a spherical body of mass M it is bent through an angle

$$\alpha = 1.75 (M/M_\odot)(R_\odot/R) \text{ arcsec.}$$

A galaxy of mass $10^{11}\, M_\odot$ and radius 10 kpc will therefore produce a bending of a few seconds of arc. It is possible that the light from a quasar can be focussed into several distinct images by the extended mass distribution in a galaxy, although some of these may not be resolvable. The separate images will have identical spectra and redshifts but not necessarily the same variability, since there is also a general relativistic time delay which can differ for the different images by up to a few years. The number of candidates for such systems now exceeds 10 and is steadily increasing; a number of them appear to be well established. In several cases the imaging galaxy is seen in a high-redshift cluster. In these cases, at least, a major part of the quasar redshift must be cosmological.

Although not so directly related to the determination of distances, we should nevertheless mention an alternative possibility in which the central engine is

The sizes of active nuclei

gravitationally focussed by a single star in an external galaxy along the line of sight to a quasar. Indeed, it has been suggested that BL Lac objects are the result of this so-called microlensing of quasars, although this appears to conflict with the different radio structures of the two classes of objects.

Henceforth, we assume that the redshifts of all active nuclei are cosmological and we assume the standard interpretation of the cosmological redshift as a distance indicator according to (7.1) and (7.2).

7.3 The sizes of active nuclei

Upper limits to the sizes of active nuclei can be obtained from direct observations of optical and radio structure; tighter limits can be set indirectly in some cases by observations of the timescales over which large changes in luminosity occur.

(a) Optical structure

Balloon observations of the nucleus of the Seyfert galaxy NGC 4151 show that the optical continuum comes from a region 7 pc in diameter. The emission line region is rather larger, of order 50 pc in this source. Indirect arguments, similar to those of Section 7.2(c), can be used in general to estimate the distance of the gas responsible for the broad emission lines from the continuum source (see Section 8.6). This gives 1–10 pc, so the optical continuum comes from a region typically no larger than this. Line variability (Section (8.6)) suggests sizes of order 0.1–1 pc.

(b) Radio structure

In a few cases the radio cores of active nuclei are resolved and in others VLBI measurements allow upper limits to be placed on the size of the compact cores. Where structure is resolved the lengthscale depends on the wavelength of observation indicating that structure is present on all accessible scales and hence, probably, also on smaller, currently inaccessible ones. At the relevant redshifts, relativistic corrections can be neglected and the size–angular size relation is simply $l = \theta cz/H_0$, where H_0, the Hubble constant, is of order 100 km s^{-1} Mpc^{-1}. The smallest angular structures resolved are a few milliarcseconds corresponding to a linear scale of order 1 pc at a distance of 100 Mpc or a redshift $z \sim 0.03$.

(c) Variability arguments

In general, a system cannot be observed to undergo substantial changes in structure on a length scale l in a time shorter than l/c. For any changes must appear smeared out over the difference in time it takes light to reach us from various parts of the system. In general, therefore, observations of variability on a timescale t_{var} provide an upper limit on the size of the variable emission region, $l \leqslant ct_{\text{var}}$. Note that the size of the region responsible for the variability need not be the same as that of a steady underlying source. Note also that the limit refers to the ultimate energy source; reflecting boundaries, for example, could be arranged to have an arbitrary separation.

The shortest reported variability is an approximately 11 min timescale in BL Lac which corresponds to a lengthscale $l < 2 \times 10^{13}$ cm. X-ray flaring has been claimed with similar timescales (e.g. in NGC 4151). The X-ray quasar OJ 169 varied by a factor 2 in a few hours which gives a lengthscale $l \lesssim 10^{14}$ cm. The Lacertids show optical flickering on a timescale of days and some (e.g. OJ 287) erupt violently over a few months ($l \lesssim 3 \times 10^{17}$ cm). Many quasars change in luminosity by large factors over a few years.

If the source motion is relativistic, as is suggested by the apparent superluminal expansion in a number of objects (Section 9.3), then the observed variability timescale is shorter than the timescale in the rest frame of the source. There is some evidence (Section 9.3) that this does indeed lead to an underestimate of the size of the most rapidly variable sources.

We conclude therefore that for rapid variability ct_{var} is an underestimate of the source size, while other more direct observations have not resolved the central engine, particularly if we interpret the non-variability of many sources as a result of reprocessing. A compromise of around $10^{15}(M/10^8 \, M_\odot)$ cm is often taken as a reasonable estimate of the size of a central engine.

7.4 The mass of the central source

If a source of luminosity L has a lifetime Δt, and if the conversion of matter to energy is accomplished in the source with efficiency η then an amount of mass

$$M = L\Delta t/\eta c^2 \tag{7.4}$$

must have accumulated somewhere in the system. This provides a method of estimating the masses of the sources in active nuclei if we can estimate Δt and η. For η it is usual to assume a fairly generous limit of $\eta \sim \frac{1}{10}$ (see Section 7.8). Limits on Δt are more convincingly obtained for the radio lobes of double sources than directly for the nuclei, but this gives the nuclear mass if we assume the lobes are powered by the central engine. At the very least we must have $\Delta t > D/c$, where D is the distance to the lobes. But the lobes are probably not separating at anywhere near the speed of light. In fact, the natural interpretation of those double sources in which the parent galaxy lies away from the line joining the lobes, which are found mainly in clusters, is that the parent galaxy is moving through the intracluster medium. The beams are then bent by the ram pressure of the medium. If the density in the beam is not too far different from that in the medium then, in order that the beam be bent significantly, the velocities must be comparable. For typical giant ellipticals this velocity is of the order of a few hundred kilometres a second, the same order as the escape velocity, which is the minimum possible speed for the jet material. More detailed investigations have given speeds in the range 500–10000 km s^{-1} so let us take 10^3 km s^{-1} for our estimate. Then for Cyg A, with a lobe separation of 80 kpc, we get $\Delta t \sim 4 \times 10^7$ yr, and (7.4) gives $M \sim 10^8 \, M_\odot$.

For sources which are not double radio sources we need an alternative way of estimating Δt. It is possible that all galaxies go through an active phase and, since about 1 % of galaxies are currently active, this would give $\Delta t = 0.01 \times t_{gal} \sim 10^8$ yr, where $t_{gal} \sim 10^{10}$ yr is a galaxy lifetime. For the more luminous sources, with $L \gtrsim 10^{47}$ erg s^{-1}, this would give a central mass of order 10^{10} $M_\odot$.

In a variant of this argument, instead of taking an estimated lifetime we take the luminosity per unit volume from quasars directly from observation; over the age of the Universe this gives the energy output per unit volume, and hence the mass density of reprocessed material. The results are consistent with a quasar phase for every galaxy and an average 'dead' mass of 10^8 $M_\odot$ per galaxy, but, of course, alternative solutions are possible with a larger mass in a fraction of galaxies which undergo a quasar phase. Strictly speaking, in all of these cases where Δt is not obtained directly it is not necessary that the quasar phase should be a single event, nor need it relate to a single massive object. As far as these arguments are concerned, the nuclear activity could result from N independent events lasting $\Delta t/N$ yr and involving masses M/N.

Several independent arguments lead to similar results for the mass of the central source:

(a)   Alignment of radio sources

Narrow sources, such as Cyg A, have cores aligned to within a few degrees of the outer lobes though the lengthscales involved may differ by factors up to 5×10^6, although there are definite fine scale misalignments of a few degrees. If we assume the core activity is recent, rather than a memory of an initial outburst, as seems to be the only reasonable assumption in view of the recent activity implied in variable compact sources, the implication is that the compact core possesses a memory of direction over the source lifetime. For some of the largest sources, even the light travel time from core to lobe is of order 10^7 yr and this is the minimum time over which alignment must be maintained.

Now a rotating object of mass M can change its axis of rotation by accreting a mass $\Delta M \sim (l_{object}/l_{acc}) M$, where l_{object} and l_{acc} are the specific angular momenta of the accreting object and the accreting gas. We have $l_{object} > l_{acc}$, unless the object is rotating at break-up speed, since material with high specific angular momentum cannot fall in. The alignment timescale Δt_l therefore exceeds

$$\Delta t_l > L/\eta \Delta M c^2$$

and we have

$$M > L/\eta c^2 \Delta t_l \sim 10^8 \, M_\odot.$$

(b)   The Eddington limit

In Section 1.2 we showed that the maximum luminosity generated by spherical accretion on to a mass M is (equation (1.2b))

$$L_{Edd} = 1.3 \times 10^{38} (M/M_\odot) \text{ erg s}^{-1}.$$

To power active nuclei by spherical accretion therefore requires masses in the range

10^6–10^{10} $M_\odot$ or larger, but, as we pointed out in Section 1.2, this limit can be violated for short times, or for a non-spherical geometry, and, in particular, in view of the discussion in Section 9.3, if there is relativistic beaming (see also Chapter 10).

(c)   Variability

Large variations in light output cannot occur on less than a light-crossing time; this time is related to the mass by

$$t \geqslant R/c \geqslant 2GM/c^3 = 0.98 \times 10^{-5}(M/M_\odot) \text{ s.} \tag{7.5}$$

For strictly periodic variations, we get a somewhat stronger condition if we assume the limiting timescale to be an orbital period, $t > t_K = 2\pi(R^3/GM)^{1/2}$, which gives $t_K \gtrsim 8.8 \times 10^{-5}(M/M_\odot)$ s. Either case gives an upper limit to the black hole mass. Combined with the Eddington limit restriction, $L < L_{\text{Edd}}$, (7.5) gives a timescale–luminosity relation which is, in principle, observable:

$$t \gtrsim 3 \times 10^{-10}(L/L_\odot) \text{ s.}$$

Interestingly enough, the reported variability in BL Lac, $t \sim 11$ min, with $L \gtrsim 2 \times 10^{47}$ erg s^{-1}, violates even this weaker inequality. This could be explained if the Eddington limit is violated, by beaming for example (see Chapter 10).

Two further approaches to obtain a mass for the central source depend on the way in which such a mass would influence the gas and stars in the galactic environment:

(d)   The stellar distribution

The discussion of the distribution of stars moving in their own gravitational field and that of a central compact mass is rather complicated. There is a net outward flux of angular momentum and an inward flux of stars governed by a diffusion equation in angular momentum and energy. The steady-state distribution of the number density of stars turns out to be

$$n(r) \propto r^{-7/4} \quad \text{for } r > r_{\text{min}}$$

where r_{min} equals either r_{coll}, the collision radius at which stellar encounters can no longer be assumed to be elastic, or the radius at which the diffusion approximation breaks down as low angular momentum stars are irretrievably captured by the black hole. This leads to an expected cusp in the surface brightness distribution in the galaxy.

Optimism over the observability of the effect has so far proved unfounded and at present this method has yielded only rather large upper limits. For example, the mass of a compact object in the core of M 87 is less than 2×10^9 $M_\odot$.

However, direct observation of the velocity dispersion of stars in the nuclei of nearby galaxies can be used to estimate the mass distribution. For example, for a spherical distribution of mass M within a radius r the virial theorem gives $\overline{v^2} \approx GM/r$. More detailed mass models show that the mass to light ratio in M 31 and

Models of active nuclei

M 32 rise sharply within a few parsecs of the centre. The results are incompatible with a normal old stellar population. The simplest interpretation in each case is a compact central object with a mass that is somewhat model dependent but is certainly within an order of magnitude of $10^7 \, M_\odot$.

(e)   Activity in normal galaxies

We have noted a certain continuity of properties of systems ranging from high-luminosity quasars to Seyfert galaxies and NELGs (Section 7.2(b)). There is a similar continuity ranging between broad line radio galaxies and optically inactive radio sources. In particular some otherwise normal E and S galaxies have been observed to possess weak extended radio structure on scales of a few arcseconds closely aligned with milliarcsecond compact radio structure in the galactic nucleus. The luminosities involved are of order 10^{39} erg s^{-1}, a factor 10^6 less than typical radio galaxies. It is therefore possible that activity observed in the radio in normal galaxies is indicative of remnant or 'dead' quasars. Close to the centre of our own Galaxy there is a compact, non-thermal, radio source, SgrA*, of size $\leqslant 3 \times 10^{14}$ cm, luminosity 2×10^{34} erg s^{-1} which could be a weak active nucleus. The central 0.1 pc of the Galaxy is complex, containing a number of infrared sources but it is not clear if any of these is associated with SgrA* or which source, if any, is coincident with the dynamical centre. There is also an intermittent emission of the 511 keV e^+-e^- annihilation gamma ray line; this line comes from a source about 45' from the galactic centre which might again be indicative of an active nucleus.

The total mass distribution in to ≈ 0.1 pc is obtained from the dynamics of interstellar gas clouds, determined from HI 21 cm and infrared emission lines, assuming that the gas motions are dominated by gravity (and not, for example, magnetic fields). Additional data comes from the velocities of stars derived from infrared absorption lines. The stellar mass can be estimated from the 2.2 μm infrared continuum assuming a constant mass to (infrared) light ratio. The two results are compatible if there is an additional mass of about $3 \times 10^6 \, M_\odot$ within 0.01 pc. Thus the evidence for a central black hole of $\approx 10^6 \, M_\odot$ is substantial but not completely compelling.

7.5   Models of active nuclei

From the foregoing arguments active nuclei must involve the generation of energy in a compact region around a massive object, or objects. In this section we consider various proposed models for the source of energy in this massive core.

(a)   Compact star clusters (without black holes)

A supernova is the one established way in which at least a small galaxy can be temporarily outshone from a region of stellar size. In this case, the available energy is the difference between the gravitational binding energy of the neutron star remnant and the stellar precursor, which is of order $GM^2/R \sim 10^{53}$ erg for a solar mass neutron star. The available energy is therefore of order 10^2-10^3 times the energy released in the outburst in the form of radiation, implying efficiencies of order 0.1–1%. Thus, one proposed model for the energy source of an active nucleus

184   *Active galactic nuclei*

is a dense cluster of massive, hence rapidly evolving, stars undergoing many supernova outbursts. However, there are several objections to this model:

(i) Fluctuations: the problem here is that for a population of N independent stars, to which Poisson statistics would be applicable, fluctuations in the supernova rate should decrease with population size, $\delta N/N \propto N^{-1/2}$. Large variability by factors $\gtrsim 2$ in timescales $\lesssim 1$ yr in the brighter sources must be attributed to cooperative processes in which one supernova can initiate another, so that stars do not behave independently. But then the less luminous active nuclei should be the more variable, contrary to observation.

(ii) Systematic motion: while in some cases the radio variability of compact cores might be interpreted as the brightening of fixed components, we shall see (Section 9.3) that in other cases there is definite evidence for increasing source separation, which, like the absence of apparent source contractions, is incompatible with a model involving random outbursts. The variant *slingshot model*, in which random gravitational encounters lead to a large fluctuation in energy and the consequent ejection of a compact object from the nucleus, is difficult to reconcile with the stability of the axis in the case of multiple outbursts.

(iii) Evolutionary argument: evolution of the mass segregation occurs in a star cluster, through random gravitational encounters, whereby the small stars escape from the cluster carrying with them all the angular momentum, and the more massive stars fall into the centre, dissipate energy in collisions and evolve to a central black hole on a timescale of about 10^7 yr. Thus, the supernova phase could represent only a short period of quasar history.

A variation on this theme attempts to explain some part of the observed energy output as contributed by hot evolved massive stars which have passed through a mass-losing phase. This is clearly important in relating starbursts and active nuclei (through the so-called 'warmers' for example) but cannot be the whole story.

(b)   Supermassive stars

The well-known upper limit of about $50 \, M_\odot$ for normal stars to radiate below their Eddington limits depends on the standard mass–luminosity relation $L \propto M^3$. It is clear that a $10^8 \, M_\odot$ star radiating $\lesssim 10^{46}$ erg s^{-1} would not violate the Eddington limit, so could, in principle, exist. For $M \gtrsim 10^8 \, M_\odot$, nuclear reactions could not provide the heating necessary to maintain pressure support. However, such a star would not be stable: normal stars are stable to small radial pulsations, since a small contraction increases the central temperature, hence increases the rate of energy generation, and provides the pressure necessary to restore equilibrium through subsequent re-expansion; but, for supermassive stars Newtonian gravity is insufficiently accurate and the effects of general relativity must be taken into account. According to general relativity, the pressure in a star makes an additional contribution to its gravitational field. Thus, a small contraction of a supermassive star leads to a pressure increase, the gravitational effect of which causes further contraction. A supermassive star supported by gas pressure is therefore unstable on a dynamical timescale, which turns out to be of order 10^7 yr. Other means of support against gravity have been suggested: in a *magnetoid* a turbulent magnetic field provides a magnetic pressure and in the *spinar* model the star is supported by rapid rotation, except along the axis, and is assumed to radiate by a scaled-up version of the pulsar mechanism (see Chapter 9). But neither model can circumvent the general relativistic instability. There is, in principle, no reason why supermassive stars should not re-form many times to power longer-lived galactic nuclei. But if general relativity – or something like it – is the correct theory of gravity, a supermassive star would evolve to a black hole.

(c) Black holes

Black holes are stable and can therefore power an active nucleus by accretion as long as there is a supply of gas from which energy can be extracted with sufficiently high efficiency. A black hole model, in which fuel is supplied either by a local star cluster or by the host galaxy globally, has therefore become the current orthodoxy. The possible existence of black holes is guaranteed in general relativity, which is supported in favour of all but rather *ad hoc* alternative theories by observational tests. But these involve relatively weak gravitational fields. Direct evidence would come from the detection of stellar mass black holes for which, as discussed in Section 6.7, Cyg X-1, LMC X-3 and A0620-00 provide probable, although not conclusive, support. The problem of the gas supply is taken up in the next section, and the efficiency of energy extraction in Section 7.8. In subsequent chapters we shall discuss active nuclei solely in the context of the black hole model.

7.6 The gas supply

Most models of active galactic nuclei require large amounts of fuel to power the central engine. Accretion on to a supermassive black hole which, as we saw in Section 7.5, is the most widely accepted model, needs mass supply rates $\approx 1-100\ M_\odot$ yr^{-1} in order to power the brightest quasars. Other models of activity which do not involve a black hole nevertheless require that large amounts of gas be channelled to the nucleus where, for example, it can turn into a cluster of massive stars which undergo supernovae.

The problem of fuelling has two parts: the origin of the gas and the mechanism by which this gas can be delivered to a tiny nuclear region at the rate required. Understanding these problems is likely to contribute to questions such as: What makes a galaxy potentially active? and What turns the activity on? We discuss two classes of models. In the first the black hole is supplied by a 'local' or nuclear source – a dense stellar cluster of radius less than 10 pc. In the alternative 'extranuclear' models the gas comes from scales $\geqslant 1$ kpc, from the main body of the host galaxy, or from infall of intergalactic gas, or from galaxy interactions.

In models in which the black hole is fuelled by a dense star cluster, it is assumed that some 10^9 stars have been brought together by stellar dynamical and/or gas dynamical processes to within ~ 10 pc from the centre of the host galaxy. This assumption in fact circumvents the basic problem of fuelling which is how to dispose of the excess angular momentum of the material in a timescale compatible with observations. Once the cluster has formed the angular momentum problem has been solved. Even so, supposing such a cluster did form, the rate at which gas is liberated by a variety of processes such as normal stellar evolution, tidal disruptions, ablation, and physical collisions between stars, is insufficient, or its retention in the cluster is uncertain, or both. Let us take a closer look at each of these processes.

The time- and mass-averaged rate at which stars liberate gas during their normal stellar evolution is $\sim 0.01\ M_\odot$ yr^{-1}. This is insufficient to power the brightest sources, and furthermore the gas liberated is likely to escape the cluster because of heating by supernovae. Attempts to use peculiar stellar mass functions and appeal to the effects of a strong radiation field to enhance mass loss remain inconclusive.

Stars whose orbits lead them too close to the massive black hole at the centre of the cluster are torn apart by tidal forces. Two fundamental timescales control the evolution of such a cluster: the dynamical time $t_{\rm dyn}$ and the two-body relaxation time $t_{\rm R}$. The rate at which gas is liberated by tidal disruptions is limited by the rate at which stellar orbits can diffuse in energy and angular momentum into an inward opening 'loss-cone' of semi-aperture $\theta_{\rm crit} \sim (t_{\rm dyn}/t_{\rm R})$ around the radial direction in velocity space. The maximum attainable fuelling rate for spherical clusters with near-isotropic velocity distributions is of the order of the cluster mass per relaxation time or approximately one stellar mass per dynamical time. This rate of disruption proves to be too low unless one pushes the cluster parameters to an extreme regime where physical collisions between stars would liberate even more mass than tidal disruptions. Another limitation of this mechanism is that once the black hole mass exceeds some $10^8\ M_\odot$, the tidal forces are no longer capable of tearing the star apart before it plunges irreversibly into the hole with relatively little electromagnetic radiation loss. For recent work on tidal disruptions and their observable consequences see Rees (1988).

Collisions between stars become disruptive when the typical stellar velocity exceeds a few times the escape velocity from individual stars. If *all* the debris of stellar collisions were accreted by the black hole, it would be possible to power quasars with this process. The rate of gas production by collisions is greater than the tidal disruption rate by a factor $\sim (v_c/v_*)^4$, where v_c is the cluster velocity dispersion and v_* is the stellar escape velocity. The fate of this gas is however uncertain since it is released at high speeds and could escape the cluster.

Finally, ablation of the outer layers of stars which move at speeds exceeding v_* through an interstellar medium has also been suggested. But it turns out that this process can become dominant only if the density of this hypothetical interstellar medium exceeds the distributed density of the cluster stars. In other words, when gas is *already* present in sufficient quantities then no production mechanism is necessary. Variants of this idea with stars whose outer layers are bloated through environmental effects have also been considered but without clear success.

It is worth noting that the dense clusters referred to here cannot form as a result of two-body relaxation processes since they would lead to an increasingly small core of high velocity stars which would destroy themselves by collisions. If gas dynamical

The gas supply

processes are involved in first concentrating the gas into a small volume from an initially larger supply, then the angular momentum problem has to be addressed satisfactorily. Since this is the main weakness of local mechanisms, one is driven to take the bull by the horns and examine other processes operating on larger scales which naturally have to deal with the angular momentum problem.

Since observations reveal that outside the galactic nucleus large amounts of fuel exist in the form of an interstellar medium, one might ask how long it would take for this gas to flow in driven by a turbulent (α-) viscosity of the type thought to operate in accretion discs in mass-transferring binaries (cf. Chapter 5). A simple estimate shows that, for reasonable assumptions about the physical state of the gas, and with $\alpha \leqslant 1$, the inflow time exceeds 10^9 years for gas beyond 10 pc. Thus, a standard accretion disc cannot steadily supply the required fuel. Furthermore, thin accretion discs tend to be gravitationally unstable at large radii where they are likely to be cool and tend to break up into lumps. If these lumps cool and contract, they decouple rapidly leading to even slower accretion. A survey of plausible sources of energy leads to the conclusion that a hot accretion flow cannot be maintained since it would tend to flow away in a wind. To summarize: in the language of accretion discs, what is required to fuel active galactic nuclei is a mechanism with an effective $\alpha \gg 1$. A few mechanisms have been suggested which could in principle yield a large effective α by coupling fluid at widely different radial scales. We shall briefly talk about two possibilities: large (galactic) scale magnetic fields and non-axisymmetric gravitational instabilities.

If a large scale poloidal magnetic field existed in the galaxies hosting a black hole, the magnetohydrodynamic torques generated by twisting this poloidal field would couple the gas at differing radial scales and effectively transport angular momentum outwards and matter inwards. This possibility is attractive because the toroidal fields generated in the process would build up a magnetic pressure tending to collimate any outflow towards the axis of rotation and thus provide a natural explanation for jets. The observational evidence for organized, large-scale magnetic fields in our own Galaxy is very good (e.g. loops near the Galactic centre) and in external galaxies it is suggestive. But it is not clear at this moment whether these magnetic fields are of the intensity required for fuelling.

Let us now turn to non-axisymmetric gravitational instabilities such as bars, and other disturbances caused by tidal interactions and mergers with passing galaxies. The observational evidence for a correlation between barred spiral galaxies and Seyfert activity goes back to 1977 and has been recently extended to include the effects of close companions. The observations suggest that most Seyferts possess some morphological evidence of transport mechanisms: large-scale stellar bars, other signs of non-axisymmetric perturbations such as oval discs and rings, and somewhat weaker evidence of excess of companion galaxies. Also galaxies with an abnormally large rate of star formation in their nuclear regions – the starburst galaxies – seem to arise more frequently in galaxies with distortions and/or companions.

Stimulated by the early evidence of a link between barred morphology and activity, a number of numerical calculations were carried out in the 1980s showing that gas, moving in a fixed (non-evolving) bar-like potential, would drift towards the centre in a time ~ 10 rotation periods. The gas is assumed to be dynamically unimportant at this stage. One can understand this as follows: the bar exists as a global $m = 2$ distortion (i.e. with azimuthal angular dependence $\cos 2\phi$) of the stellar distribution with some pattern speed Ω_p. The gas rotates in the galaxy with the local angular speed for nearly circular orbits $\Omega(R)$. Inside the corotation radius where $\Omega(R_{co}) = \Omega_p$, the gas rotates faster than the bar. As the gas catches up with the bar it is first accelerated towards the bottom of the potential trough and it is then decelerated and compressed as it overtakes the bar. Thus the gas shocks at the leading edge of the bar, loses angular momentum to the bar and flows in as a consequence. A similar inflow would be generated by the bar-like disturbance induced by a close encounter with another galaxy.

The gas does not flow all the way to the centre because the effects of the bar become weaker as the gas flows in. Typically these processes cease to drive inflow at scales ~ 1 kpc, where ordinary viscous processes are still incapable of bridging the gap to the black hole. If, however, the mass of gas so accumulated is large enough to be dynamically important, an $m = 2$ bar-like gravitational instability now in the *gaseous* disc may set in, redistributing angular momentum, and driving the gas further inwards. Once the gaseous disc has gone unstable, a significant fraction of the disc mass could flow in if two competing conditions are met: (a) coupling over different radial scales is maintained by gravitational or other torques; and (b) the instability efficiently sweeps the mass at every radial scale down to a smaller scale. These conditions are in principle contradictory since inflow results in higher mass concentrations, hence higher angular velocities, and gravitational coupling between non-axisymmetric features that rotate at widely different speeds is poor. The obvious solution is some sort of homologous inflow which might satisfy both constraints. The viability and details of such flows still remain to be studied with the help of numerical simulations.

The processes discussed above may also operate in elliptical hosts at radial scales < 1 kpc. Direct infall of the interstellar medium in high luminosity ellipticals whose stars are supported essentially by random motions but with $\sim 1\%$ rotational support, is likely to lead to a gaseous disc at scales $\lesssim 1$ kpc, unless other mechanisms contribute to angular momentum loss during infall. Thus, even in active ellipticals with little rotational support, a mechanism involving dynamical instabilities may be required to transport material further in to fuel a giant black hole.

In this picture, where dynamical instabilities operate at different radial scales, differences in initial conditions and/or the efficiency of mass transport may lead to different observable outcomes. For example, disc galaxies possessing a large-scale stellar bar, but whose swept-up gaseous material fails to go unstable at $\lesssim 1$ kpc scales, may turn into a starburst. Under certain dynamical conditions – e.g. in the presence of resonances – a gaseous ring forms which can turn into stars as perhaps is the case in the Seyfert galaxy NGC 1068.

7.7 Black holes

In order to discuss the efficiency of energy extraction we need to consider black holes in a little more detail than has been given so far. A theory of gravity provides equations for the motion of particles subject to the gravitational influence of massive bodies. In Newtonian theory, the radial position $r(t)$ of a particle in orbit about a mass M is given by the energy equation:

$$\tfrac{1}{2}\left(\frac{dr}{dt}\right)^2 + V(r) = E \tag{7.6}$$

where E = constant is the total energy of the particle per unit mass, and

$$V(r) = h^2/2r^2 - GM/r \tag{7.7}$$

is called the *effective potential* for a particle of constant angular momentum h per unit mass. Particles of zero angular momentum, $h = 0$, fall radially into the centre, while particles of any non-zero angular momentum escape if $E \geqslant 0$ (provided that they do not collide with the mass M). Particles remain bound to the mass M if $E < 0$; a bound particle moves between potential barriers at radii determined by $dr/dt = 0$, $V(r) = E$. The particular case of a circular bound orbit is possible if $dr/dt \equiv 0$ is a solution of (7.6); hence if $V(r) = E$ = constant. Thus circular orbits occur at radii at which $V(r)$ has a stationary point, $\partial V/\partial r = 0$.

In general relativity all energy contributes to the gravitating mass of a system, including the gravitational potential energy, so for strong gravitational fields the Newtonian inverse square law is significantly modified. A gravitational field is strong if $GM/rc^2 \sim 1$; this implies $GMm/r \sim mc^2$ for a body of mass m, i.e. its gravitational potential energy is of the order of its rest mass energy, and in such circumstances special relativity must also be taken into account. Thus the energy equation which replaces (7.6) must involve proper time s, in place of coordinate (absolute) time t, and this must be supplemented by a relation between s and t. In particular, for a spherical body of mass M in a vacuum (the *Schwarzschild solution*) it turns out that

$$\frac{1}{c^2}\left(\frac{dr}{ds}\right)^2 + V^2(r) = E^2, \tag{7.7a}$$

$$\frac{dt}{ds} = E(1 - 2GM/rc^2)^{-1}, \tag{7.7b}$$

and the effective potential is given by

$$V^2(r) = (1 - 2GM/rc^2)(1 + h^2/r^2), \tag{7.8}$$

where h = constant is again the (relativistic) angular momentum per unit mass of the orbiting body (Fig. 58).

As $r \to 2GM/c^2 = R_{\text{Schw}}$ equations (7.8) and (7.7a) show $V \to 0$ and $dt/ds \to \infty$ and the theory appears to break down. For the Sun we have $2GM/c^2 \sim 3$ km, so the

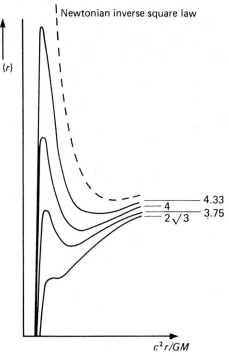

Figure 58. The effective potential $V(r)$ for various values of the specific angular momentum $c^2 h/GM$. (Adapted from C. W. Misner, K. S. Thorne & J. A. Wheeler (1973), *Gravitation*, Freeman.)

orbiting particle encounters the surface of the gravitating mass before it can approach $r = R_{\text{Schw}}$ and the problem is of no significance. However, it is relevant for a sufficiently compact body. In such a case, essentially by tracing light rays, it can be shown that $r = R_{\text{Schw}}$ represents the surface, or *horizon* of a black hole, and demarcates a region of space from the inside of which nothing can escape. This can be made plausible by an argument discovered independently by Michell and Laplace. Namely, we note that the expression for R_{Schw} is equivalent to $V_{\text{esc}} = (2GM/R_{\text{Schw}})^{1/2} = c$, so the Newtonian escape velocity from the horizon equals the velocity of light. Thus to escape from a black hole a particle would have to exceed the speed of light, which is impossible. Note also, however, that this argument is internally inconsistent, since there is no restriction on the speed of objects in Newtonian physics, and is inapplicable since Newtonian theory is certainly invalid for strong gravitational fields. The result is properly obtained from a general relativistic analysis.

Since we must have $(dr/ds)^2 \geq 0$ in (7.7a), motion is again possible only for $V(r) \leq E$ and the particle moves between the turning points. The crucial difference between the Newtonian orbits and the general relativistic ones arises from the facts that (i) $V(r)$ in (7.8) has a maximum as a function of r, as well as a minimum, if $h \geq 2(3)^{1/2} GM/c^2$ and (ii) there is no turning point at all if $h < 2(3)^{1/2} GM/c^2$, not just, as in Newtonian theory, if $h = 0$. The effects of the strong gravitational field therefore are (i) particles with sufficient energy overcome the centrifugal potential barrier and fall into the gravitating mass whatever their angular momentum, and (ii) particles with low (not just zero) angular momentum are captured by the hole.

Circular orbits where $dr/ds \equiv 0$ are possible, as in Newtonian gravity, at radii such that $\partial V/\partial r = 0$, but this now includes maxima as well as minima. However, at a maximum a small inward displacement causes the particle to fall in while an outward perturbation causes it to move out. Therefore the maxima represent *unstable* circular orbits. *Stable* circular orbits are possible at $r = (GM/2c^2)[H^2 + (H^4 - 12H^2)^{1/2}]$, where $H = c^2h/GM$, if $h \geqslant 2(3)^{1/2}GM/c^2$, and the innermost of these stable circular orbits occurs at $r_{\min} = 6GM/c^2$. This is the effective surface for the maximum feasible extraction of energy from infalling particles; beyond this the infall time must be short compared with a radiative lifetime. From an unstable orbit a particle carries its energy with it into the hole. A crude Newtonian analysis, which one might hope would be accurate to a factor ~ 2, gives a maximum efficiency of energy generation

$$\varepsilon_{\max} = \text{(maximum available gravitational potential energy)/(rest mass energy)}$$

$$= \text{(maximum binding energy)/(rest mass energy)}$$

$$= \frac{(GMm/2r_{\min})}{mc^2} = \tfrac{1}{12}.$$

A proper relativistic calculation of the binding energy gives a maximum efficiency of 6% (see equation (7.10)).

If the black hole itself possesses angular momentum the details of this discussion are somewhat altered quantitatively but not qualitatively. A rotating black hole (the *Kerr solution*) is characterized by a mass M and angular momentum per unit mass H, usually expressed in terms of parameters with the dimensions of lengths: $m = GM/c^2$ and $a = H/c$. For the Kerr black hole the horizon occurs at $r = m + (m^2 - a^2)^{1/2}$. Here the effective potential for motion in the equatorial plane can be *defined* as the minimum value of the energy per unit mass $E_{\min}$ for which motion is possible at each radius r, and is given by

$$V(r) \equiv E_{\min}(r) = [(r^2 - 2mr + a^2)^{1/2} \{r^2 h^2 + [r(r^2 + a^2) + 2a^2m] r\}^{1/2}$$
$$+ 2ahm][r(r^2 + a^2) + 2a^2m]^{-1}.$$

We now have a system of curves for various values of h for each value of a/m; for $a = 0$ we regain (7.8), of course. For $a \neq 0$ the curves are qualitatively similar with both a maximum and a minimum on each curve, representing unstable and stable circular orbits, provided h is large enough. The innermost stable circular orbit occurs at

$$r_{\min} = m\{3 + A_2 \mp [(3 - A_1)(3 + A_1 + 2A_2)]^{1/2}\} \tag{7.9}$$

(Bardeen (1973)) where $A_1 = 1 + (1 - a^2/m^2)^{1/3}[(1 + a/m)^{1/3} + (1 - a/m)^{1/3}]$, $A_2 = (3a^2/m^2 + A_1^2)^{1/2}$, and where the upper sign refers to particles orbiting in the same sense as the rotation of the hole, while the lower sign occurs for counter-rotating particles. Equation (7.9) reduces to $r_{\min} = 6m$ if $a = 0$. The last stable orbit in the equatorial plane can be shown to correspond to a maximum efficiency of energy extraction

$$\varepsilon_{\max} = 1 - [r_{\min} - 2m \pm (am^{1/2}/r_{\min}^{1/2})][r_{\min} - 3m \pm (2am^{1/2}/r_{\min}^{1/2})]^{-1/2} \quad (7.10)$$

or approximately 40% for a particle corotating with a hole having the maximum allowed angular momentum $a = m$.

So far, our black hole solutions are characterized by at most two parameters and are either exactly spherically symmetric or exactly axially symmetric. Thus the gravitational fields are of a highly symmetrical and restrictive form. In general, the irregular Newtonian gravitational field of a real body requires for its specification a whole range of parameters corresponding to an analysis into monopole ($1/r^2$), dipole ($1/r^3$), quadrupole ($1/r^4$), ... components. It is necessary to ask what happens to these irregularities when a real body collapses in on itself with no means of support. Indeed, one might suspect that the formation of a black hole could be an artefact of the symmetry. However, the *singularity theorems* show that provided the initial configuration is reasonable, a gravitational singularity of some sort must arise. The *cosmic censorship* hypothesis conjectures that the singularity is always hidden behind a horizon, as a black hole is formed. There is no general proof of this, but there are arguments that make it plausible. In particular, for small departures from spherically symmetry, hence in astrophysically reasonable cases, *Price's theorem* shows that higher moments of the gravitational field (quadrupole, etc.) are radiated away as gravitational waves leaving only the monopole (mass) and dipole (angular momentum) terms appropriate to the Kerr solution. In other words, the body tends to become more symmetrical as a result of collapse. This is just as required because the *Israel–Carter–Robinson theorem* shows that there is a unique (electrically neutral) black hole for a given mass and angular momentum, which therefore must be just that given by the Kerr solution. The endpoint of the catastrophic collapse of a massive body in a realistic astrophysical context can therefore be taken to be the formation of a Kerr black hole.

7.8 Accretion efficiency

We have seen that a high efficiency of conversion of gravitational potential energy to radiation energy is an essential requirement of a quasar model if we are to avoid an excessive accretion rate and the accumulation of an exceedingly massive black hole over a quasar lifetime. In disc accretion material cannot fall into the hole without dissipating its binding energy, so the efficiency must be given by the binding energy per unit mass at the last stable orbit, and this exceeds 6% even for a Schwarzschild (non-rotating) black hole. Thus, in a disc the efficiency presents no problem, provided only that a suitable source of viscosity exists to ensure that material falls in. In this section we shall show that much the same claim can be made for spherical accretion. Consequently other arguments are required to distinguish between the two possibilities. We shall also consider the limits placed on accretion efficiency from observations of variable sources. The following discussion of the

Accretion efficiency

efficiency of spherical accretion is taken from McCray (1979). Suppose that the black hole accretes optically thin gas and that the infall timescale is short compared with the radiative timescale due to bremsstrahlung, both somewhat unpromising assumptions for maximum efficiency. Within the accretion radius (equation (2.33))

$$r_{\text{acc}} = 3 \times 10^{18} M_8 / T_8 \text{ cm} \tag{7.11}$$

the flow velocity v approaches free fall

$$v \sim (2GM/r)^{1/2};$$

conservation of mass,

$$4\pi r^2 N v = \text{constant},$$

implies

$$N = N_0 (r_{\text{acc}}/r)^{3/2}. \tag{7.12a}$$

From the assumption that the infall is approximately adiabatic we have, using (7.12a),

$$T \propto N^{\gamma-1} \sim T_0 (r_{\text{acc}}/r) \tag{7.12b}$$

if $\gamma = \frac{5}{3}$ as appropriate to a fully ionized plasma at a moderate temperature (compare (2.21) and (2.22)). In fact, the temperature probably becomes high enough that the electrons, but not the ions, become relativistic, at which stage $\gamma = \frac{13}{9}$ is correct, but we shall ignore this detail save to note that it changes the results by only a small factor.

The emitted luminosity is

$$L = 4\pi \int_{R_{\text{Schw}}}^{r_{\text{acc}}} 4\pi j_{\text{br}}(T) r^2 \, dr$$

where $j_{\text{br}}(T)$ is the bremsstrahlung emissivity. The lower limit of integration can be taken as the Schwarzschild radius of the hole, since, of course, beyond this the accreting material cannot emit. The upper limit is the accretion radius beyond which the gas is relatively uninfluenced by gravity. Since $r_{\text{acc}} \gg R_{\text{Schw}}$ and the integrand is a decreasing function of r, we can let the upper limit tend to infinity. Then

$$L \sim 2.5 \times 10^{-27} R_{\text{Schw}}^{-1/2} r_{\text{acc}}^{7/2} N_0^2 T_0^{1/2} \text{ erg s}^{-1}$$

and substituting $R_{\text{Schw}} = 2GM/c^2$ for a spherical hole (Section 7.7), and using (7.11), we obtain

$$L = 4 \times 10^{32} M_8^3 T_8^{-3} N_0^2 \text{ erg s}^{-1}. \tag{7.13}$$

By definition, the efficiency is $\eta = L/\dot{M}c^2$ where $\dot{M}$ is the Bondi accretion rate given by equation (2.31). Therefore from (2.31), which is valid in the limit $\gamma \to \frac{5}{3}$, whence $[2/(5-3\gamma)]^{(5-3\gamma)/2(\gamma-1)} \to 1$, we obtain

$$\eta = 9 \times 10^{-3} (L/L_{\text{Edd}})^{1/2}. \tag{7.14}$$

For accretion from the interstellar medium by a solar mass black hole, (7.13) gives a very low luminosity ($L \lesssim 10^{-16} L_{\text{Edd}}$) and hence a low efficiency. But for luminosities approaching L_{Edd}, it appears that the efficiency increases to a reasonable value.

However, it is impossible to achieve $L \sim L_{\text{Edd}}$ in this model without violating the assumptions. Consider first the effects of a non-zero optical depth. We have

$$\tau = \sigma_T \int_r^\infty N_e \, dr = \sigma_T \int_r^\infty N_0 (r_{\text{acc}}/r)^{3/2} \, dr = 2\sigma_T N_0 r_{\text{acc}}^{3/2} r^{-1/2} \tag{7.15}$$

for the optical depth from r to infinity due to electron scattering. But this is

$$\tau \sim \frac{\sigma_T \dot{M} c^2}{4\pi G M c m_p} \left(\frac{GM}{c^2}\right)^{1/2} \frac{1}{r^{1/2}} = \frac{1}{\eta} \frac{L}{L_{\text{Edd}}} \left(\frac{R_{\text{Schw}}}{r}\right)^{1/2}, \tag{7.16}$$

where the intermediate step arises from eliminating $c_s(\infty)$ from (2.25) and (2.31) in the case $\gamma = \frac{5}{3}$ and substituting for r_{acc} in (7.15). Thus $L \sim L_{\text{Edd}}$ at $r \sim R_{\text{Schw}}$ implies $\tau \sim 10^2$ from (7.14). This would have the effect of extracting energy from outward flowing high-energy photons in Compton scattering off the thermal electrons, thereby reducing the flux above an energy $h\nu = kT(r_1)$, where r_1 is the radius at which $\tau = 1$. It also implies that not all the radiation produced can escape. For in order to do so photons must random walk their way to regions of low optical depth at a speed in excess of that with which the infalling material would carry in trapped radiation. This latter is of the order of the free-fall velocity. After n scatterings of a random walk with mean free path λ_{mfp} a photon has travelled a path length $n\lambda_{\text{mfp}}$ in a time $n\lambda_{\text{mfp}}/c$ and moved a distance $n^{1/2}\lambda_{\text{mfp}}$ equivalent to an optical depth $\tau = n^{1/2}\lambda_{\text{mfp}}/\lambda_{\text{mfp}} = n^{1/2}$. Thus the outward random walk velocity is $n^{1/2}\lambda_{\text{mfp}}/(n\lambda_{\text{mfp}}/c) = c/\tau$. The radius at which this equals the infall velocity is called the *trapping radius* r_t and is given by

$$c/\tau(r_t) = (2GM/r_t)^{1/2}, \tag{7.17}$$

or using (7.16)

$$r_t = \frac{1}{\eta} \left(\frac{L}{L_{\text{Edd}}}\right) R_{\text{Schw}}.$$

If we try to make $L \sim L_{\text{Edd}}$ we find $r_t \sim 10^2 R_{\text{Schw}}$ and most of the high-frequency radiation is convected into the hole.

The second assumption of the model, that the flow is approximately adiabatic, will be justified provided the infalling gas cannot radiate its gravitational potential energy on an infall timescale t_{in}. In our case $t_{\text{in}} \sim t_{\text{ff}} \sim (GM/r^3)^{1/2}$, and the ratio of the cooling timescale for bremsstrahlung (the time taken by the gas to radiate its internal energy) to the infall timescale is

$$\frac{t_{\text{rad}}}{t_{\text{in}}} \sim \frac{NkT}{4\pi j_{\text{br}}} \left(\frac{r^3}{GM}\right)^{-1/2} \sim 10^2 T_8^{1/2} \eta \frac{L_{\text{Edd}}}{L} \left(\frac{r_{\text{acc}}}{r}\right)^{1/2},$$

where the final estimate follows on using equations (7.12a), (7.12b), (7.11), (7.13),

Accretion efficiency

(2.31) and the definitions of the efficiency and of the Eddington limit. So it is inconsistent to postulate a hot gas ($T_8 \sim 1$) at $r = r_{\rm acc}$ for $L \sim L_{\rm Edd}$ and assume adiabatic inflow. High efficiencies cannot be produced in this way.

Since $t_{\rm rad}/t_{\rm ff} \propto \eta/L \propto \dot{M}^{-1}$ we have essentially two distinct cases. For sufficiently low accretion rates $t_{\rm rad} > t_{\rm ff}$ and the adiabatic assumption is valid. For high accretion rates $t_{\rm rad} < t_{\rm ff}$ and the temperature is determined instead by balancing gains and losses of energy in the gas: in a stationary state the rate of loss of energy by radiation (the 'rate of cooling') in an element of volume fixed in space must equal the sum of the rate of energy input into the volume (the 'heating rate'), the rate at which pressure does work on the gas, and the net rate of gain by advection of internal energy into and out of the volume element (the 'enthalpy' flux, cf. (2.5)). In all cases, there is a general minimum value for the temperature, whatever the cooling mechanism, given by the temperature of a blackbody radiating at the same rate. Thus $L(r) = 4\pi r^2 \sigma T_{\rm min}^4$ and

$$T(r) \gtrsim 2.3 \times 10^4 [L(r)/L_{\rm Edd}]^{1/4} (300 R_{\rm Schw}/r)^{1/2} M_8^{-1/4} \ {\rm K}. \tag{7.18}$$

It follows that if $L \sim L_{\rm Edd}$ then T must exceed 10^4 K at $r < 300 R_{\rm Schw}$. The condition also shows that if $L \sim L_{\rm Edd}$ and if the gas reaches 10^8 K at $r \sim R_{\rm Schw}$ then it is radiating at the blackbody limit and so must be optically thick, and this is consistent with our previous discussion. Of course, the discussion does not imply that these conditions can be attained: in order for the gas to achieve thermal equilibrium the electrons and protons must be able to come into equipartition on a timescale short compared with the heating timescale, a condition which may well not hold in X-ray emitting active nuclei.

The maximum efficiency of extraction of energy will be obtained if the heating and cooling rates balance ($t_{\rm rad} = t_{\rm ff}$) at the horizon, $r = R_{\rm Schw}$. This follows because a lower cooling rate means that energy that might have been radiated is instead advected into the hole. On the other hand, a higher cooling rate at $r = R_{\rm Schw}$ implies a reduced gas temperature and pressure and hence a reduction in the work that can be done on the gas ($\int P dV$) during infall; as a consequence, the infalling kinetic energy is less effectively converted into internal energy and radiated. It follows that the inclusion of other possible cooling mechanisms, in addition to bremsstrahlung, does not necessarily lead to a high efficiency, although it may alter the ambient conditions of temperature and density far from the hole under which the maximum efficiency is obtained. For example, it is probable that the accreted material is threaded by magnetic field so synchrotron emission will become important if the electrons reach high temperatures (or if the other acceleration mechanisms operate). Detailed calculations show the efficiency is still low even at optimum conditions.

In order to improve the efficiency, in addition to an efficient radiation mechanism, we need a source of heating of the infalling gas which is more effective than adiabatic compression. Thus we need to invoke dissipative processes to transform the ordered infall into heat which can be radiated. Realistic models involve heating by dissipation of turbulent velocities in the infalling gas, by reconnection of magnetic fields, or by acceleration in shocks.

Alternatively an improvement in efficiency might be obtained by increasing the inflow timescale. Rather surprisingly there turns out to be a natural way of achieving this. Solutions exist for spherical accretion in which protons downstream of a standing spherical shock are heated to temperatures at which the e–p collision timescale is longer than the free-fall timescale. The electrons and protons in this region therefore behave as individual particles in orbit around the black hole diffusing slowly inwards on an e–p collision timescale.

Consider accreting gas that, in addition to thermal energy, carries with it magnetic and turbulent energy. In the interstellar medium these are in approximate equipartition:

$$Nc_s^2 \sim Nv_t^2 \sim \left[\frac{4\pi}{\mu_0}\right]\frac{B^2}{8\pi} \equiv \tfrac{1}{2}v_M^2 N$$

where v_t is the mean turbulent velocity and v_M the Alfvén velocity (Section 3.7). It is therefore reasonable to assume that the black hole accretes material that is in equipartition outside the accretion radius. This internal energy is then available in principle to heat the gas. For turbulent energy this could be achieved by dissipation through shear viscosity or hydromagnetic shocks from the collision of turbulent eddies. This turbulence will also act to amplify frozen-in magnetic fields (Section 3.7) to energy densities at which they could locally dominate the flow, and this would lead to currents in the plasma, and Joule heating. The timescale for dissipation of magnetic energy is the Alfvén timescale, l/v_M where l is the lengthscale of enhanced field (Section 3.7). For equipartition at the accretion radius r_{acc}, $v_t = v_M = c_s = (2GM/r_{acc})^{1/2}$, so if $l \sim r_{acc}$ we expect the dissipative heating to occur on a free-fall timescale, $(r_{acc}^3/2GM)^{1/2} = t_{ff} \lesssim t_{in}$.

We assume that the dissipative heating rate per unit volume $H(N_e, T)$, is of the order of the rate of loss of gravitational potential energy, hence

$$H(N_e, T) \sim (GMNm_p/r)/t_{in}. \tag{7.19}$$

At $r = r_{acc}$ this follows from equipartition, with $t_{in} = t_{ff}$ according to the previous argument, since the energy to be dissipated is $\rho v_M^2 \sim \rho v_t^2 \sim \rho c_s^2 \sim GM\rho/r_{acc}$; the assumption consists of extending (7.19) to $r \leqslant r_{acc}$. If advected energy can be neglected and the system is in a steady state, the temperature of the material, obtained by balancing energy gains and losses, is given by

$$4\pi j \equiv N_e^2 \Lambda(T) \sim H(N_e, T), \tag{7.20}$$

since we are assuming dissipative heating dominates adiabatic compression. In (7.20), $\Lambda(T)$ is the standard *radiative cooling function* defined in terms of the usual emission coefficient by $4\pi j \propto \rho^2 \Lambda$ (p. 99). Whether this balance between heating and cooling does in fact hold is governed by the parameter

$$Q = \frac{\text{rate of dissipative heating}}{\text{rate of radiative cooling}}. \tag{7.21}$$

If $Q \lesssim 1$ cooling is efficient and the gas falls in without a large temperature rise. However, in this case the efficiency with which energy is extracted is not forced to

Accretion efficiency

be low if dissipation feeds sufficient energy into the gas, which is then radiated. If $Q \gtrsim 1$ the gas temperature will rise to a value at which cooling does become efficient. Such peaks in the cooling function $\Lambda(T)$ occur at 10^4 K, when the line cooling operates effectively, and above about 10^8 K when relativistic cooling processes come into play. Thus if $Q \gtrsim 1$ at 10^4 K a rise to efficient cooling at $\gtrsim 10^8$ K is generally assured and a high efficiency of energy extraction virtually guaranteed, provided, of course, that our initial assumption of a steady state is realized.

For example, suppose for simplicity that the dominant cooling mechanism is bremsstrahlung, so

$$t_{\rm rad} = N_{\rm e} kT/4\pi j_{\rm br} \sim N_{\rm e} kT/2 \times 10^{-27} T^{1/2} \langle N_{\rm e}^2 \rangle \text{ s} \tag{7.22}$$

where the averaging over $N_{\rm e}^2$ allows for some clumpiness in the accreting gas. The accretion rate is

$$\dot{M} \sim 4\pi r^2 m_{\rm p} N_{\rm e} r/t_{\rm in} \sim M N_{\rm e} t_{\rm ff}^2 / t_{\rm in}, \tag{7.23}$$

where we have assumed $N \sim N_{\rm e}$. Therefore, using equations (7.20)–(7.23),

$$Q \sim \tfrac{1}{3} \left(\frac{L_{\rm Edd}}{L} \right) \eta \left(\frac{kT}{m_{\rm p}c^2} \right)^{-1/2} \left(\frac{R_{\rm Schw}}{r} \right) \left(\frac{t_{\rm ff}}{t_{\rm in}} \right)^2 \frac{N_{\rm e}^2}{\langle N_{\rm e}^2 \rangle}.$$

Unless the flow is very lumpy ($\langle N_{\rm e}^2 \rangle \gg N_{\rm e}^2$) or the infall timescale is long compared with free fall (as in disc accretion), or the overall efficiency η very small, then even for $L \sim L_{\rm Edd}$ we have $Q > 1$. Thus the gas is heated by dissipation and, if we can neglect radiation losses during the heating so that $kT \sim GMm_{\rm p}/r$, relativistic temperatures are reached at $r \sim 10^3 R_{\rm Schw}$. At this point synchrotron, inverse Compton, or perhaps electron–positron pair production processes radiate the infall energy with an efficiency that can approach unity.

The efficiency problem in spherical accretion can therefore be overcome if sufficient dissipation occurs. As a final point, we should consider the spectrum of the emerging radiation, since at first sight the observed power law spectra might appear to present a difficulty. In fact, it is only too easy to produce approximate power law spectra either directly or by reprocessing of the emergent radiation. For example, in the optically thin bremsstrahlung model with which we started we have

$$L_\nu = \int 4\pi j_\nu \, dV \underset{\sim}{\propto} \int N_{\rm e}^2 T^{-1/2} e^{-h\nu/kT} r^2 \, dr \underset{\sim}{\propto} \int r^{-1/2} e^{-h\nu/kT} \, dr$$

where we have used (7.12a) and (7.12b). Defining $x = h\nu/kT \propto \nu r$ and substituting for $r \propto x/\nu$ gives $L_\nu \propto \nu^{-\alpha}$, $\alpha = \tfrac{1}{2}$. A similar result is obtained for the spherical shock model with a slightly different $\alpha (\approx 0.6$–$0.7)$ arising from the different radial dependence of the temperature. Alternatively, reprocessing by Compton scattering in a hot plasma can produce a power law spectrum (see e.g. Rybicki & Lightman (1979) and Section 9.4).

Finally we turn to an observational constraint on efficiency (Fabian (1979)) that, at least for spherical sources, is largely model-independent. Suppose the luminosity

L is radiated by a variable source over a timescale Δt. If the source has a significant electron scattering optical depth $\tau_{es} \geq 1$, the observed timescale of variation must satisfy

$$\Delta t \geq (1 + \tau_{es}) R/c.$$

This follows from the discussion leading to equation (7.17) above where it was shown that the effective diffusion velocity of radiation in the source is c/τ; hence the additional time delay due to scattering is $R/(c/\tau_{es}) = \tau_{es} R/c$. If the density of material in the source is n the mass involved in the outburst is $M \leq \frac{4}{3}\pi R^3 n m_p$ and the luminosity is $L = \eta M c^2 / \Delta t$. So, since $\tau_{es} = n \sigma_T R$, we have

$$\eta = \frac{3L\Delta t}{4\pi c^2 R^3 m_p n} \geq \frac{3L\sigma_T}{4\pi c^4 m_p \Delta t}$$

where we have eliminated n and R and used $(1 + \tau_{es}) > 1$. Hence, if L is in erg s^{-1} and Δt in seconds, we get

$$\eta > \frac{L}{2 \times 10^{42} \Delta t}.$$

Clearly the argument will yield a limit on the efficiency for a general source of opacity, not only for electron scattering.

Some of the more violent variations observed grossly violate this constraint by implying $\eta > 1$. In these cases relativistic effects have to be invoked to lengthen the timescale in the frame of the source and to reduce the intrinsic change in luminosity.

8

Accretion discs in active nuclei

8.1 The nature of the problem

At first sight an accretion disc appears to be an unlikely candidate for the central engine of an active nucleus. Scaling the thin disc equations (5.45) to typical parameters for active nuclei gives

$$\Sigma = 5.2 \times 10^6 \, \alpha^{-4/5} \dot{M}_{26}^{7/10} M_8^{1/4} R_{14}^{-3/4} f^{14/5} \text{ g cm}^{-2}$$

$$H = 1.7 \times 10^{11} \, \alpha^{-1/10} \dot{M}_{26}^{3/10} M_8^{-3/8} R_{14}^{9/8} f^{7/5} \text{ cm}$$

$$\rho = 3.1 \times 10^{-5} \, \alpha^{-7/10} \dot{M}_{26}^{11/20} M_8^{5/8} R_{14}^{-15/8} f^{11/5} \text{ g cm}^{-3}$$

$$T_c = 1.4 \times 10^6 \, \alpha^{-1/5} \dot{M}_{26}^{3/10} M_8^{1/4} R_{14}^{-3/4} f^{6/5} \text{ K}$$

$$\tau = 3.3 \times 10^3 \, \alpha^{-4/5} \dot{M}_{26}^{1/5} f^{4/5}$$

$$v = 1.8 \times 10^{18} \, \alpha^{4/5} \dot{M}_{26}^{3/10} M_8^{-1/4} R_{14}^{3/4} f^{6/5} \text{ cm}^2 \text{ s}^{-1}$$

$$v_R = 2.7 \times 10^4 \, \alpha^{4/5} \dot{M}_{26}^{3/10} M_8^{-1/4} R_{14}^{-1/4} f^{-14/5} \text{ cm s}^{-1}$$

with $f = 1 - (6GM/Rc^2)^{1/2}$, where the inner radius R_* is taken to be the last stable orbit (Section 7.7) for a Schwarzschild black hole. The effective temperature of the disc photosphere is, from equation (5.42),

$$T = 2.2 \times 10^5 \, \dot{M}_{26}^{1/4} M_8^{1/4} R_{14}^{-3/4} \text{ K}.$$

The characteristic temperature is therefore 10^5 K in the inner disc where most of the energy is released. The spectrum (Fig. 20) can be described as a rather broadened blackbody curve peaking in the soft X-ray region, clearly nothing like the observed power law extending to hard X-rays. The above formulae are valid for a disc where Kramers' opacity dominates (Section 5.6). From equation (5.51) Kramers' opacity is more important than electron scattering if

$$R \gtrsim 5.4 \times 10^{17} \dot{M}_{26}^{2/3} M_8^{1/3} f^{8/3} \text{ cm},$$

so in the inner region we expect electron scattering to dominate. This flattens the spectrum but does not demolish the argument. Gas pressure will be greater than radiation pressure at radii

$$R \gtrsim 5.2 \times 10^{14} \, \alpha^{8/30} \dot{M}_{26}^{14/15} M_8^{1/3} f^{56/15} \text{ cm}$$

from equation (5.53). In terms of the critical accretion rate (5.55) this is

$$R \gtrsim 0.8 \times 10^{14} \left(\frac{\dot{M}}{\dot{M}_{\mathrm{crit}}}\right)^{14/15} \alpha^{8/30} \eta^{-14/15} M_8^{19/15} f^{56/15} \text{ cm},$$

so for efficient subcritical accretion the disc is largely gas pressure dominated. For supercritical accretion radiation pressure is important and the thin disc approximation breaks down (Section 5.6). Other considerations then apply (Chapter 10).

If we take this discussion seriously (and in the next section we shall see why perhaps we should not) then the objective becomes not to explain the observed spectrum in terms of an accretion disc but rather to find where the disc emission is hiding.

Let us look in more detail at a representative spectrum of an active nucleus (Fig. 59, cf. Fig. 57). In section 7.1 we looked at the typical spectrum from the point of view of the characteristic flatness of the νF_ν distribution over many decades of frequency. Now we consider some of the bumps in this spectrum (after due allowance for emission from the host galaxy). One or both of two features can often be found in the UV – a bump around 3000 Å and a turn-up at shorter wavelengths (as measured in the rest frame of the source, of course). A large fraction of X-ray sources show an excess of soft X-ray emission around 0.5 keV relative to an extrapolation from the 2–10 keV band of a power law ($\nu F_\nu \propto \nu^{0.3}$) modified by absorption. The putative bump which fills in the spectrum between the upturn in the UV and the steep spectrum in soft X-rays is called the 'big blue bump'. This is to distinguish it from the 3000 Å bump, also called the 'little blue bump' with which it was confused in some of the early literature. Some authors also refer to an infrared bump relative to an extrapolation of the radio spectrum.

Despite early attempts to ascribe it to blackbody emission from the inner regions of an accretion disc it is now agreed that the 3000 Å feature arises from overlapping Fe II lines and Balmer continuum associated with the broad line region (Section 8.6). On the other hand the 'big blue bump', by which we mean the bump remaining after subtraction of any Fe II and Balmer continuum contribution, is widely regarded as the signature of an accretion disc. It can be reasonably fitted by a standard thin disc spectrum, particularly if the effects of an electron scattering atmosphere to the disc are included. However, as we shall see in the next section, the argument is by no means conclusive.

We should mention here two arguments that have been raised as objections to the disc model. The first is the absence of an absorption edge in the spectra of active nuclei at the Lyman limit, which should result from a cool disc atmosphere. The observation is a difficult one since in high redshift objects the UV spectrum is confused by the 'Lyman α' forest of absorption lines due to intervening material (Section 7.3); and at low redshifts the Lyman α photons are absorbed in the Galaxy. However, where the edge has been looked for it is not observed. Electron scattering in a hot corona will broaden such features in the observed spectrum, perhaps beyond the limit of detectability.

A second objection is the polarization of radiation from a disc surface. When radiation is scattered by an electron the cross-section for the two states of linear polarization depends on the scattering angle. Depending on inclination the disc

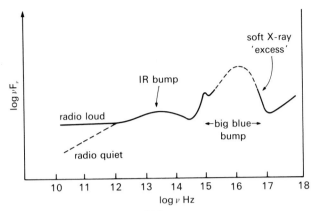

Figure 59. Schematic AGN spectrum.

radiation should therefore be partially polarized. In fact, the observed net polarization should be substantial with the electric vector perpendicular to the axis of the disc if the axis is close to the plane of the sky i.e. except for close to face-on discs. In radio sources where a big bump is observed one would expect the blue light in the bump to be substantially polarized perpendicular to the radio axis. In fact it is weakly polarized parallel to that axis. Nevertheless, polarization is not observed in cataclysmic variables (CVs) where we know a disc is present. One suggested resolution of the problem is that the disc surface is irregular so that the scattering angle to the observer spans a wide range thereby reducing the net polarization to a random residual.

In subsequent sections therefore we shall consider what circumstantial support the disc picture gains from observational features it can be invoked to explain. Even though one cannot (yet) use the arguments as compelling evidence in their favour, accretion discs emerge as 'good-buy' models.

8.2 Radio, millimetre and infrared emission

The remarkable alignment of radio structure in classical double radio sources which must be maintained over 10^8 years suggested early in the history of the subject the presence of net angular momentum in active nuclei. From this the idea that the angular momentum is in an accretion disc which, moreover, could be responsible for accelerating and collimating the relativistic outflows of jet plasma developed into the theory of thick discs (Chapter 10). Whatever the merits of thick discs, the idea that they are in themselves responsible for highly relativistic jets has now been almost universally abandoned. We are left with the thought that angular momentum may provide the orientation in radio sources but without any conclusive evidence that the angular momentum is in a disc rather than the black hole (or the galaxy). Indeed, in BL Lacs where we are presumably exploring the relativistic jet on the smallest scales, there is no evidence for a big bump in soft X-rays which could be attributed to a disc.

In those quasars where radio emission is not dominated by a compact core, and even possibly where it is, the millimetre emission does not interpolate smoothly between the radio and infrared. This suggests that the radio emission is a separate component from the infrared–optical spectrum, and supports a unified picture in which radio emission in all cases comes from a separate jet component viewed at different angles.

On the other hand the near infrared is related statistically in an apparently remarkable way to the 2–10 keV X-ray emission: the X-ray flux can be predicted to within a factor 2 from the near infrared slope in some sample sets of sources. The connection is all the more remarkable in view of the fact that in the handful of cases where both are measured the timescales of variability appear to be very different, the large amplitude short term fluctuations in X-rays being entirely absent from the infrared. This suggests reprocessing of the primary X-rays at large radii (in a disc?). But this cannot be the whole story because the far infrared is not well correlated with X-rays, if at all.

We conclude that this region of the spectrum does not provide much evidence, one way or the other, on the role of accretion discs.

8.3 X-rays and discs

The features of the X-ray spectrum to be considered are (i) rapid variability, (ii) the canonical 2–10 keV power law, (iii) the soft excess and (iv) the iron emission.

The rapid variability (Section 7.1) implies that the variable X-ray component is emitted from a region of size $\leqslant 10^{15}$ cm at most and within 10^{13} cm in low luminosity sources. It is not yet clear how the 2–10 keV emission and the soft excess are related. Neither are produced in a standard accretion disc around a massive black hole, although suggested mechanisms (Section 9.4) are compatible with a disc geometry within a region of this size.

As we remarked in Section 8.1, some authors attempt to identify the big blue bump with a modified disc spectrum. There is, in fact, no direct evidence that the UV and X-ray soft excess are connected; one can certainly be found without the other. Nevertheless, where variability is observed, albeit at different times and in different sources, the UV is found to harden as the luminosity increases whereas the soft X-rays appear to get softer. If such behaviour were seen in a single object it would be suggestive of the variation of a single component connecting the two wavebands.

The real problem however, and one which we warned in Section 8.1 should caution us from taking this picture too seriously, is the disc spectrum. Despite the widespread assumption to the contrary, the spectrum almost certainly cannot be represented locally by a blackbody in anything but the crudest of approximations, and certainly not for the purpose of fitting observed spectra. Real disc spectra are not available; for illustration therefore Fig. 60 shows the spectrum of a white dwarf having an effective temperature and surface gravity comparable to the inner part of an accretion disc. The large departures from the blackbody curve at the same temperature as the star arise because at high frequency the reduced opacity means one is seeing deeper into the atmosphere.

Finally here, we should note that the disc interpretation for the origin of the big

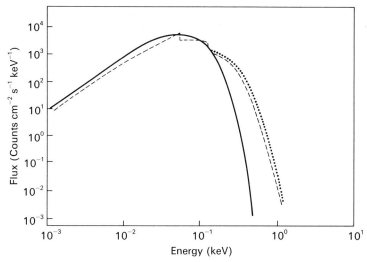

Figure 60. Model white dwarf atmospheres computed for solar abundances of H, He, C, N, O, a temperature of 2.32×10^5 K, and two values of surface gravity $\log g = 8.11$ (dashed line) and $\log g = 9.31$ (dotted line) by M. Barstow (unpublished). A black body curve at the same effective temperature is shown for comparison (solid line).

bump is by no means unique. From equation (7.18) we see that $T \approx 10^5$ K is a characteristic thermal temperature within about 30 Schwarzschild radii. Thus any sufficiently dense gas in this region will radiate at this temperature. A disc provides perhaps the simplest way to confine this gas, but clouds, either optically thick or optically thin, and confined perhaps by a magnetic field, have also been proposed.

A more promising link between X-rays and discs comes from the observation of an X-ray line at about 6.4 keV. This is identified as a fluorescence line of iron. Such X-ray line photons are produced when an atom, or ion, of a heavy element, is left in an excited state following ejection from an inner K- or L-shell by an incident X-ray photon of sufficient energy. The ion may return to a lower energy state by emitting an electron from a higher shell (the 'Auger effect') or by a radiative transition. The relative probability of a radiative transition is referred to as the fluorescence yield. For K-shell electrons the fluorescence yield increases with atomic number; the largest product of element abundance and yield, by a factor of about 5, occurs for iron. The L-shell fluorescence yield is comparatively small. The actual yield depends on the stage of ionisation. For neutral iron it is 34%, but it varies between 11% and 75% for higher states of ionization.

The energy of the K-α line also depends on the number of electrons present. For Fe XXVI which has only one electron, the levels are hydrogenic and the K-α line (the analogue of Lα in hydrogen) is at 6.9 keV. This energy falls to 6.45 keV in Fe XVII and then more slowly to 6.4 keV in neutral iron. The photoelectric absorption cross-section for the removal of the K-shell electrons is approximately independent of ionization stage, but the threshold energy varies between 7.1 keV for neutral iron to 9.3 keV for Fe XXVI. Thus the identification of this line at

6.4 keV is evidence for the presence of relatively cold ($T \leqslant 10^6$) material irradiated by a sufficient flux of energetic X-rays and emitting into our line of sight.

Furthermore, the equivalent width (i.e. the line flux divided by the continuum flux at the line energy) of the observed line is relatively large. This means the cold material must cover a large fraction of the source of hard X-rays without at the same time obscuring it. Reprocessing of hard X-rays reflected from the surface of an accretion disc undoubtedly meets these conditions. At the same time it gives a hard X-ray tail, variations in which should lag those in the primary power law continuum; this is at least compatible with the observed spectral changes. Detailed modelling and observations of variability and line profiles in the future appears to be a promising way of identifying the signature of an accretion disc in active nuclei.

8.4 The broad and narrow, permitted and forbidden

We turn now to the second aspect of the problem, namely the information to be obtained from the line spectra. We have already mentioned an observationally quite clear distinction between the broad and narrow emission lines. The principal broad lines, with their wavelengths in angstroms, are, in the ultraviolet, Mg II $\lambda 2798$, C III] $\lambda 1909$, C IV $\lambda 1549$, N V $\lambda 1240$, and Lα $\lambda 1216$ and blends of Si IV plus a narrow [O IV] line at $\lambda 1400$, and O VI + Lβ $\lambda 1030$, and in the optical the Balmer lines of hydrogen, Hα $\lambda 6562$, Hβ $\lambda 4861$ and Hγ $\lambda 4340$. Helium lines He II $\lambda 1640$, He II $\lambda 4686$ and He I $\lambda 5876$ are commonly distinguishable (in spectra that sample the appropriate range of redshifted wavelengths, of course) as are numerous bands of Fe II emission, and Pα $\lambda 18751$ has been observed in the infrared.

The absence of a square bracket distinguishes a line as *permitted*, which means that the atomic transition giving rise to it has a non-zero probability at the lowest ('electric dipole') order of quantum-mechanical perturbation theory, and therefore occurs rather rapidly after an atom has been excited (typically after less than 10^{-8} s). The square brackets, as in [O IV] indicate that the transition is *forbidden* in first order perturbation theory, so depends on higher order effects and is therefore slower to occur. In fact, if there is a high density of free electrons in the emitting plasma, forbidden line emission will tend to be *relatively* depressed because a bound excited electron will more readily lose its excitation energy in a collision with a free electron rather than by making a forbidden radiative transition. We can define a critical free electron density N_e^*, above which this effect is important, by balancing the radiative transition rate from the upper to lower state A_{ul} with the rate of collisions $N_e R_{ul}$ at which an electron in the upper state is collisionally de-excited in a medium of free electron density N_e. Thus

$$N_e^* = A_{ul}/R_{ul}. \tag{8.1}$$

For typical forbidden lines this is in the range 10^4–10^7 cm^{-3}. Note that the higher free electron density also implies that more bound electrons will be excited by collisions, so there will be some forbidden line emission at any density. However, at high densities it will be depressed relative to the permitted lines to the point where it becomes negligible in comparison, and hence not detected.

A final technicality is the single square bracket in C III] λ1909 which indicates that this is a so-called *intercombination line* involving a change in electron spin and having a transition rate intermediate between those for permitted and forbidden lines. This turns out to be important because the line would be depressed at free electron densities above about 10^{10} cm^{-3}. From the ubiquitous presence of a prominent C III] λ1909 line in quasar spectra covering the appropriate redshifted spectral range we can therefore deduce that the electron density in the broad line region is less than 10^{10} cm^{-3}.

On the other hand, no broad forbidden lines are seen in the spectra of active nuclei: therefore the electron density in the broad line region must exceed about 10^7 cm^{-3}.

The principal narrow lines are those of [O III] λ5007, [O III] λ4959, [O III] λ4363, [Ne III] λ3869, [O II] λ3727, [Ne V] λ3426, as well as the permitted lines of hydrogen. Since the narrow lines include both permitted and forbidden lines it follows that they are formed separately from the broad lines in a region, or regions, with electron densities below about 10^4 cm^{-3}.

To produce an emission line region requires an input of energy both to ionize and to excite the gas. The ionization can be produced either by sufficiently energetic photons or electrons, or other fast particles, and excited ions can be obtained either by the recombination of electrons and once further ionized species to an excited state, or by collisional excitation of ions in their ground state. An ion in an excited state can usually be taken to decay since the relative probability of further excitation or ionization is in most circumstances negligible. Models in which the ionization is achieved by ultraviolet photons are called *photoionization models*, independently of whether the lines are then formed by radiative transitions following recombination or collisional excitation. (Successful models require both.) These models are characterized by the assumed abundance of the elements, the density of the photoionized gas and the spectrum of the ionizing radiation from which they yield ratios of emission line intensities. The various models in the literature differ in the number of more complicated physical processes taken into account and in the nature of the assumptions made in order to obtain tractable systems. Other models have been discussed in which particles provide the energy input via, for example, shocks or relativistic electrons and there are circumstances, for example in the forbidden line emission associated with 'knots' of radio emission in jets, in which these alternatives could be relevant. However, we shall confine our attention to photoionization models for both the narrow and broad line regions.

8.5 The narrow line region

We begin with the physical conditions in the more accessible narrow line regions for which the arguments and results are similar to those for planetary nebulae. We assume the line emitting material is photoionized by ultraviolet radiation from a central source, an assumption consistent with the resemblance between narrow line spectra and those of planetary nebulae, for which the photoionization model is established. The lines themselves are produced mainly by the radiative decay of *collisionally excited* electrons. In these circumstances the relative numbers of ions in two widely spaced excited levels depends sensitively on the temperature, and this fact can be used to provide an estimate of the temperature. For a three-level atom in which the levels $i = 1, 2, 3$ have statistical weights g_i and populations N_i the emission coefficients for the lines $3 \to 2$ and $2 \to 1$ are related by

$$\frac{j_{32}}{j_{21}} = \frac{N_3 h\nu_{32} A_{32}}{N_2 h\nu_{21} A_{21}}$$

where the As are the Einstein coefficients. To illustrate the method we assume local thermodynamic equilibrium (LTE, see the Appendix) in order to eliminate the relative populations, $N_3/N_2 = (g_3/g_2) \exp[-(E_3 - E_2)/kT]$, and we obtain

$$\frac{j_{32}}{j_{21}} = \frac{A_{32} g_3 \nu_{32}}{A_{21} g_2 \nu_{21}} \exp[-(E_3 - E_2)/kT]. \tag{8.2}$$

For a medium optically thin to the lines, the ratio of emissivities is also the observed ratio of intensities and the remaining quantities other than T in (8.2) are atomic constants. Applied to [O III] $\lambda 4363$ and [O III] $\lambda 5007$ ($+ \lambda 4959$ which has approximately the same lower level and which cannot in practice be separated) this yields temperatures in the range $(1-2) \times 10^4$ K. Of course, detailed applications of the method take into account departures from LTE and possible collisional de-excitation of the upper levels.

The density can be obtained more precisely than the upper limit in Section 8.4 from lines corresponding to closely spaced upper levels. These are clearly insensitive to temperature but depend on the extent of collisional de-excitation. If this is not important then every excited electron gives rise to a line photon: the ratio of emissivities is therefore equal to the ratio of collisional excitation rates from the ground state and for $E_3 - E_2 \to 0$ this depends only on g_3/g_2. Thus

$$j_{31}/j_{21} = g_3/g_2.$$

For example, for [S II] $\lambda 6716$, $\lambda 6731$, which arise from transitions from $^2D_{5/2}$ and $^2D_{3/2}$ states respectively, we have

$$\frac{j(6716)}{j(6731)} = \frac{(2J+1)_{J=5/2}}{(2J+1)_{J=3/2}} = 1.5.$$

On the other hand, if collisional de-excitation is important the emissivities are given by an equation similar to (8.2) with $E_3 - E_2 \sim 0$:

$$\frac{j_{31}}{j_{21}} = \frac{g_3 A_{31}}{g_2 A_{21}}.$$

For the pair of lines in our example this ratio is 0.35. Thus, if the observed intensity ratio, which equals the ratio of emissivities for a medium optically thin to the lines, is near 0.35 both the upper levels must be collisionally de-excited, and the electron density must be greater than the critical densities for both levels (equation (8.1)). In this case a density greater than $10^{4.5}$ cm^{-3} would be implied. If the ratio is found to be near 1.5 the density must be less than critical for both levels, which here means less than $10^{3.5}$ cm^{-3}. Again, detailed computations can make this more precise. Observed values lie in the range 10^3 cm^{-3} to 1.6×10^4 cm^{-3}, but these are averages over the whole narrow line region. More detailed work, for example in NGC 4151, suggests a somewhat higher-density high-ionization zone (e.g. O III), and a lower-density low-ionization region (e.g. O II, S II).

It is now easy to argue that the neglect of *collisional ionization* is justified. To get oxygen predominantly as O III rather than O II by collisional ionization would require $T \sim 10^6$ K, inconsistent with the temperature determined from the line ratios. Our assumption that collisional excitation rather than radiative recombination must be responsible for populating the excited levels of the forbidden lines can be shown to follow from the relatively strong [O III] lines compared with, say, the narrow components of the Balmer lines. Such efficient production of [O III] lines could not be obtained from radiative recombination.

The flux in a given line depends on the product of the electron density with the density of the atomic species producing the line and the volume of emitting material; hence, for fixed element abundances and a given ionization state, the flux is proportional to the emission measure $\int N_e^2 dV$ with a factor that depends on the particular line. Consequently, if N_e is known the volume and mass of emitting material can be obtained from absolute measurements of the line fluxes. Typical masses are found to be of order $10^5 M_\odot$.

The emitting material turns out to occupy a volume much smaller than that available in the narrow line region, $V \sim 10^6$–10^9 pc^3, by a factor, the *filling factor*, $f \sim 10^{-3}$ or less. This is in agreement with indications of cloud-like spatial structure and with the existence of structure in the line profiles. If we take the widths of the line profiles to indicate radial velocities of order 10^3 km s^{-1} the travel time across the narrow line region is greater than 10^5 yr and hence the mass flow rate through the region is less than $1 M_\odot$ yr^{-1}.

To try to determine whether this represents a mass outflow or infall we can turn to details of the line profiles. The majority of measured profiles are asymmetric in the sense that there is more emission to the blue of the peak than to the red. The more asymmetric examples also have larger Hα:Hβ ratios consistent with absorption by dust, which would 'redden' the lines by preferentially absorbing the shorter wavelengths, as a source of the asymmetry. Suppose, for example, that the lines are produced in *outflowing* clouds which are embedded in a dusty medium. The greater optical depth in the medium to redshifted emission from receding clouds on the far side of the nucleus would give rise to the observed asymmetry. However, it could also be produced if dusty *infalling* clouds in a transparent medium were themselves opaque to the Lyman continuum, hence ionized only on the side facing the nucleus – because redshifted line photons would be absorbed on passing through the nearside clouds to the observer whereas blueshifted photons from the far side would be unaffected. One should note that a realistic picture might be more complex than either of these models. Although the line profiles are fairly smooth at the resolution available, they do sometimes appear to consist of distinct components, in particular of a blueshifted narrow core superimposed on a broader base.

There are also interesting correlations between line asymmetry on the one hand and critical density and ionization potential on the other. On the basis of these correlations a case can be made for dusty *infalling* clouds, if the asymmetries are indeed produced by dust extinction and radial cloud motion; this model can produce the asymmetries for lines of both high and low ionization potential with a single gas to dust ratio.

However, where there is direct evidence of gas motions in the NLR this appears to favour outflow, or, at least, to be against purely radial infall. For example, blueshifted radio recombination lines are observed in Mkn 668; since this emission results from induced transitions in which the emitted photon must be travelling away from the source of ambient photons, the emitting gas must lie between us and the active nucleus. Hence the motion must be outflow.

Nevertheless, we know that the narrow line widths are not correlated with the luminosity of the active nucleus but do correlate with the galactic nuclear light, hence with the mass. Indeed, the velocity widths are of the same order as the velocity dispersion of stars in the nuclear bulges of early type spiral galaxies. This argues in favour of an external origin for the NLR gas, hence favours infall. We cannot avoid the conclusion that we do not know what the gas motion is at the outer boundary of the active nucleus; the Space telescope may yet provide a verdict.

Surprisingly it is line emission from still further from the nucleus that provides some insight into the structure of the nuclear region. Spatially resolved spectra of some nearby Seyferts and radio galaxies show an extended narrow line region ('ENLR' or 'EELR', where the second 'E' stands for emission) of a few kpc. This can be interpreted as gas in rotational motion in the galactic disc illuminated by ionizing radiation from the active nucleus. From the flux in the lines, usually [O III] or Hβ, we can estimate the mass of gas involved, and hence the photon flux required to keep it ionized. It turns out that the number of ionizing photons incident on this gas is up to ten times the flux we see directly. We therefore conclude that the radiation from the central source is anisotropic in these sources. This may be a consequence of the emission process or a result of reflection or of absorption along the direct line of sight.

If the ionizing flux is assumed to be the UV–soft X-ray big bump there is approximate agreement with the statistics of soft X-ray excesses. The fact that about half of sources in which a bump might be expected on the basis of hard X-ray do not show a detectable feature could be a result of the anisotropy of the radiation field.

It may also tie in with observations of nearby Seyfert 2 galaxies which, in polarized light, are seen to have weak broad emission lines. This has been interpreted as light scattered into the line of sight from a broad line region that is obscured from direct view. Presumably the weakness of the hard X-rays in Type 2 Seyferts – they are detected only in NGC 1068 – can be attributed to the same obscuration. The obscuring material must lie between the broad line and narrow line regions and must be absent along the radio axis; it must therefore have some sort of toroidal geometry. In this picture, viewing over the top of the torus, the observer would see a Seyfert 1 nucleus, as would the ENLR gas.

8.6 The broad line region

The most compelling evidence that the broad emission lines, like the narrow ones, arise from gas photoionized by the ultraviolet component of the central source comes from the absence of any C III $\lambda 977$ emission in those cases where it would fall within the observed spectral band. The presence of strong C III] $\lambda 1909$ in these cases shows that carbon is certainly predominantly in its doubly ionized state. The upper level for $\lambda 977$ emission lies at an energy $\Delta E \sim 5$ eV above that for $\lambda 1909$, so collisional excitation would populate the former significantly unless $\exp(-\Delta E/kT) \ll 1$, i.e. unless $T \lesssim 7 \times 10^4$ K. But to produce C III requires an ionization energy of $\Delta E' \sim 24$ eV. To obtain significant C III by collisional ionization we can estimate roughly that we would need $\exp(-\Delta E'/kT) \gg 1$, or $T \gtrsim 3 \times 10^5$ K. This version of the argument is somewhat crude, because we do not have local thermodynamic equilibrium, but it is in essence correct. In fact, photoionization is the only way in which high ionization states can be produced at low temperatures (less than a few times 10^4 K), and the evidence for such temperatures severely limits the energy that could be deposited in the line-emitting region by fast particles or shock heating or turbulence.

Further evidence for photoionization comes from a general overall correlation between the strengths of lines relative to the continuum (the *equivalent widths*) and the overall level of the continuum flux (extrapolated to the ultraviolet where this is not observed directly). In individual sources, time variations of the continuum do not always correlate well with line variability, even allowing for light travel time delays, but this may be taken to indicate a complex dependence of the line-emitting gas on the continuum power.

The broad lines are observed to come from within the narrow line region. For example, in NGC 4151 direct observation shows that the broad line region lies within 0.025 pc of the central source. However, the broad line region cannot be resolved so we have no direct information on the spatial distribution of gas; this must be determined by indirect arguments.

Note first that we observe strong Lα emission lines but that Lβ, if present at all, is much weaker. Now in a medium optically thick to Lβ absorption there is a tendency for Lβ photons to be destroyed, whereas Lα photons are not. To see this, observe that only absorption from the ground state is important, since the radiative lifetimes of excited levels of hydrogen are very short. Then a Lα absorption must be followed by a Lα emission; but the absorption of a Lβ photon, to the second excited state, can be followed either by a Lβ emission or by Hα followed by Lα with relative probabilities 7.5:1. Thus Lβ can come only from regions of low optical depth and conversely a large Lα:Lβ ratio implies large optical depths in the Lyman lines. From this it does not follow that there must be a large optical depth in the Lyman continuum, since $\tau(L\alpha) \sim 10^4 \, \tau$ (Ly limit), a relation which follows directly from the ratio of the cross-sections involved. However, Mg II $\lambda 2798$ is a prominent quasar line which cannot be produced efficiently at a temperature of around 10^4 K by recombination; it must therefore be produced by collisional excitation of Mg II. But the ionization potential of Mg II to Mg III is only ~ 15 eV, very close to the ionization potential of hydrogen. If the medium were optically thin to the Lyman continuum then magnesium would be primarily Mg III. On the other hand, if the medium is optically thick there will be a transition zone from largely ionized hydrogen to largely neutral hydrogen at the point where the supply of Lyman continuum photons is exhausted, and in this region the collisional excitation of Mg II will be possible. We conclude that the broad line gas must be optically thick to the ionizing continuum; it cannot consist of a mist of optically thin droplets with a total optical depth greater than unity.

Early observations showed, however, that only in a fairly small fraction ($\sim \frac{1}{10}$) of cases was Lyman continuum absorption seen in the spectrum! For example, OQ 172 and OH 471 have the same emission line strengths but one has a Lyman continuum jump and the other does not. This led to the idea of a *cloud model* in which the emission line gas is arranged in clouds that do not completely cover the continuum source. The Lyman continuum absorption would then appear only if a cloud were in the line of sight to the central source. This would be consistent with observations if the *covering factor* ε, the fractional area of the quasar sky as viewed from the centre that is covered by clouds, were of order $\frac{1}{10}$. However, this argument is not, as it stands, conclusive. We know that the 'Lα forest' in high-redshift quasars is associated with absorption in intervening clouds which are not related to the broad line region. This material will also absorb in the ultraviolet continuum. Analysis of high-redshift quasars shows that *all* of their Lyman continuum absorption could be due to such clouds, i.e. that $\varepsilon \sim 0$ within the observational errors. At best then, the argument leads to a rather weak upper limit on ε.

To obtain a better estimate, consider the fate of each Lyman continuum photon in the so-called case B of nebula theory, in which the medium is completely optically thick to the higher Lyman lines (Lβ, Lγ, etc.). Each Lyman photon ionizes one hydrogen atom and this is followed by a recombination. If recombination directly to the ground state occurs another ultraviolet photon capable of ionizing a hydrogen atom is produced and there is no net change. Otherwise a series of radiative transitions to the ground state ends with the emission of a Lyman line photon. If this is Lα it is resonantly scattered until it escapes. If it is a higher Lyman line then it will be degraded to, amongst other things, exactly one Lα photon by further absorptions and re-emissions, according to our case B assumption. Thus, if there is no collisional excitation of Lα,

$$\varepsilon = \frac{\text{number of L}\alpha \text{ photons emitted}}{\text{number of Ly continuum photons}}.$$

The neglect of collisional excitation is unwarranted, but, provided this is not excessive, the argument yields a reasonable estimate of ε. Furthermore, the number of continuum photons available for absorption must be estimated from the observed continuum. It is not sufficient to use, for example, the crude approximation $F(\nu) \propto \nu^{-1}$ since this gives $\varepsilon > 1$! The result is $\varepsilon \sim 0.08$, somewhat fortuitously close to our first estimate above. But, in addition to the cloud model already mentioned, this constraint can be satisfied if the broad lines arise from the photoionized atmosphere of an accretion disc.

The gross physical properties of the broad line region can now be obtained fairly

The broad line region

readily. This will enable us to introduce what has become known as the standard model of the broad line region, although everyone recognizes it to be untenable. We shall explain the difficulties and why these lead some workers to think that at least part of the broad line spectrum comes from a disc. We can also reverse the argument: if a disc is present then it must make some contribution to the line emission. One must therefore consider the extent to which the disc could contribute and yet remain hidden.

As we discussed in Section 7.2, the relative strengths of lines from different ionization species are controlled by the ionization parameter. There are several definitions of this parameter current in the literature, all simply related. Here we shall use U, the ratio of the number density of photons to the number density of hydrogen,

$$U = \frac{Q}{4\pi R^2 c} \frac{1}{n},$$

where $Q = \int (L_\nu/h\nu) d\nu$ and the integration is over the ionizing continuum, L_ν; $\nu > \nu_{Ly}$. The following equivalent parameters are also used: (i) $\Xi = U\langle h\nu\rangle/kT$, the ratio of the pressure of ionizing radiation to the pressure of hydrogen (here $\langle h\nu\rangle$ is the mean energy of ionizing photons, so $L = \int L_\nu d\nu = \langle h\nu\rangle Q$); (ii) $U_1 = L/nr^2$; (iii) $\Gamma = L_{\nu_{Ly}}/8\pi R^2 hcn$; and (iv) $\xi = 2c\Gamma$. Balancing recombinations and photo-ionizations for a hydrogen plasma we have

$$n_{H^+}n_e \alpha = n_H \int_{\nu_0}^{\infty} \frac{L_\nu}{h\nu} \frac{1}{4\pi R^2} e^{-\tau_\nu} \sigma_\nu d\nu \tag{8.3}$$

at distance R from the ionizing source at an optical depth $\tau_\nu = \int \sigma_\nu n_H dz$ into the plasma. Provided $n_e \cong n_{H^+}$, i.e. the plasma is highly ionized, we have, to order of magnitude

$$\frac{n_{H^+}}{n_H} = \frac{Q}{4\pi R^2 c n_e} \frac{1}{\alpha} \frac{\sigma_{tot} c}{\alpha} \sim 10^5 U.$$

A similar expression for the C IV/C III ratio yields an estimate of U from the observed line ratios. Detailed models originally gave $\log_{10} U \approx -2$. In fact, later work showed that this ratio is double valued as a function of U; the higher value of $\log_{10} U \approx -0.5$ gives better agreement with the relatively strong lines of O VI and N V. From this we can estimate the size of the broad line region as in section 7.2 with the result

$$r_{BLR} \sim 10^{18} \left(\frac{Q_{56}}{N_9}\right)^{1/2} \text{cm} \sim 0.3 \left(\frac{Q_{56}}{N_9}\right)^{1/2} \text{pc} \tag{8.4}$$

where $N_9 = N_e/10^9$ cm^{-3} and $Q_{56} = Q/10^{56}$ photons s^{-1}. The use of the present high value of U brings down the radii calculated by this method towards those estimated from the lag between variability in the continuum and in the lines. (If the line fluxes follow the continuum with a lag time of Δt, the characteristic scale of the broad line region is of order $c\Delta t$.)

The luminosity in the Hβ line, if we assume this arises entirely from recombination rather than collisional excitation, is

$$L_{H\beta} = (\tfrac{4}{3}\pi r_{BLR}^3 f) N_e^2 \alpha_{H\beta} h\nu_{H\beta}. \tag{8.5}$$

Here $\alpha_{H\beta}$ is the 'Hβ recombination coefficient' (the rate per unit volume at which electron recombination leads to the production of an Hβ photon), and f is the filling factor which, as we have seen, gives the fraction of the available volume within r_{BLR} that is filled with emitting material. We have assumed a fully ionized pure hydrogen medium, so $N_e = N_p$. Taking $L_{H\beta} \sim 10^9 L_\odot$ as found for luminous Seyferts, we obtain

$$f^{1/3} r_{BLR} = 1.5 \times 10^2 N_9^{-2/3} \text{ pc}.$$

and a mass of line-emitting material of order $30 N_9^{-1} M_\odot$, a surprisingly small amount. Using (8.4) we deduce

$$f \sim 3 \times 10^{-8} L_{46}^{-3/2} N_9^{-1/2}.$$

We shall see that Hβ emission is probably boosted from its recombination value by a factor of about 10, so (8.5) should have this factor on the right hand side. But this makes little difference to the general conclusion of a relatively small mass of gas filling a small fraction of the available volume and covering a small fraction of the source sky. From the covering factor $\varepsilon \sim \tfrac{1}{10}$, the mass of broad line gas, $M_{BLR} \sim 30 N_9^{-1} M_\odot$, and the size of the broad line region (equation (8.4)) we can deduce the total number of clouds involved. If there are N_c spherical clouds each of radius r_c then

$$\varepsilon = N_c \pi r_c^2 / 4\pi r_{BLR}^2$$

and

$$M_{BLR} = \tfrac{4}{3}\pi r_c^3 N_c m_H N_e.$$

So for $N_e = 10^9$ cm^{-3}, $L = 10^{46}$ erg s^{-1}, we deduce $N_c \sim 10^{16}$. This is a surprisingly high number, even though it is sensitive to the parameters of the broad line region ($N_c \propto \varepsilon^3 r_{BLR}^6 / M_{BLR}^2$). Certainly the cloud structure could not be responsible for the irregularities in the line profiles.

In the standard model these conditions are realized by assuming the broad line region to consist of a uniform distribution of clouds of density around 10^9 cm^{-3}, and column density around 10^{23} cm^{-2}. The ionization parameter is assumed constant across the region. The clouds are assumed to be confined by a high temperature, low density gas at the same pressure as the clouds, hence at the same value of the ionization parameter Ξ. Let us consider if this is a possible solution to the problem of confinement.

In the original picture of Krolik, McKee & Tarter (1981) there is a range of ionization parameters in which cool clouds at 10^4 K can be in pressure equilibrium with a hot intercloud medium at 10^8 K. This can be seen in a plot of the equilibrium

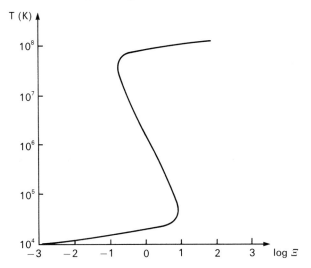

Figure 61. Schematic curve of equilibrium gas temperature as a function of ionization parameter for a relatively hard input spectrum with Compton temperature 10^8 K.

temperature of a photoionized cloud against Ξ shown schematically in Fig. 61. In a certain range a line of constant Ξ intersects the S-shaped curve in three points. The central point is an unstable equilibrium (because a slight decrease in temperature at this point leads to a net cooling and an increase in temperature gives a net heating). The outer points correspond to the two stable phases. However, existence of multiple solutions depends on the spectrum of ionizing radiation. The S-shaped curve arises for relatively hard spectra; in the presence of a big bump in the UV–X-ray flux the curve probably does not bend back on itself and a two-phase equilibrium does not exist. In any case, the confining medium must have a column density which would have been detected by X-ray absorption. It appears that this picture can only be saved if the confining medium is somehow heated to mildly relativistic temperatures. Alternative proposals for solving the confinement problem include magnetically confined clouds and line emitting gas gravitationally confined to the bloated atmospheres of externally irradiated high-mass stars.

A further set of difficulties relates to the dynamics of the clouds. Early work on line broadening by repeated Compton scattering of line photons by free electrons at 10^4 K in the emitting plasma failed to produce sufficient line widths. Turbulence, involving random subsonic motions in the emitting gas, is likewise ruled out as the principal broadening mechanism. Thus the line widths are generally ascribed to Doppler shifts from large differential velocity components along the line of sight in the photoionized gas. Note, however, that electron scattering in a *hot* ambient plasma at 10^7 K is sometimes postulated to account for the broad *wings* of the lines (i.e. the widths at zero intensity). If the clouds are in motion with velocities of order 10^8 cm s^{-1} then the crossing time of the broad line region is of order 10^9 s. The problem is now basically this: in this time it is impossible to form the clouds (at least as cooling density enhancements in the flow) and it is equally difficult to stop them breaking up (by various fluid dynamical instabilities or mass outflow from the flanks). Thus the clouds must be injected already formed throughout the broad line region.

A further set of problems arise when we consider the structure of a photoionized cloud in more detail. The presence of hard X-rays in the ionizing spectrum makes this quite different from a standard H II region. This can be seen by returning to the ionization balance equation (8.3) for a pure hydrogen plasma. If $L_\nu \sigma_\nu/\nu$ is peaked around a frequency ν_p the ionized hydrogen column is cut off sharply at around a depth $\tau_{\nu_p} \approx 1$. If L_ν has a power law form, then, since $\sigma_\nu \propto \nu^{-3}$ the integral in (8.3) gives $n_{H^+}/n_H \propto \tau_0^{-s}$, for some $s > 0$. So there is no sharp edge to the ionized region. We can therefore distinguish two regions in a photoionized broad line cloud: the normal highly ionized H II region, and an extended H I/H II boundary in which the ionized fraction varies between $\frac{1}{2}$ and a few per cent. Lines produced predominantly in the former (e.g. Lα, C IV, C III, N V, O VI) are referred to as high ionization lines and those coming from the extended boundary region (e.g. Mg II, Ca II, Fe II, Balmer lines) are called low ionization lines.

The problem is now this: the energy input into the high ionization lines is that absorbed in the H II region from the UV part of the spectrum; the energy input into the H I/H II boundary is the hard X-ray band. Thus, the ratio of luminosity in high to low ionization lines is determined by the spectrum. The observed ratio of about unity is an order of magnitude smaller than required by this argument.

One special case which was known from the early days of quasar spectroscopy is the Lα/Hβ ratio. This cannot be solved by reddening (the effect of dust absorbing more strongly in the UV than in the optical) since this would make the required intrinsic Hα/Hβ ratio impossibly small, and would imply a highly super-Eddington intrinsic continuum. The problem must be solved by enhancing the Balmer emission. This can be achieved by assuming high densities in the low ionization regions ($n \geq 10^{11}$ cm^{-3}) in the following way.

If a Lα photon is produced at a large optical depth its probability of escaping from the nebula is small, and its probable absorption will excite a neutral hydrogen atom from its ground state to its first excited state. After a short time the excited atom decays again with the emission of a Lα photon, but because the atom has a random thermal velocity the emitted photon will be Doppler shifted in frequency relative to the absorbed photon. Thus, we can picture the Lα photon as executing a random walk in frequency space, in a region of relatively small spatial extent, until it enters the wings of the line where the optical depth is sufficiently small that it can leave the nebula in a single leap. As a result, there is a relatively high density of trapped Lα photons in such a plasma and therefore a high density of atoms in the first excited state. Now if the electron density is sufficiently high a significant proportion of such atoms can be excited by collisions to higher states, where they provide an extra contribution to Balmer line emission (in addition to that from radiative recombinations). This can be important at the usual nebula temperature of around 10^4 K when collisional excitation directly from the ground state can be neglected.

In a standard H II region, photoionized by the ultraviolet continuum of a hot star, line emission from large optical depths cannot, in fact, occur, because the ultraviolet continuum cannot penetrate there. The crucial difference in active nuclei is the existence of the X-ray continuum. This can ionize the gas at large optical depths to Lα (and the Lyman continuum) because the continuum absorption cross-section falls off strongly with increasing frequency ($\sigma_\nu \propto \nu^{-3}$). The broad emission lines in active nuclei therefore probably come from a region quite unlike the standard H II nebula. Detailed computations support the conclusion that the Balmer lines can indeed be enhanced by collisional excitation to yield Lα/Hβ ratios in agreement with observations and hence that this anomaly is not a fundamental objection to the photoionization model.

One cannot, however, enhance the other low ionization lines in this way. These require a large column density (10^{25} cm^{-2}) to absorb the hard X-ray energy. To enhance Fe II sufficiently relative to Hβ in strong Fe II emitters, mainly radio sources, even this may not be sufficient and an additional heating mechanism may be required. In any case, clouds with density 10^9 cm^{-3}, column density $\sim 10^{25}$ cm^{-2} would be larger than the broad line region in lower luminosity sources. One can appeal to density stratification in the clouds but this is, in effect, the same as saying that the broad line must come from two distinct regions: a low density, high column density, low ionization region, and a higher density, high column density, low ionization region. This brings us to what for our purposes is the main point of this discussion: it is natural to identify this latter region as the surface of an accretion disc.

In these circumstances one may wonder whether a pure disc model would work. Indeed, in radio sources there is some evidence that line width is correlated with viewing angle as would be expected for a flattened emission region orthogonal to the radio axis. However aside from the problem of line ratios, the line profiles provide further constraints. If a disc is illuminated with a $1/r^2$ photoionizing flux (at a radius r in the disc) then, assuming constant emissivity, equal steps in radius contribute equal fluxes to a line. The line profile will therefore be characterized by the velocities at the inner and outer radii of the disc emission. Instead of a single peak to the line one obtains a profile with a central dip and two peaks separated by the smaller of these two velocities. However, in at least some of those few cases where such features are even present the two peaks do not vary simultaneously in response to variations in the continuum as would be expected for a disc geometry (since the approaching and receding material in the disc is equidistant both from the central source and from the observer).

There is also a general argument, due to Shields (1978), called the [O III]-problem.

To understand this, observe first that the variations in profile between different lines in a given object are often not large, even though in some cases significant differences certainly exist. As we have explained (above and Section 7.2) the ratio of line intensities from a given location depends only on the parameter U at that point. Thus, if the line profiles of different lines match when plotted as functions of the Doppler velocity $v/c = (\lambda - \lambda_c)/\lambda_c$, U must be constant over the regions responsible for the different velocities. Hence the ionizing flux F must be proportional to the electron density N_e in the broad line plasma; if $F \propto r^{-2}$ then

$N_e \propto r^{-2}$, or $r \propto N_e^{-1/2}$. Now compare the radius of the [O III] emission region from gas of density $N_e \sim 10^6$ cm^{-3} with that of the broad lines from a density $N_e \sim 10^{10}$ cm^{-3}. We find

$$r_{\text{[O III]}} \sim (10^{10}/10^6)^{1/2} r_{\text{BLR}} \sim 10^2 r_{\text{BLR}}.$$

For a Kepler disc we have $v \propto r^{-1/2}$. Hence

$$v_{\text{[O III]}} \sim 10^{-1} v_{\text{BLR}}$$

which implies that the [O III] lines should be at least an order of magnitude broader than observed. This problem also occurs for inflowing clouds in free-fall, since then $v \propto r^{-1/2}$ also.

Both of these problems can be circumvented if the disc illumination falls off faster than $1/r^2$. The smaller contribution to the lines from large radii reduces, or removes completely, the double peaks. Indeed, one can choose the illumination to give profiles with power law wings, as in the Seyfert galaxy 3C 120 or the logarithmic wings, $F_v \propto \log|v - v_0|$, where v_0 is the line centre, which have been popular fits to the observed line profiles. At first sight this makes the [O III]-problem worse by implying even broader wings, but the more rapid decrease in illumination means that the flux in the wings will be negligible. Nevertheless, the fact that the high ionization lines are often blueshifted with respect to the low ionization lines, are generally broader, and show different time lags in response to continuum changes makes a disc origin for all of the lines unlikely. On the other hand, the fact that line profiles do not show large differences, in particular that they do not exhibit a two component character, means that the velocity fields of the low and high ionization lines must be linked.

In these circumstances it is difficult to believe that *ad hoc* modelling of the broad line data is going to tell us very much: every model fits those aspects its inventor considers important. Rather we should regard the broad line data as a diagnostic for physical models of the region producing the emission lines. To show how this works we present one such model (which just happens to contain a disc). This is the 'duelling winds' model (Smith & Raine (1988), Raine (1988)).

The one assumption of this model, in addition to the presence of a disc and a central continuum source of UV to X-rays, is that there is also a highly supersonic spherical wind. This wind scatters the ionizing continuum on to the disc. This has two effects. First it gives an illuminating flux $F \propto r^{-2.7}$ (from an approximate fit to a numerical simulation) which gives rise to a disc chromosphere from which the low ionization lines can be emitted. Second, it produces a hot corona at 10^7–10^8 K, depending on the assumed irradiating spectrum. At large radii ($\sim 10^{18}$ cm) the corona is no longer bound to the disc and material tries to leave the surface (see also Section 9.5). Under certain circumstances the material cannot leave the surface as a steady wind: were it to do so its pressure would be raised to the stagnation pressure of the nuclear outflow; this would give an ionization parameter too low for the wind to remain hot. The material therefore leaves the disc in cool clouds. The clouds are unstable and break up in a few sound crossing times, but not before they can be accelerated to the order of their rotation velocity. This produces a flattened

The broad line region

cloud system, stratified in density and velocity, which emits the high ionization lines. If the inner part of the cloud system on the far side of the disc is partially obscured by the disc, lines predominantly produced in these higher density clouds, such as C IV $\lambda1549$, will be blue shifted more than those coming mainly from larger radius, such as C III] $\lambda1909$ and Mg II $\lambda2800$. There will also be the observed decrease in velocity width from the former to the latter. Furthermore, the velocity fields of the high and low ionization lines are linked, so the profiles can be similar. Unfortunately, detailed fits to observation depend on how the clouds break up. In principle, however, we see that a study of the broad emission lines may eventually provide evidence of the velocity field of gas in this region and in particular may enable us, even indirectly, to see the presence of a disc.

9

Accretion power in active nuclei

9.1 Introduction

In these final two chapters we come to the mechanism by which the gravitational potential energy of material accreting on to a supermassive black hole is extracted as radiation. We start with the radio emission on kpc–Mpc scales which, where it occurs, is probably the best understood feature. To power the extended radio lobes the central engine in sources with such large-scale radio structure must turn accretion energy into directed bulk relativistic outflow. It is universally accepted that the power law radio spectrum from the lobes is synchrotron radiation from relativistic electrons. At the other extreme the X-ray emission appears to be produced on scales down to tens of Schwarzschild radii, providing the deepest possible probe of conditions near the black hole. However, the X-ray power law spectrum seems to admit any number of explanations. There is a problem here in distinguishing the primary radiation from any reprocessed components.

Since, therefore, we are not certain of either the geometry of the central source or the emission mechanism responsible for any part of the spectrum from this region, a discussion of accretion power in active nuclei contrasts sharply with our previous consideration of binary star systems. We shall present a range of partial theories each focussed on a different aspect of the problem. The thick discs expected at super-Eddington accretion rates to be considered in Chapter 10, and the electrodynamic disc theories discussed here in Sections 9.6–9.8 are primarily concerned with the production of bulk relativistic outflows. In contrast, the synchrotron–Compton theories and the electron–positron cauldron models (Section 9.4) are directed to producing the observed radiation spectrum under various assumptions about possible geometries. It is clear that a more complete picture will emerge as theories and observations are able to go beyond discussions of power law slopes over restricted wavebands.

9.2 Extended radio sources

Extended radio sources generally consist of a pair of lobes, typically 50 kpc across, separated from a central elliptical galaxy by up to a few Mpc, but more typically by a few hundred kpc. They are classified into Fanaroff–Riley (FR) classes. FR I sources are low luminosity, having a power at 14 GHz, $P_{14} < 10^{25}$ W Hz^{-1}, with a brightness that fades gradually with increasing distance away from the central object ('limb darkened'). FR II sources have $P_{14} > 10^{25}$ W Hz^{-1} and are limb-brightened often showing bright spots (or 'hot spots') a few kpc across. Radio galaxies of the FR II class are also referred to as 'classical doubles'. The central galaxy itself is also very often a radio source. In fact such systems are important here for several reasons: (i) they are similar to the so-called quasi-stellar sources in which the parent galaxy is replaced by a quasar (the FR II class includes double lobe quasars and 'one-sided' radio quasars); (ii) the total energies involved in giant radio galaxies are not very different from those estimated for quasars; (iii) a radio galaxy can be taken to be at the distance indicated by the galaxy redshift; (iv) there are, as we shall see, physical links between the radio lobes and the central galaxy or quasar which must provide the ultimate source of power for the lobes; (v) furthermore, the continuum emission from the radio lobes, which appears to be well understood, shares many features with the emission from active nuclei – principally its approximate power law spectrum and polarization – and this can be used as a clue to the nature of the mechanism of energy production in active nuclei. In this section, therefore, we look at extended radio sources in some detail.

The radio spectrum generally has an approximate power law form, $F_\nu \propto \nu^{-\alpha}$ with $\alpha \sim 0.6$ typically, but with a steepening of the power law, (i.e. α increasing) away from the hot spot and towards the parent galaxy. The emission away from the hot spot is significantly linearly polarized. Possible deviations from this picture include the absence of a dominant hot spot, and the bending of tails and other distortions of the lobes, which can be attributed to environmental influences from either an external medium or other nearby galaxies. Now the best studied examples of objects with power law spectra are supernova remnants and, in particular, the Crab nebula. These can be explained in terms of *synchrotron emission* from relativistic electrons spiralling round lines of magnetic field, i.e. from electrons having energies much in excess of their rest-mass energy, so that $\gamma = E/m_e c^2 \gg 1$. For we know that cosmic ray electrons, for which there is at least some evidence of an origin in supernovae, have a power law spectrum, in the sense that the number of electrons per unit volume $n(E)\,dE$ with energies between E and $E+dE$ has the form

$$n(E)\,dE = n_0 E^{-\alpha}\,dE; \tag{9.1a}$$

furthermore the synchrotron emission from such a distribution of electrons is a power law:

$$F_\nu\,d\nu \propto \nu^{-(\alpha-1)/2}\,d\nu. \tag{9.1b}$$

In the case of extragalactic radio sources there is no independent evidence for a power law input spectrum of relativistic electrons. Nevertheless the hypothesis of a synchrotron origin for the radiation here too is a reasonable one and is the universally accepted theory. Support for this view comes from observations of polarization. For a single electron, synchrotron radiation is linearly polarized with polarization integrated over frequencies $\Pi = (I_{max} - I_{min})/(I_{max} + I_{min})$ of order 75%. For a power law electron distribution, this gives $\Pi = (\alpha+1)/(\alpha+\frac{7}{3}) \sim 30\%$. The observed degree of linear polarization is much smaller, but this is attributed to inhomogeneities in the sources. As a bonus, the mean magnetic field directions can be mapped by polarization measurements since the magnetic field at each point, projected on to the plane of the sky, is perpendicular to the polarization. Results show a field perpendicular to the axis joining the lobes to the parent galaxy and rather disordered around the hot spots, and more uniform and parallel to the axis in the diffuse lobes.

The rate of loss of energy of a single electron of energy $E \gg m_e c^2$ in a magnetic field B is

$$-dE/dt = \tfrac{4}{3}\sigma_T c(E/m_e c^2)^2 [4\pi/\mu_0] B^2/8\pi$$
$$= 1.1 \times 10^{-15} \gamma^2 B^2 \text{ erg s}^{-1}. \tag{9.2a}$$

The timescale to radiate its energy in synchrotron emission is therefore

$$t_{\text{sync}} = E/(-dE/dt) = 6.7 \times 10^2 \, B^{-2} E^{-1} \text{ s}. \tag{9.2b}$$

If a synchrotron source has luminosity L the energy content in electrons must be of order $U_e = L t_{\text{sync}}$. To estimate this in terms of observable parameters note that a power law spectrum must be cut off at either high or low frequencies, depending on the slope, to avoid an infinite total luminosity. Let this cut-off be at frequency v_*. The electrons principally responsible for radiation at frequency v have energy (see the Appendix)

$$E \sim (2\pi v m_e c/eB)^{1/2} m_e c^2 [c^{-1/2}] \tag{9.3}$$

and hence at v_*

$$E_* \sim 5 \times 10^{-10} B^{-1/2} v_*^{1/2} \text{ erg};$$

therefore

$$U_e \sim 1.5 \times 10^{12} B^{-3/2} L v_*^{-1/2} \text{ erg}.$$

The total energy content of the source is obtained by adding the energy of the magnetic field $U_{\text{mag}} = V[4\pi/\mu_0] B^2/4\pi$, where V is the volume of the source, and the energy in protons U_p. If the protons are accelerated to the same average *velocity* as the electrons then $U_p = (m_p/m_e) U_e$, whereas if the protons and electrons acquire the same average *energy* $U_p = U_e$. Since the acceleration mechanism is unknown we put $U_p = kU_e$ with $1 \lesssim k \lesssim 2000$ (probably). Fortunately, the subsequent discussion will be insensitive to the value of k.

Thus we have obtained an estimate of the total energy content of a synchrotron source

$$U = (1+k)U_e + U_{\text{mag}}$$

as a function of magnetic field and observable quantities. The *minimum* energy content is obtained for a field such that $dU/dB = 0$, which gives

$$U_{\text{mag}} = \tfrac{3}{4}(1+k)U_e; \tag{9.4}$$

i.e. the minimum energy requirement corresponds to approximate equality of magnetic and relativistic particle energy or *equipartition*. For the Crab nebula, (9.4) gives a magnetic field $B_{min} \sim 10^{-4}(1+k)^{2/7}$ G, in agreement with estimates from several independent methods. For classical double radio sources such as Cygnus A, we obtain a minimum energy requirements of 10^{59} erg and a field of 5×10^{-5} G. This yields a lifetime t_{rad} of 10^7 yr at the observed luminosity of about 5×10^{44} erg s^{-1}.

To see the importance of this result we compare t_{rad} with t_{sync}. Electrons radiating at around the maximum accessible survey frequency of about 10^{11} Hz in fields of 10^{-5} G have Lorentz factors $\gamma \sim 10^5$ (from (9.3)) and lifetimes of 10^4 yr (from (9.2)). Thus for these electrons $t_{sync} \ll t_{rad}$; we could increase t_{sync} by reducing the magnetic field (equation (9.2)) but this would increase the energy requirement unreasonably. In fact, t_{sync} is of the order of the light travel time across the hot spots, and much less than the light travel time through the lobes or from the central galaxy. It follows that there must be a mechanism in the hot spots which accelerates the electrons to relativistic energies. This is consistent with the deficit of higher frequency radio emission from the lobes away from the hot spots as indicated by a steepening of the spectrum away from the hot spot observed in Cyg A. It is also consistent with a picture in which ageing electrons leak out into the lobes from the hot spot, which is continually supplied with fresh electrons to be accelerated. We shall not go into the mechanism by which this acceleration is achieved. It is assumed that some variant of the several cosmic ray acceleration mechanisms will work here too; indeed in some theories radio source lobes are supposed to be the origin of a component of cosmic rays.

In principle, the input of fresh electrons could come from the lobe itself or from an ambient medium, but the central galaxy appears the most plausible source on both observational and theoretical grounds. There are several persuasive arguments. First, the parent galaxy often exhibits radio emission from a compact nuclear region similar to that found in some quasars. Conversely, some 10% of quasars have associated double radio lobes similar to the classical doubles. Thus the central galaxy or quasar is a source of activity and hence a potential energy supplier. Second, such a continuous supply would alleviate problems associated with the large energy requirement of the lobes. But third, and most suggestively, bridges of radio emission are observed in many cases between the active nucleus in the quasar or galaxy and the lobes. Especially in those cases where the bridges are long and narrow, they are referred to, somewhat presumptuously, as *jets*.

For a long time the one-sided jets observed in the quasar 3C 273 and in the elliptical galaxy M 87 were thought to be exceptional. Improvements in resolution through the use of very long baseline interferometry (VLBI) and in the dynamic range of radio telescopes (the ability to distinguish low surface brightness features in the neighbourhood of high-intensity sources) have shown that jets or bridges are perhaps the rule rather than the exception. They are observed on scales that range from hundreds of kpc down to pc scale jet-like features in the core. The power law spectrum of the radio emission, with typical spectral index $\alpha \sim 0.6$, and the observed polarization in the range 10–40% are most readily interpreted as synchrotron emission. Polarization measurements can again be used to map the direction of the magnetic field at each point along the jet. Near the core the field is predominantly parallel to the jet axis while further out the field near the axis of the

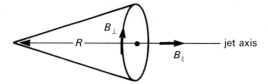

Figure 62. Geometry of a conical jet.

jet becomes orthogonal to the axis. This suggests that magnetic field is being convected in the jet with conservation of flux. For, approximating the jet as a cone, conservation of flux at a distance R from the apex of the cone implies $\pi R^2 B_\parallel$ = constant. Taking $B_\perp$ to be the component of magnetic field encircling the jet axis, the line integral of $B_\perp$ round the axis (Fig. 62) gives $2\pi R B_\perp$ = constant. Thus $B_\parallel$ falls off faster than $B_\perp$ which must dominate at large radii.

There are, however, several problems with this picture; we mention two. (a) The asymmetry problem: especially in the more powerful sources jets are seen on one side of the nucleus only, even in cases of symmetric lobes. It is not clear whether this involves an intrinsic asymmetry, perhaps a switching from one side to the other, or an effect of preferred orientation with a beam pointing almost along the line of sight towards the observer being brightened by the Doppler effect. (b) The confinement problem: a minimum energy density U_{min} and corresponding magnetic field in the synchrotron emitting plasma of the jet can be estimated as for the lobes. Fields of between 10^{-4} and 10^{-6} G are obtained, implying magnetic pressures up to 10^{-10} dyne cm^{-2}. Since the minimum energy occurs for equipartition, the internal gas pressure must be of the same order. To keep the jet collimated, one can then invoke an external medium, which, if it is to exert a pressure of 10^{-10} dyne cm^{-2}, must have a density N and temperature T satisfying $NT = 10^{-10}/k = 10^6$ cm^{-3} K. In the cores of clusters of galaxies, X-ray observations yield $N \sim 10^{-3}$ cm^{-3} and $T \lesssim 10^8$ K; a plausible intercluster medium contributing significantly to the X-ray background could have $N \sim 10^{-7}$ cm^{-3}, $T \sim 10^8$ K. In neither case is the medium anywhere near capable of confining the more powerful jets. If a jet is *not* confined it will have an opening angle θ, which we assume to be small, given by the ratio of transverse velocity to velocity along the jet axis: $\theta = v_\perp/v_\parallel$. If the transverse velocity arises from internal pressure P, we have $P \sim \rho v_\perp^2$ so the ram pressure in the jet is $\rho v_\parallel^2 \sim P/\theta^2 \sim u_{min}/\theta^2$, which amounts to 4×10^{-2} dyne cm^{-2} if $\theta \sim \frac{1}{20}$. There appears to be no possibility that such a jet could be stopped by the intergalactic medium to give the observed radio lobes. However, many observed jets are not straight, but show multiple bends and even, in some cases, appear to curve back on themselves. The forward momentum here could be destroyed by many small angle collisions with density inhomogeneities in the intergalactic medium. An alternative possibility is that jets are confined magnetically by fields generated by currents in them.

The conclusion, then, is that at least some active nuclei emit relativistic particle beams as well as radiation. Further support for this view comes from the variability arguments to be considered in Section 9.3. Observations have shown radio features in Seyfert galaxies resembling the classical double structure but confined within the

Compact radio sources

galaxy on a much smaller scale. This suggests that the appearance of extended radio structure may be connected more with the environment of the active nucleus than with the nature of the central source.

9.3 Compact radio sources

We now turn to the radio emission from the nucleus, with the hope that the large quantity of detailed observational data will provide clues to the overall nature of the central engine, despite the relatively small contribution to the total energy output in most cases. One should not be unaware, however, that the radio emission could represent a minor side-effect, unrelated to the main problem.

With increased detector sensitivity many of the classical double radio sources reveal *compact* radio cores with angular sizes much less than one arc second, the conventional definition of 'compact'. In addition, many active nuclei lacking extended radio emission have compact radio cores. The core characteristics do not seem to depend strongly on the presence or absence of extended emission. However, the core emission of at least the radio quiet and the lobe dominated radio loud quasars shows a distinct break in the spectrum in the millimetre region indicating that the radio emission is indeed a separate component.

A simple characterization of the specific intensity at a given observed frequency is the temperature of a blackbody producing an equal intensity at that frequency. This is the *brightness temperature* T_B, of the source at the given frequency, and is given by

$$I_\nu = \frac{2h\nu^3}{c^2}[\exp(h\nu/kT_B) - 1]^{-1}.$$

Note the distinction between T_B and the blackbody temperature T_b (equation (1.6)) in that the latter refers to an average over frequency. Clearly, T_B is a function of frequency unless the emission is blackbody, in which case T_B is the source temperature. For $h\nu \ll kT_B$ on the Rayleigh–Jeans part of the spectrum, the case in which the brightness temperature is most commonly used, we have

$$kT_B = (c^2/2\nu^2)I_\nu. \tag{9.5}$$

For compact radio sources, T_B is in the range 10^{10}–10^{12} K. Since blackbodies at such temperatures do not emit a significant fraction of their power in the radio, this tells us that the radio sources cannot be thermal.

The observed spectra fall into two classes. Some cores have power law spectra rather like those from the lobes of classical doubles (*steep spectrum sources*). Others are flat (or, in some cases, inverted) up to several hundred GHz and a turnover to a power law $F_\nu \propto \nu^{-0.25}$ can sometimes be observed at millimetre wavelengths (*compact sources*). It is reasonable to attribute the steep spectrum sources to synchrotron emission, and, indeed, it appears possible that they could be classical doubles with relatively weak cores viewed end-on. The inverted non-thermal spectra in compact sources suggest *self-absorbed* synchrotron emission, the main points of which we shall briefly recall.

At sufficiently high densities, or, it turns out, at sufficiently low frequencies, the absorption of radiation by relativistic electrons in the presence of a magnetic field, the process inverse to synchrotron emission, can become important. If the same electrons are responsible for producing the radiation in the same magnetic field, we refer to this as synchrotron self-absorption. Even for theoretically infinite optical depths in this plasma, the resulting emission will *not* be blackbody, because the electrons do not have a thermal distribution and therefore the absorption coefficient, μ_ν does not satisfy Kirchhoff's law, i.e. $\mu_\nu \neq j_\nu/B_\nu$ (see the Appendix). The observed flux is obtained by solving the appropriate radiative transfer equation (see for example Rybicki & Lightman (1979)). For a simplified discussion, imagine each electron to emit and absorb only at the peak frequency related to its energy by equation (9.3). A proper analysis then shows that we obtain the emitted intensity by replacing the mean energy per particle for a thermal source ($\sim kT$) by just the energy given in equation (9.3). Thus in the Rayleigh–Jeans limit, which is appropriate here, we obtain for an optically thick synchrotron source

$$I_\nu = \frac{2\nu^2}{c^2} E_\nu \sim 10^{-30} \nu^{5/2} B^{-1/2} \text{ erg s}^{-1} \text{ cm}^{-2} \text{ Hz}^{-1} \text{ sr}^{-1}. \tag{9.6}$$

At a frequency ν_m the observed intensity equals this limit: below ν_m (9.6) should hold; above ν_m the system becomes optically thin and the plasma radiates less efficiently. Thus, if F_ν is the observed flux and $\Omega \sim \theta^2$ the solid angle of the source, we have $F_\nu = \Omega I_\nu$ and (9.6) gives

$$\nu_m \sim 10^{12} B^{1/5} F_{\nu_m}^{2/5} \theta^{-4/5} \text{ Hz}. \tag{9.7}$$

An optically thin source cannot be smaller than

$$\theta \gtrsim 2 \times 10^{-3} F_{-26}^{1/2} B^{1/4} \nu_8^{-5/4} \text{ arcsec} \tag{9.8}$$

where $\nu_8 = 10^{-8}\nu$ and $F_{-26} = 10^{26} F_\nu \text{ erg s}^{-1} \text{ cm}^{-2} \text{ Hz}^{-1}$ is the flux measured in the standard unit used by radio astronomers, the milli-Jansky (1 Jy = 10^{-23} erg s^{-1} cm^{-2} Hz^{-1}). Equality holds in (9.8) if F_{-26} is the flux at the turnover frequency ν_m.

Applying this theory to the observations of the compact sources yields typical values for field strengths and total energies of

$$B \sim 10^{-3} \text{ G}, \quad U_{\text{particles}} \sim 10^{54} \text{ erg}, \quad U_{\text{mag}} \sim 10^{49} \text{ erg},$$

well away from equipartition; but there are large uncertainties since errors in θ get magnified to large powers. Of course, there is an obvious objection to applying this theory directly to observed sources, since the spectra are often flat and never as steep as $\nu^{5/2}$. However, inhomogeneous source models with either smooth variations in properties (field, electron density, etc.) or discrete components can produce agreement with the observed spectra.

For compact sources in which the self-absorption, and the angular size can be estimated, the synchrotron lifetime of the relativistic electrons can be obtained. The observed brightness temperature, $T_B \sim 10^{12}$ K at ν_m, gives the energy and hence Lorentz factor of electrons radiating at this frequency;

$$\gamma = E_m/m_e c^2 \sim kT_B/m_e c^2 \sim 100.$$

Compact radio sources

The magnetic field B must be such that electrons with this Lorentz factor $\gamma = \gamma_m$ do indeed radiate at the turnover frequency ν_m, (equation (9.3)). The synchrotron lifetime (9.2) is

$$t_{\text{sync}} \sim 5 \times 10^{21} \gamma^3 \nu_m^{-2} \text{ s} \qquad (9.9)$$
$$\sim 200 \text{ yr}$$

for $\nu_m \sim 1$ *GHz*, $\gamma = \gamma_m \sim 100$. Since this is much less than the source lifetime, which must exceed $U/L \sim 10^3$ yr at the very least and is almost certainly a lot longer (in excess of U/L for the extended lobes), there must be an essentially continuous input of relativistic electrons. This is similar to the conclusion we reached for the radio lobes but the constraints on the nature of the central source, at least for some objects, can be considerably tightened.

We should ask what happens to the electrons after they lose their energy and cease to be relativistic. Any non-relativistic electrons will lose their remaining energy by cyclotron emission. The timescale is still given by (9.9) but with $\gamma \sim 1$ and ν_m replaced by the cyclotron frequency $\nu_{\text{cyc}} \sim \nu_m / \gamma_m^2$. The resulting loss time is somewhat longer than the age of the source, and such electrons would eventually be thermalized by collisions. However, there are limits on the amount of such thermal plasma in the radio source. One approach would be to consider optical depth effects – the smoothing of variability by electron scattering for example. Another is the observation of the variable polarization of the radio emission during source outbursts on timescales of months to years. This is possible only if Faraday rotation of the plane of polarization of radiation crossing the source is too small to smear out the polarization changes. Thus we require the rotation $\psi = R_m \lambda^2 \lesssim 1$ radian, where $R_m \sim 10^6 N_e BR$ m^{-2} for a source of size R pc, electron density N_e cm^{-3}. Relativistic electrons contribute with a factor $1/\gamma^2$ since $R_m \propto \omega_p^2 \omega_{\text{cyc}} \propto m_e^{-2} \gamma^{-2}$, and so can be neglected. For $B \sim 10^{-3}$ G and $\lambda \sim 0.1$ m we require $N_e R \lesssim 10^{-1}$ cm^{-3} pc. Compact radio sources vary from less than one to more than ten pc in radius; for $R \sim 3$ pc, say, we find $N_e \lesssim 1/30$ cm^{-3}. We can compare this with the relativistic electron density $N_{e,\text{rel}} \sim u_e/\gamma m_e c^2$; for $u_e \sim U_{\text{particles}}/R^3$ and $\gamma \sim 100$ we obtain $N_{e,\text{rel}} \sim 10$ cm^{-3}, which is much greater than the non-relativistic component. An even more stringent constraint is obtained from the detection of variable Faraday rotation during an outburst. We conclude that the synchrotron power cannot be maintained by the repeated injection of fresh electrons into the magnetic field, but that the *same* electrons must be continuously re-accelerated. An alternative is a plasma of relativistic electrons and positrons for which the net Faraday rotation vanishes.

Indirect evidence of the radio structure comes from observations of variations in radio power. A growing number of compact radio cores are interpreted to have a pair of components which are separating faster than light. In addition, there is direct evidence in 3C 273 from VLBI observations that the source does indeed contain a pair of components moving apart and does not consist of fixed components flaring independently – which would in any case produce contractions, which are not observed – nor result from observations of varying emission at a fixed observing frequency. In fact, the extrapolation of the component separation to zero

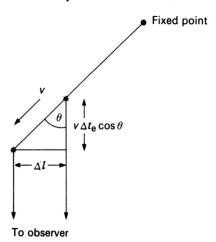

Figure 63. Geometry of superluminal expansion: a blob of emitting material moves at relativistic speeds at a small angle θ to the line of sight.

often coincides with an observed outburst in the core. This superluminal expansion has a simple *kinematic* explanation in the *relativistic beam model* quite independent of details of the emission mechanism.

To see how this model works, consider a blob of emitting material ejected from a fixed emitting core with velocity v at an angle θ to the line of sight. It makes little difference if both regions of emission are assumed to be moving, since the result depends only on the fact that the approaching source makes a small angle with the line of sight. The true motion is not known (except in the case of 3C273), since only relative positions can be measured with sufficient accuracy at present. The apparent velocity of separation is the ratio of the difference in *observed* positions to the *observed* time-interval (Fig. 63):

$$v_{app} = \Delta l / \Delta t_o$$
$$\Delta l = \Delta t_e v \sin \theta,$$

where

$$\Delta t_o = \Delta t_e (1 - v/c \cos \theta),$$

allowing for the decrease in light travel time as the object approaches. Thus

$$v_{app} = v \sin \theta / (1 - v/c \cos \theta) \qquad (9.10)$$

is the required expression for the apparent velocity as a function of v and θ. For v fixed, the maximum apparent velocity v_{max} occurs for $\sin \theta = 1/\gamma$ and is $v_{max} = \gamma v$. For θ fixed, the maximum occurs for $v = c$. So

$$v_{max} < v_{app}(v = c) = \sin \theta / (1 - \cos \theta) \sim 2c/\theta \quad \text{as } \theta \to 0. \qquad (9.11)$$

It is clear that an arbitrarily high v_{app} can be achieved for sufficiently small θ and large γ, but at the expense of decreasing probability of such alignment. In fact, the probability $P(\Theta)$ that a source axis makes an angle $\leqslant \Theta$ to the line of sight is the

Compact radio sources

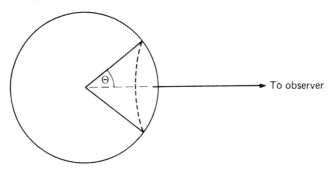

Figure 64. Source axes at angles less than Θ to the line of sight cut out a conical sector on the hemisphere of all possible directions.

ratio of the area of a conical sector of a sphere to the area of a hemisphere (Fig. 64); thus

$$P(\Theta) = 1/2\pi \int_0^{2\pi} \int_0^{\Theta} \sin\theta \, d\theta \, d\phi = (1 - \cos\Theta), \qquad (9.12)$$

and for γ large, $\sin\Theta \sim \Theta \sim 1/\gamma$, this is $P(\Theta) \sim 1/2\gamma^2$. Therefore for $\gamma \sim 5$, as required by the observed velocities, the probability of alignment to within an angle $\leqslant 1/\gamma$, at which relativistic effects would be important, is of order 2%.

This interpretation in terms of specially oriented relativistic outflow is qualitatively consistent with the interpretation of core dominated radio sources as 'end-on' radio doubles in a model in which only orientation effects affect the radio properties of quasars: up to one half and possibly all of the compact sources, which are themselves a subset of the 10% of radio loud quasars exhibit superluminal expansion or milder forms of 'over rapid' variability. In this case we should say that the parent population of the beamed (superluminal) sources are the radio loud quasars. An alternative picture which for some reason has come to be known as '*the unified model*' postulates that radio galaxies are the parent population and all radio loud *quasars* the specially oriented, beamed subset. In this picture all quasars would host a superluminal source at some level. The idea has some success in accounting for the distribution of the separation of the radio lobes. There is also some suggestion that it can be extended to radio quiet quasars where luminous infrared galaxies would be the unbeamed population.

A further argument for relativistic bulk motion comes from observations of variable sources, since some of the shorter timescales present problems for non-relativistic source models. In particular, the high energy density implied by the small volume and high luminosity leads to what has become known as the *inverse Compton catastrophe*, which, as implied by the name, was historically considered to be more than a mere problem. It arises because the synchrotron emitting electrons are also bathed in the radiation they are producing. The energy of a typical radio photon of this radiation field is clearly much less than the energy of a typical electron so that the net effect of scattering is a transfer of energy to the radiation which is referred to as the inverse Compton effect. If the radiation field were too intense the electrons would lose all their energy by inverse Compton scattering with the catastrophic result that they would cease to be synchrotron emitters. If the electron energy were maintained by suitable external forces they would further scatter these and subsequent inverse Compton photons with the catastrophe of an infinite energy density as a result. This puts a limit on the brightness temperature of a synchrotron source. The inverse Compton power of a relativistic electron of energy E in a radiation field of energy density U_{rad} is

$$-dE/dt \sim \tfrac{4}{3}\sigma_{\text{T}} c(E/m_e c^2)^2 U_{\text{rad}}. \tag{9.13}$$

Comparison with (9.2a) shows that synchrotron losses will dominate if

$$U_{\text{B}} \gtrsim U_{\text{rad}}. \tag{9.14}$$

The brightness temperature then satisfies

$$T_{\text{B}} \lesssim \frac{c^3}{2v^3 k}\left[\frac{4\pi}{\mu_0}\right]\frac{B^2}{8\pi}, \tag{9.15}$$

where we have used (9.5) in (9.14) and estimated I_v by $cU_{\text{rad}}/4\pi v$. For an equipartition magnetic field of somewhat less than 10^{-3} G, and at the peak of brightness at frequency $v = v_{\text{m}} \sim 10^9$ Hz, we find $T_{\text{B}} \lesssim 10^{12}$ K. The limit on T_{B} provided by equation (9.15) cannot be increased by postulating a stronger magnetic field, since measurement of T_{B} at v_{m} gives a value for B by putting $E_{\text{m}} = kT_{\text{B}}$ in (9.3). Where T_{B} is obtained from direct measurements of angular size, the inequality (9.15) is found to be satisfied and the inverse Compton catastrophe is avoided.

Consider, however, those variable sources in which the brightness temperature is inferred. The condition that synchrotron emission is not self-absorbed, $I_v = F_v/\Omega < 2v^2 kT_{\text{B}}/c^2$ (equation (9.6)), can be written

$$T_{\text{B}} \geq 10^{12}\left(\frac{\Delta E}{3\times 10^{44}}\right)\frac{1}{v_9^3}\left(\frac{3\times 10^6}{t_{\text{var}}}\right)^3, \tag{9.16}$$

with equality at $v = v_{\text{m}}$, where the variability on a timescale t_{var} involves an energy $\Delta E \sim vL_v t_{\text{var}}$, measured in ergs, and we have assumed a source size $l < ct_{\text{var}}$; v_9 is the observation frequency in units of 10^9 Hz. Thus at 1 GHz the limit (9.15) on T_{B} is violated by variability over less than a month involving an energy in excess of 3×10^{44} erg or more. Observed violations amount to factors 10–10^3 in T_{B}. Conversely, in order to avoid the inverse Compton catastrophe, sources have to be larger than the limit $l \lesssim ct_{\text{var}}$ by factors of order 10. Other evidence appears to support this conclusion. For example, all sources observed with VLBI are at least partially resolved indicating sizes for the non-varying source component up to $100 ct_{\text{var}}$. Furthermore, there is no evidence of the scintillations at radio frequencies caused by inhomogeneities in the interplanetary medium which should occur for sources of the sizes indicated by the variability lengthscale.

We can circumvent the light crossing time limit on the source size if either the brightness temperature or the observed timescale t_{var}, or both, do not refer to the rest frame of the source. Thus, if the radiation is beamed towards us from a source moving at relativistic speed and having Lorentz factor $\gamma_B \gg 1$, close to the line of sight, then the timescale in the source frame is larger than that observed by a time dilation factor $t_{var} = \gamma_B t_{obs}$ and the emitted intensity is smaller by a factor γ_B^{-3} (since I_ν/ν^3 is invariant under Lorentz transformations). Note that γ_B refers to the bulk motion of material towards the observer, not to be confused with the relativistic particle velocities in the moving source. Quite modest values γ_B would resolve the inverse Compton catastrophe, and would be consistent with the beaming suggested by the superluminal sources ($3 \lesssim \gamma_{min} \lesssim 10$), and, at least for quasars, would be statistically consistent with the view that the compact sources are end-on doubles ($\gamma_B \sim 5$). Note that a preferred source axis is required in all of these cases: relativistic outflow in a spherical shell would not produce the time dilation effect because light travel time delays from different parts of the shell would smear out the outburst over a longer observed timescale. Whether or not the γ_B^3 enhancement of intensity in the approaching beam, an effect usually referred to as *Doppler favouritism*, can account for the observed jet asymmetry (Section 9.2) is unclear at present.

9.4 The nuclear continuum

The Compton catastrophe appeared historically as a constraint on the source structure imposed by limits to the Comptonized flux. With Lorentz factors of $\gamma^2 \sim 10^8$ envisaged in radio synchrotron models the Comptonized flux will emerge, scattered by a further γ^2, in X-rays. Given that X-rays are a ubiquitous feature of active nuclei we can turn the discussion round and ask if Comptonization is a viable way of producing the X-ray flux, and, by implication, a significant contribution, at least, to the IR–optical–UV flux. If the same electrons are taken to be responsible for the Comptonization and synchrotron emission the models are referred to as synchrotron-self-Compton models (SSC). The initial objective is to obtain the canonical power law spectrum in X-rays, $F_\nu \propto \nu^{-0.7}$. The relation between the X-ray flux and the infrared emission outlined in Section 8.6 has been cited both for and against models of this kind.

An implication of the radio source models is that the synchrotron emission may be associated with a jet. In many BL Lacs in particular, a reasonably strong case can be made that the X-rays are either a straightforward extension of the direct synchrotron emission or result from Comptonization. In these cases it is appropriate to model the sources as jets. However, both spherical and disc geometries have also been considered for SSC models; the qualitative discussion we give here is essentially independent of the assumed geometry.

In the simplest SSC models synchrotron radio photons are Compton scattered by relativistic electrons of Lorentz factor γ from frequencies ν_R to $\gamma^2 \nu_R$. The final radiation spectrum is therefore wholly a consequence of the input electron spectrum. If the input is a burst of relativistic electrons with a power law energy distribution the output will be a power law photon spectrum with a break at a frequency for which the electron lifetime equals the age of the source; by adjusting the input one can obviously obtain more complex outputs to match the observations.

An alternative version is the synchrotron inverse Compton model. In this an extra external soft photon input, corresponding to the 'big bump', is included since SSC models do not produce this as an output. The Compton scattering of the radio is taken to give the infrared emission; further scattering gives soft X-rays, while electrons with the same γ^2 ($\sim 10^4$) scatter UV photons to hard X-rays. Such models can give quite impressive fits to steady state spectra. They appear to depend on the unknown acceleration mechanism to give the right electron spectrum, but, against this, one can argue that the mechanism is invoked in the same way in which it is known to be operating in the extended radio structures and in synchrotron emitters such as the Crab nebula. Some modification of the simplest models is required to take account of electron–positron pair production (see below).

At the opposite extreme one can try to construct theories that yield the observed spectra independently of the input electron spectrum. In fact, Compton scattering also provides several ways of doing this. First, if the escape time from a source is much greater than the cooling time and the electron distribution reaches a steady state, one finds that the number of electrons with Lorentz factors between γ and $\gamma + d\gamma$ is related to the rate of input of fresh electrons at γ, $Q(\gamma)$, by

$$n(\gamma) \propto \dot{\gamma}^{-1} \int Q \, d\gamma$$

where $\dot{\gamma} \propto dE/dt \propto -\gamma^2$ (9.2a). Hence $L_\nu \propto \nu^{-0.5}$ from equation (9.1b). If there is some further reprocessing that softens the spectrum (scattering on cool pairs for example – see below) then one may obtain something close to $L_\nu \propto \nu^{-0.7}$ in X-rays.

We can also consider the effect of multiple Compton scattering. The general argument here is due to Zeldovich. In a single Compton scattering on average a photon undergoes a relative change in frequency $\langle \Delta \nu / \nu \rangle$ given by $\Delta = \langle \Delta \nu / \nu \rangle \sim \langle E \rangle / m_e c^2 = \langle \gamma \rangle$, where E is the mean electron energy. After n scatterings we have $\langle \Delta \nu \rangle \propto \Delta^n \langle \nu \rangle$; if n or ΔE, is large, and if $\Delta \nu \gg \nu$, then we can write this in terms of initial and final mean frequencies as $\nu_f = \Delta^n \nu_i$. Thus, the probability of an input photon at ν_i emerging as a Comptonized photon at ν_f is equal to the probability of n scatterings, i.e. to P_1^n, where P_1 is the probability of a single scattering and $n = \log(\nu_f/\nu_i)/\log \Delta$. Thus

$$L_{\nu_f} = \nu_f P_1^{\log(\nu_f/\nu_i)/\log \Delta} \frac{L_{\nu_i}}{\nu_i} = \nu_f \left(\frac{\nu_f}{\nu_i}\right)^{(\log P_1/\log \Delta)} \frac{L_{\nu_i}}{\nu_i};$$

so if we input photons in a relatively narrow band about ν_i we get a power law output with energy slope $1 + \log P_1/\log \Delta$.

The argument applies provided (i) $\Delta \nu \gg \nu$; (ii) $\langle \Delta \nu / \nu \rangle \sim$ constant, independent of ν, hence provided that we can ignore the energy dependence of the Klein–Nishima cross-section; (iii) the number of scatterings is large. The first requirement is satisfied if the mean electron energy is much greater than the photon energies attained. This is also necessary in order that we can ignore down scattering. The second condition implies a break in the spectral slope in γ-rays where the fall off in the Klein–Nishima cross-section becomes important. The final condition is controlled by the so-called *y*-parameter which measures (energy change per scattering) × (number of scatterings). For large optical depth, τ, photons random walk out of the source. To random walk through a source of size R with a mean free path λ required $(R/\lambda)^2 = \tau^2$ scatterings; so if τ is large, $y \sim (\Delta E/m_e c^2)\tau^2$. If $\tau < 1$ the relevant parameter is $y \sim (\Delta E/m_e c^2)\tau$: since only single scatterings are relevant, the mean number of scatterings equals the mean fraction of scattered photons which is equal to τ. Note that y may be a function of frequency through τ if absorption is also important. The argument applies in particular to a thermal electron distribution for which $\langle \Delta E \rangle/E = \langle \Delta v \rangle/v = 4kT/m_e c^2$ if the electrons are non-relativistic and $\langle \Delta E \rangle/E = 4kT/m_e c^2 + 16(kT/m_e c^2)^2$ in the relativistic case. For multiple scattering of a relativistic distribution we get a power law photon spectrum for $y \gg 1$ and $y \ll 1$. This gives rise to many variants of synchrotron–Compton models.

In all of these models there are various reasons why electron-positron pairs may be present. One possibility is that the acceleration of protons in shocks may be more effective than that of electrons. Relativistic protons will lose energy by pion production (rather than Compton scattering, since the Compton scattering cross-section is smaller by a factor of $(m_e/m_p)^2$ compared with that for electrons). The pions are unstable and decay to electrons and positions. In principle inelastic e–p or e–e collisions could produce $e^\pm$ pairs. But the other important possibility turns out to be photon–photon interactions (the inverse of $e^\pm$ annihilations). For this we require high energy γ-rays ($h\nu \gg m_e c^2$) and a high density of low energy photons. The latter may be provided by the 'big bump'. The former are observed to be emitted by a small number of sources, 3C 273 for example. Not all active nuclei can be γ-ray sources otherwise the upper limits on the γ-ray background would be violated. But the idea is that in most active nuclei the high energy γ-rays would be present but would generally not escape: most would be reprocessed to $e^\pm$ pairs.

The relevant parameter for the importance of photon–photon interactions is the 'compactness parameter', l. This is the soft photon energy expressed in units of $m_e c^2$, intercepted per Thompson cross-section in a photon crossing time:

$$l = \frac{L}{m_e c^2} \cdot \frac{\sigma_T}{4\pi R^2} \cdot \frac{R}{c} = \frac{L\sigma_T}{4\pi R m_e c^2}.$$

If γ-rays are present and if $l > 10$ then $e^\pm$ pairs will be produced in significant numbers. We can rewrite l as

$$l = \frac{1}{2}\left(\frac{m_p}{m_e}\right)\left(\frac{L}{L_{\text{Edd}}}\right)\left(\frac{R_{\text{Schw}}}{R}\right).$$

so $l > 10$ even for $L < L_{\text{Edd}}$ near R_{Schw}. The presence of $e^\pm$ pairs modifies the source opacity. At high enough density $e^\pm$ annihilation becomes important producing γ-rays and a self-sustaining system, the so-called $e^\pm$ cauldron. In most circumstances this does not occur; $e^\pm$ pairs modify the radiation spectrum but do not determine it. However, for $30 < l < 300$ a self-sustaining system can be obtained in which

Compton scattering on cooled pairs gives rise to a spectrum somewhat steeper than $v^{-0.7}$ without violating the γ-ray background constraint. (Further reprocessing can then give the required 0.7 slope as mentioned in Section 8.3.)

9.5 Applications to discs

There are two ways in which the above considerations impinge on disc models. First there is the extent to which some of the above processes might operate in a disc geometry to produce a radiation spectrum from the disc that is closer to that observed. Certainly the similarity of the variability timescales of at least the soft X-ray end of the big bump and of the hard X-rays suggests an origin in similarly sized regions. This would obviously be satisfied if, given that the big bump comes from the disc, the hard X-rays were to be produced in the disc also. Second, one must consider the influence of the disc in producing the bulk relativistic outflows associated with jets, spectral considerations being relegated to second place after the geometry has been fixed. The former question is taken up here and the latter in the remainder of the book.

It is beyond the scope of this book to discuss in detail the modified disc models that have been proposed to account for various aspects of the observed spectra. We shall restrict ourselves to showing where the uncertainties lie that make such deviations from the orthodox thin disc possible or necessary.

In Section 5.8 we saw that a viscous instability develops unless $d\mu/d\Sigma > 0$. For the inner region of a thin accretion disc, where radiation pressure dominates gas pressure and the main source of opacity is electron scattering (Section 5.6) this condition is not satisfied. In fact, solving (5.68) under these circumstances and with $\tau = \Sigma\kappa_{\text{es}}$ gives $\mu \propto 1/\Sigma$. Thus a steady thin disc solution is not possible. Now, one can suppress this instability by modifying the *ad hoc* viscosity prescription. For example, we might use the sound speed in the gas, $c'_s = (P_{\text{gas}}/\rho)^{1/2}$, instead of $(P/\rho)^{1/2}$ (with P given by equation (5.39)), to construct the viscosity in (4.30). In this case the viscosity is proportional to the gas pressure, rather than the total pressure, and the resulting thin disc turns out to be stable. However, since we are seeking to obtain a spectrum more in agreement with observations than that of the standard thin disc, let us use the instability as an opportunity to depart from the standard model.

To obtain X-rays requires a high temperature; the maximum temperature is T_{th} (Section 1.3), which, at the surface of a Schwarzschild black hole, $R = 2GM/c^2$, is of the order of the proton rest energy, $m_\text{p} c^2/k \sim 10^{12}$ K. Now, if the disc really is responsible for the emitted radiation we can estimate the ratio of cooling time to infall time (i.e. to heating time) for the electrons and the protons. If we assume that we are going to be led eventually to a stable disc structure then we must take this to be optically thin, in order that the radiation pressure does not dominate the gas pressure. Then, using the non-relativistic Compton cooling time from (A9) and the timescales from Section 5.3, for electrons we have, roughly,

$$\frac{t_{\text{Compt}}}{t_{\text{infall}}} \sim \frac{\alpha}{\mathcal{M}^2}\left(\frac{\dot{M}_{\text{crit}}}{\dot{M}}\right)\eta\left(\frac{R}{R_{\text{Schw}}}\right)^{1/2}\left(\frac{m_\text{e}}{m_\text{p}}\right);$$

for protons there is a further factor $(m_\text{p}/m_\text{e})^3 \sim 10^9$. Thus, if the coupling between

Applications to discs

electrons and protons is weak, we expect a *two temperature disc* with the electrons cooled by Compton scattering to a lower temperature than the protons.

Suppose the only coupling between electrons and protons is Coulomb collisions i.e. we ignore possible collective plasma effects. Then the energy exchange timescale, $(m_p/m_e)t_d(e-i)$, from Section 3.4 can be compared with the infall timescale. It turns out that the temperature difference between electrons and ions will be maintained down to $3R_{\text{Schw}}$ if the density in the disc is less than of order 10^9 cm^{-3}. Of course, we do not yet have a consistent disc structure to know whether this condition can hold. We obtain a consistent structure by taking the pressure in the two-temperature disc to be

$$P = nk(T_i + T_e).$$

Thus we replace T by $\frac{1}{2}(T_i + T_e)$ in the disc equations, assuming electron scattering opacity and an optically thin medium with gas pressure dominating radiation pressure. It turns out that this is indeed a consistent picture at intermediate accretion rates $(\dot{M}/\dot{M}_{\text{crit}}) < 50(v_R/v_{\text{ff}})^2/\eta$. Note this leaves open the possibility of thick discs at supercritical accretion rates – see Chapter 10. Strictly, if the ion temperature reaches T_{th} it is also inconsistent to ignore thick disc effects in the two-temperature disc. This model was originally proposed for Cyg X-1 where it gives a good fit to the X-ray data. The two-temperature disc is also referred to as an ion torus.

There are numerous variations that can now be played on this theme. By threading the disc with magnetic field one can include synchrotron emission; if the viscosity parameter is high one can imagine that the implied supersonic turbulence leads to shocks and non-thermal particle injection. Thus, all of the emission mechanisms of the previous section become relevant. In particular, one must consider the effects of electron–positron pairs. The indications from the present approximate models are that the disc structure is not much affected by the presence of pairs but that the spectrum may well be.

There are two ways in which these considerations may necessitate *further* modifications of the standard thin disc picture. First, we can define a new temperature, T_{esc} at which material becomes gravitationally unbound to the black hole by

$$\frac{3}{2}kT_{\text{esc}} \sim \frac{1}{2}m_p v_{\text{esc}}^2 = \frac{GMm_p}{R}.$$

Thus, $T_{\text{esc}} \sim T_{\text{th}}$, so disc material at its maximum temperature is close to becoming gravitationally unbound. In addition, if pairs are present, the Eddington limit is reduced: if the pair density (electrons and positrons) is $n_\pm$ and the density of ionization electrons is n_e then the mass per electron is reduced from m_p to $m_p(1+n_\pm/n_e)$ and the Eddington limit is lowered by the same factor. Thus, we must allow for the loss of mass and angular momentum in a radiatively driven wind from the disc surface in the inner regions.

In fact, mass loss may also be important in the outer regions of the disc. This will occur if the disc surface intercepts a significant amount of the radiation flux from the inner region, either because it is flaring or by scattering of the radiation off ambient material above the disc. This external heating gives rise to a thermally driven wind wherever the disc surface temperature is raised sufficiently close to T_{esc}. If the dominant process is Compton scattering the disc surface is raised by a flux F_ν to the Compton temperature $T_c = \langle h\nu F_\nu\rangle/4k\langle F_\nu\rangle$ at which Compton heating and cooling balance. If the disc is exposed to a typical quasar spectrum then $T_c \sim 10^7$ K, so this process is important only for radii greater than $10^{19} M_8$ cm.

If the disc extends to even larger radii (10–100 pc) other considerations become important. The standard model predicts a very low surface temperature so the opacity is dominated by dust grains. Self-gravity in the disc cannot be neglected since it gives rise to the Jeans instability. (If the thermal energy becomes less than the self-gravitational potential energy on some scale then material will tend to form into gravitationally bound clouds (or stars) on that scale.) However, at these radii the disc is no longer powered by accretion: the energy balance in the disc is dominated by *external* heating by radiation from the central source and from starlight.

The second modification is more fundamental. We expect most of the accretion energy to be extracted near the black hole. (Precisely where depends on the disc model but radii 5–$10 \times R_{Schw}$ are typical.) In this region Newtonian gravity is inapplicable and we must therefore expect a departure from Keplerian rotation. Since $v_R \to c$ at R_{Schw} we must have $v_R > c_s$ at some radius in the disc, and the thin disc approximation breaks down. We can still have $H < R$ for $\dot{M} < \dot{M}_{crit}$ so the disc is still thin in some sense; but we cannot neglect the radial motion and the departure of u_ϕ from Keplerian form. Thus in the vertically averaged equation of motion, obtained from the Euler equation (2.3), we retain the inertial terms and the radial pressure gradients to get

$$v_R \frac{dv_R}{dR} - \Omega^2 R = -\frac{1}{\rho}\frac{dP}{dR} - \Omega_K^2 R.$$

In addition, the energy equation must allow for the fact that the dissipated energy need not be radiated locally but can be advected with the flow. Thus the flux from the disc is reduced by a term of order $(\dot{M}T/4\pi R)(dS/dR)$, where S is the entropy per unit mass on the equatorial plane.

Disc structures calculated with these modifications have come to be known as 'slim discs'. The effects, on the emitted spectrum for example, become important as $\dot{M}$ approaches $\dot{M}_{crit}$.

9.6 Magnetic fields

The inclusion of magnetic fields in models of accretion powered sources has been most widely discussed in relation to the generation of collimated bulk outflows or jets. Originally the primary concern was the dynamics of the outflow from the disc. In the simplest case (electrodynamic models) we consider the motion of individual charges under the action of magnetic fields and gravity. At greater density it is necessary to treat the outflowing material as a fluid (magneto-hydrodynamic models). The basic idea is that collimation occurs via a form of the

Magnetic fields

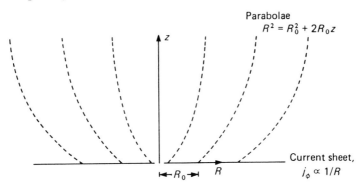

Figure 65. Field lines due to a current sheet $j_\phi \propto R^{-1}$.

'pinch' effect: a current flow along the jet generates a toroidal field, B_ϕ, the magnetic pressure of which, $(B_\phi^2/8\pi) \times [4\pi/\mu_0]$, increases away from the axis and so confines the current flow. Such a jet is therefore self-collimating.

More recently the structures of magnetized discs have been discussed, particularly to show that such discs can join smoothly on to the outflow. A steady state disc model must take into account the diffusion of magnetic field to counteract the rotational winding up of the field-lines. A detailed treatment is beyond our scope.

Outside the disc one can imagine the outflowing plasma particle moving along field-lines which, projected into a meridional plane, point away from the disc axis (as in Fig. 65). The material moves up and out along these field-lines at constant angular velocity until it reaches the Alfvén velocity (Section 3.7). Since this can occur at a radius much greater than the radius at which a given field-line leaves the disc, a small mass loss can be accompanied by a large angular momentum loss. In fact, this can be the dominant mechanism by which loss of angular momentum drives accretion, dispensing with the need for an anomalous α-viscosity altogether.

In the remainder of this chapter we shall look at energy extraction in an electrodynamic disc.

The basic idea of the electrodynamic model is that large-scale magnetic fields tied to the disc extract rotational energy from the disc which is resupplied by the infall of material in the disc. The mechanism is taken to be essentially a scaled up version of pulsar emission. In addition, the rotational energy of the accreting black hole can be extracted by electromagnetic effects and provides a roughly equal contribution to the emitted power for typical parameter values. The model has several clearly desirable features. Thus, in contrast to non-electrodynamic disc models there is no need to postulate *ad hoc* sources of viscosity: the angular momentum of infalling material is removed in a natural way by the magnetic field (Section 9.7). The energy is expected to appear initially as a directed beam of particles, organized by the magnetic field and readily, in principle, accelerated to relativistic energies. The Faraday rotation problem (Section 9.3) is avoided because these particles radiate in the same field that accelerates them, so there is no accumulation of non-relativistic material in the emission region. However, there are also some difficulties, in practice if not in principle. For, since there is no definitive pulsar model, details of this scaled-up version are uncertain, and, in particular, the generation of and radiation from the relativistic beam has yet to be worked out. Our aim here is to describe the model and to estimate the luminosity it might produce. For detailed considerations we refer the reader to the literature cited in the bibliography.

The motion of an ionized gas with overall charge neutrality is governed by the Euler equation (2.3) with, in general, a force density

$$\mathbf{f} = -\rho\mathbf{g} + q\mathbf{E} + [c]/c\mathbf{j} \wedge \mathbf{B} \qquad (9.17)$$

(cf. equation (3.2) and Section 3.7); here q is the charge density. In order to obtain solutions it is necessary to make various limiting approximations. If the net force density is sufficiently small the velocities induced will be small and the inertial terms (the mass × acceleration terms on the left hand side of the Euler equation) can be neglected. For a sufficiently tenuous plasma with large electric fields the gravitational terms can be neglected. But in the vicinity of a black hole, the case of interest to us here, the conditions under which this leads to a stationary magnetosphere (a region in which magnetic fields are important) turn out to be rather artificial. In the opposite extreme of small electromagnetic fields, gravity would dominate and material would be accreted; but this would drag in the frozen-in magnetic field and probably enhance it to the point where electromagnetic forces were no longer small (Section 3.7). Thus the relevant limits would appear to be the *magnetostatic case*, in which strong magnetic fields balance gravity, and the *force-free case*, in which the charges in the magnetosphere are assumed to be sufficiently mobile to allow them to act as sources of field. The force-free condition

$$q\mathbf{E} + [c]/c\mathbf{j} \wedge \mathbf{B} = 0 \qquad (9.18)$$

then follows if one assumes that large field strengths are set up which dominate the other terms in the force density. Note that this condition is distinct from the perfect magnetohydrodynamic assumption, for which $\mathbf{E} + [c]/c\mathbf{v} \wedge \mathbf{B} = 0$ and which is a consequence of perfect conductivity irrespective of the forces dominating the dynamics. The two are equivalent only if $\mathbf{j} = q\mathbf{v}$, which is rarely satisfied in magnetohydrodynamics, since the currents arise from the net motion of positive and negative carriers in a neutral medium, rather than from a net space charge moving with the fluid. On the other hand, the force-free condition requires a non-zero space charge density q since $q\mathbf{E} = -\mathbf{j} \wedge \mathbf{B} \neq 0$. Overall neutrality can be maintained if there are regions of opposite net charge. Thus charges must be sufficiently mobile that they can flow as needed if the force-free condition is to be satisfied. The magnetostatic case is probably important in determining the structure of a magnetized accretion disc, and may be important above the disc if there is a large amount of plasma there. The force-free condition is appropriate to pulsar magnetospheres and in the Blandford–Znajek model below.

9.7 Newtonian electrodynamic discs

We turn from these general considerations to an outline of the specific example of electrodynamic energy extraction from a disc in a Newtonian gravitational field. The simplifying assumption here comes not so much in the choice of a theory of gravity as from the fact that one does not have to worry about the boundary conditions for the magnetic and electric fields on a black hole horizon.

Consider first the field of currents flowing in circular paths in the plane $z = 0$. This is a standard, if somewhat difficult problem in electrodynamics, but we can see what must happen as follows. The magnetic field will, of course, be steady, and in cylindrical coordinates (R, ϕ, z) we will have $B_\phi = 0$, but B_R and B_z non-zero. If the surface current in the disc falls off rapidly with increasing R the solution will approach that of a circular loop of current with an approximately dipole field. Near the axis the field-lines will open out. If the current falls off more slowly towards the outer edge of the disc, so that loops at larger radii dominate as one moves away from the disc near the axis, we might expect the field-lines to bend towards the axis; this would provide some hope of carrying charges away in a directed jet (Fig. 65). In fact $j_\phi \propto R^{-1}$ gives parabolic field lines (Blandford (1976)). There is no radiation of energy in this system because the axisymmetry implies a stationary field.

Consider now how the picture changes when we replace the current loops by a magnetized disc. The magnetic energy density in the disc must be small compared with the thermal energy density and much less than the kinetic energy density of rotation, since otherwise the disc would be disrupted. Furthermore, we would expect the dense fully ionized disc to have a high electrical conductivity. In these circumstances, the magnetic field will be frozen into the disc, to a good approximation, and the field will be swept round with the disc material. We distinguish between the *toroidal* component of a vector, here the magnetic field, $\mathbf{B}_T = (0, B_\phi, 0)$ in cylindrical polar coordinates, and its *poloidal* part $\mathbf{B}_P = (B_R, 0, B_z) = \mathbf{B} - \mathbf{B}_T$. We can make the same decomposition of the electric field, $\mathbf{E} = \mathbf{E}_T + \mathbf{E}_P$. Since the model is stationary the components of all observable quantities are independent of time, and since it is axisymmetric they are independent of the angle ϕ. In particular, since the magnetic flux through a surface spanning a circle on the symmetry axis does not depend on time, it follows from Faraday's law that the e.m.f. round the circle is zero. Therefore, in fact, $\mathbf{E}_T = 0$, $\mathbf{E} = \mathbf{E}_P$. Furthermore, since B_ϕ is independent of ϕ, we have

$$\nabla \cdot \mathbf{B}_T = 0$$

and therefore, since $\nabla \cdot \mathbf{B} = 0$, we deduce

$$\nabla \cdot \mathbf{B}_P = 0. \tag{9.19}$$

This result is simple but important. It means we can picture unending lines of magnetic field associated *separately* with the poloidal and toroidal components. The lines of poloidal magnetic field can therefore be thought of as rotating with the disc material to which they are anchored.

An important result is Ferraro's law of isorotation, which states that, provided certain conditions are satisfied, the angular velocity of a field-line is constant along the field-line. Sufficient conditions for stationary axisymmetric fields are the perfect magnetohydrodynamic condition (3.46), the force-free condition, (9.18) or, more generally, degeneracy (3.47). In this last case the condition that a charge rotating with the field does not experience a force, hence remains at rest in the rotating frame, is that the electric field $\mathbf{E}'$ in the rotating frame vanishes; thus

$$0 = \mathbf{E}' = \mathbf{E}_\mathrm{P} + \frac{[c]}{c}(\mathbf{\Omega} \wedge \mathbf{r}) \wedge \mathbf{B}$$

where the angular velocity of the field-line $\mathbf{\Omega}$ has the form $\mathbf{\Omega} = (0, 0, \Omega_\mathrm{F})$. Therefore,

$$\mathbf{E}_\mathrm{P} = -\frac{[c]}{c}(\mathbf{\Omega} \wedge \mathbf{r}) \wedge \mathbf{B}_\mathrm{P}. \tag{9.20}$$

Using $\nabla \wedge \mathbf{E}_\mathrm{P} = 0$, which is a consequence of the time independence of $\mathbf{B}$, and $\nabla \cdot \mathbf{B}_\mathrm{P} = 0$ (equation (9.19)), we obtain

$$\mathbf{B}_\mathrm{P} \cdot \nabla \Omega_\mathrm{F} = 0.$$

This expresses the fact that Ω_F does not change along poloidal field-lines. Therefore the field-lines cannot wind up and discharge the field and a stationary solution as envisaged is possible.

Far out along the field-lines their rotational velocities will exceed the speed of light. We can therefore construct a speed of light surface on which equality holds. This conclusion is not in conflict with relativity because field-lines are purely mathematical entities, but any charged particles in the region beyond the speed of light surface have to move outwards and upwards back along lines of the B field if they are to move at less than the speed of light while remaining attached to lines of magnetic field. In other words, particles are forced to move away from the disc, rather like beads on a rotating wire. This provides a mechanism for the generation of a flux of relativistic particles, if only we can produce a supply of particles and a source of energy.

The particles are assumed to be extracted from the disc by a process analogous to that which operates in pulsar models. Inside the disc the perfect magneto-hydrodynamic assumption implies that there is an electric field $\mathbf{E} = -([c]/c)\mathbf{v} \wedge \mathbf{B}$ and hence a charge density

$$q = [1/4\pi\varepsilon_0]\nabla \cdot \mathbf{E} = -[1/4\pi\varepsilon_0 c]2\mathbf{\Omega} \cdot \mathbf{B}/c \neq 0 \tag{9.21}$$

where $\mathbf{\Omega} = (0, 0, v/R)$. This charge density gives rise to an electric field outside the disc which is available to pull charges out of the disc. For a magnetic field of 10^{12} G (10^8 T) this gives electric fields which dominate the gravity of the central black hole in the region above the disc but do not grossly violate the condition of degeneracy (3.47) (because $\mathbf{E} \cdot \mathbf{B} \sim ([c]/c)R\Omega B^2$ from (9.21) and hence $\mathbf{E} \cdot \mathbf{B} \ll [c]B^2$ if $R\Omega \ll c$). Thus we may idealize the magnetosphere as gravity free and degenerate.

If the particle density in the magnetosphere is sufficiently large the particle inertia will be important and the region above the disc will not be force-free. In this case, the particles may be accelerated along field-lines directly, extracting energy and angular momentum from the stored gravitational potential energy of the disc (Blandford & Payne (1982)). We shall not discuss this case further. Instead we assume that the particle density is sufficiently low and there is a force-free region above the disc. This cannot extend right out to the speed of light surface because relativistic effects mean that particles travelling on field-lines close to the speed of light become more massive so that inertial terms (mass × acceleration) cannot be neglected in the determination of their motion. The magnetic field-lines therefore are attached to non-force-free material in the disc and to non-force-free material some distance above the disc, where particles are accelerated to relativistic velocities. Electromagnetic energy and angular momentum flow through the intervening force-free region creating a torque on the disc. This has the same effect as the usual disc viscosity, causing material to flow in and extracting gravitational energy from the disc. The disc does not *radiate* electromagnetic wave energy, since this is impossible in an axisymmetric configuration, but it transmits a 'DC' flux. A rough analogy is the coupling of two flywheels by a viscous medium where the transmitted energy resides in the shear of the medium, not in sound waves.

The detailed implementation of this scheme is far from complete. The force-free region can be discussed in some detail, but the acceleration of particles in the non-force-free region, which, as we shall see, actually determines the efficiency of the process, is essentially informed guess-work.

9.8 The Blandford–Znajek model

The quasar model consists not of a Newtonian disc, but of a disc around a black hole. We shall see that the black hole can be an important source of energy, but we expect its direct effect on the disc to be small except near the horizon. Therefore we can estimate the disc luminosity in the Newtonian model.

We have seen that the magnetic lines of force are approximately frozen into the disc material but this freezing-in cannot be exact since there would then be no magnetic torque on the disc and no energy extraction. A small relative velocity is both necessary and sufficient to extract energy. This slippage of the field-lines is brought about because we have assumed the field to be anchored also in non-force-free material above the plane of the disc, where, as in the disc, the field is controlled by the inertia of the matter. In the intermediate force-free region the field determines the particle motions.

We have seen that the angular velocity Ω is constant on a poloidal field-line (Ferraro's law, Section 9.7). In fact, since Ω is independent of ϕ by symmetry, it is constant over the surface generated by rotating a poloidal field-line about the axis of symmetry $R = 0$. Let us call such surfaces (poloidal) *magnetic surfaces* (Fig. 66). We define the magnetic flux $\Phi(r, z)$ at a point by integrating the normal component of **B** over any surface bounded by a circle through the given point and centred on the symmetry axis. As a consequence of flux conservation ($\nabla \cdot \mathbf{B} = 0$), Φ is independent of the surface over which the integration is performed, and is constant on magnetic surfaces. Consider now a magnetic surface in $z \geqslant 0$ bounded by a circle of radius R in the disc. A current I must flow down through the interior of any such surface to generate a non-zero $\mathbf{B}_T$ in order to give a non-zero energy flux (see equations (9.22) and (9.23) below). It can be shown for both Newtonian and black hole models that the force-free condition implies that the current flow occurs along magnetic

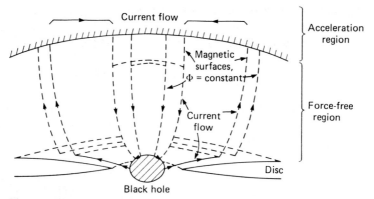

Figure 66. Magnetic surfaces and current flows in the vicinity of the black hole and disc. (Adapted from D. Macdonald & K. S. Thorne (1982), *Mon. Not. R. astr. Soc.*, **179**, 433.)

surfaces and that I is constant on these surfaces (see Fig. 66). Ampère's law applied to a circle centred on the axis gives the magnitude of $\mathbf{B}_T$ induced by a given current:

$$B_T = \left[\frac{c\mu_0}{4\pi}\right]\frac{2I}{cR}. \tag{9.22}$$

Consider now two adjacent magnetic surfaces bounded by circles of radii R and $R+dR$ in the disc, and let I and $I+dI$ be the currents flowing interior to these surfaces. In order to avoid a build-up of charge there must be a radial current I flowing between the two surfaces in the surface of the disc, and also a similar return current in the acceleration region. This is possible because these regions are not force-free so current is not constrained to flow along magnetic surfaces. There is therefore a dissipation of energy in the acceleration region, which represents the power transmitted from the disc, and a dissipation of energy in the disc. The latter is wasted energy in terms of the transmission efficiency, but does not affect the overall efficiency, since it is radiated directly by the disc. In the intermediate region it can be shown that the force-free condition implies there is no dissipation. (In the black hole case the transmitted energy suffers the usual general relativistic redshift.)

The electromagnetic power flowing through the force-free region from the disc to the acceleration region is given by the Poynting vector

$$\mathbf{S} = \left[\frac{4\pi}{\mu_0 c}\right]\frac{c}{4\pi}\mathbf{E}\wedge\mathbf{B} = \left[\frac{4\pi}{\mu_0 c}\right]\frac{c}{4\pi}\mathbf{E}_P\wedge\mathbf{B}_T \tag{9.23}$$

since $\mathbf{E}_T = 0$. Equation (9.20) gives $\mathbf{E}_P = ([c]/c)(\mathbf{R}\wedge\boldsymbol{\Omega})\wedge\mathbf{B}_P$; strictly, in the black hole case Ω should be replaced by the angular velocity of the field-lines relative to local inertial frames which, for a rotating black hole, are themselves dragged round with a non-zero angular velocity. In either case, we have

$$S \sim [4\pi/\mu_0]R\Omega B_P B_T/4\pi, \tag{9.24}$$

and in order to find S we need to know $\mathbf{B}_P$ and $\mathbf{B}_T$, which are, in principle, determined by conditions in the non-force-free regions. To obtain a numerical

The Blandford–Znajek model

estimate while circumventing this problem we assume that near the acceleration region particles move up field-lines, parallel to $\mathbf{B}_\mathrm{P} + \mathbf{B}_\mathrm{T}$, with a velocity approaching c, and a circular velocity $R\Omega$ parallel to $\mathbf{B}_\mathrm{T}$. Then a diagram of relative velocities gives

$$B_\mathrm{P}/B_\mathrm{T} \sim (c^2 - R^2\Omega^2)^{1/2}/R\Omega \sim c/R\Omega. \tag{9.25}$$

Therefore the disc luminosity is

$$L_\mathrm{D} \sim \int |S| 2\pi R \, \mathrm{d}R \sim [4\pi/\mu_0] \int B^2 c (R\Omega/c)^2 2\pi R \, \mathrm{d}R.$$

We assume that the major contribution comes from close to the minimum disc radius, which is the innermost stable orbit $R \leqslant 3R_\mathrm{Schw}$, and so we evaluate Ω as the Kepler velocity at this radius. Therefore

$$L_\mathrm{D} \sim 10^{45} (B/10^4 \, \mathrm{G})^2 (M/10^8 \, M_\odot)^2 \, \mathrm{erg} \, \mathrm{s}^{-1},$$

which is clearly large enough to be observationally interesting.

The major difference in the more realistic general relativistic model is that power can also be extracted from the rotational energy of the black hole. Magnetic field-lines are anchored on the hole by the magnetic pressure of field-lines convected in with material in the disc. (In the absence of magnetic field anchored in the disc the magnetic field of the hole would disperse on approximately a light-crossing timescale.) The black hole behaves as if currents are set up on the horizon; these *fictitious* currents are just those required to act as sinks for the current flowing down magnetic surfaces (Fig. 66) and so 'complete the circuit'. The horizon is then subject to a torque depending on the difference between the angular velocity of the field-lines and that of the hole. The efficiency η' is defined as the ratio of power extracted to the maximum power that could be extracted from material falling into a black hole, i.e. η' measures the contribution of the hole to the total power output. It can be shown that $\eta' = \Omega_\mathrm{F}/\Omega_\mathrm{H}$ where Ω_H is the angular velocity of the hole at the horizon. The angular velocity of a field-line Ω_F will vary from one field-line to another, and is determined by conditions in the acceleration region. A plausible value is $\eta' = \frac{1}{2}$, again obtained if particles move along field-lines with $v \sim c$ at large distances. We shall see that this corresponds to the maximum power output from the hole. In this case, the luminosity of the hole L_H turns out to be of order $L_\mathrm{H} \sim (a/m) L_\mathrm{D} \leqslant L_\mathrm{D}$ (Section 9.9).

These results follow strictly from a formulation of electrodynamics modified to take account of strong (i.e. non-Newtonian) gravitational effects and, in particular, to impose the appropriate boundary conditions on the electromagnetic field at the horizon of the black hole. There is one further qualitative difference from the Newtonian theory, in addition to the possibility of energy extraction from the hole, that deserves comment. Away from the equatorial plane of the disc the flow of charge is towards the hole near the horizon and away from the hole at large distances. In fact, there are now two speed of light surfaces, an inner as well as an outer one. If the model is to be consistent these currents have to be maintained even though, since there can be no current sources, they cannot be supplied by the disc.

(If, instead, the black hole were a material body there would be no inner light surface and the body could supply the necessary charges.) The mechanism postulated is the 'breakdown' of the vacuum in the sense of the creation of electron–positron pairs from high-energy photons scattering in the strong magnetic field, a mechanism that is also supposed to operate in the pulsar environment.

9.9 Circuit analysis of black hole power

The emission of energy from the black hole in the Blandford–Znajek model can be described in terms of an interesting analogy: the black hole behaves as a battery with an internal resistance; the maximum power output from the hole results when the circuit through which current flows between the hole and the acceleration region is 'impedance matched'.

To see this, note first that the detailed analysis of the boundary conditions on the horizon leads to a relation between the electric field $\mathbf{E}_H$ tangential to the horizon and the surface current density on the horizon. Let $\mathbf{J}_H$ be the current crossing a unit length on the horizon; then the total current crossing a circle of radius R is

$$I = \oint \mathbf{J}_H \cdot \mathbf{n} \, dl = 2\pi R J_H \qquad (9.26)$$

where $\mathbf{n}$ is normal to the curve and where R is a suitably defined radial coordinate in units such that the horizon is the unit sphere. (See Macdonald & Thorne (1982) for details; R is there denoted by $\bar{\omega}$). Then, it can be shown that

$$E_H = R_H J_H \qquad (9.27)$$

where $R_H = [1/4\pi\varepsilon_0](4\pi/c) = 377$ ohm is the impedance of free space.

Consider now two adjacent magnetic surfaces which intersect the horizon at latitudes λ and $\lambda + \Delta\lambda$. On the horizon we define a potential difference between these surfaces

$$\Delta V_H = E_H \Delta\lambda \qquad (9.28)$$

(recall that $\mathbf{E}$ is poloidal). If the impedance on the horizon between the magnetic surfaces is ΔZ_H, Ohm's law gives

$$\Delta V_H = \Delta Z_H I. \qquad (9.29)$$

Comparing equation (9.29) with equations (9.26)–(9.28) we get

$$\Delta Z_H = R_H \frac{\Delta\lambda}{2\pi R} = \left[\frac{1}{4\pi\varepsilon_0}\right] \frac{2\Delta\lambda}{cR}. \qquad (9.30)$$

For comparison with our previous results we need to relate $\Delta\lambda$ to the magnetic flux difference between the surfaces $\Delta\Phi$. This comes from the application of Stokes' theorem to the circuit composed of the *two* circles of latitude λ and $\lambda + \Delta\lambda$ (or the difference of two applications). We obtain

$$2\pi R \Delta\lambda B_\perp = \Delta\Phi \qquad (9.31)$$

where $B_\perp$, the normal component of magnetic field on the horizon, is the limiting value of $\mathbf{B}_P$ on the horizon. Now we can calculate the potential difference at the base of the magnetic surfaces on the horizon directly from the electric field induced by rotation of the field-lines. We have

$$E_H = -\frac{[c]}{c}(\mathbf{v} \wedge \mathbf{B}_P)_H. \tag{9.32}$$

The appropriate $\mathbf{v}_H$ introduces an important subtlety. If the field-lines were to rotate at the same rate as the hole there would be no torque, hence no power output. The appropriate velocity in (9.32) is the difference between the rotation rate of the field and of the hole $R(\Omega_F - \Omega_H)$. This arises naturally in the relativistic analysis; there $\mathbf{v}_H$ appears as the velocity of the field-lines in an inertial frame, this being dragged round with velocity Ω_H relative to stationary observers at infinity by the gravity of the spinning hole. Thus from (9.31) and (9.32)

$$E_H = -([c]/2\pi c)(\Omega_F - \Omega_H)(\nabla \Phi)_H,$$

and from (9.28)

$$\Delta V_H = [c](\Omega_H - \Omega_F)\Delta\Phi/2\pi c. \tag{9.33}$$

In a precisely similar manner we can calculate V_A at the interface of the force-free and acceleration regions. We have

$$\Delta V_A = I\Delta Z_A \tag{9.34}$$

corresponding to (9.29)

$$\Delta V_A = [c]\Omega_F \Delta\Phi/2\pi c \tag{9.35}$$

corresponding to (9.33), provided the acceleration region is sufficiently distant that the relativistic dragging of inertial frames can be neglected. We obtain for the ratio of impedances

$$\Delta Z_A/\Delta Z_H = \Omega_F/(\Omega_H - \Omega_F); \tag{9.36}$$

and for the current

$$I = (\Delta V_A + \Delta V_H)/(\Delta Z_A + \Delta Z_H) = [4\pi/c\mu_0](\Omega_H - \Omega_F)R^2 B_\perp/2$$

using (9.30) for ΔZ_H and (9.36) for ΔZ_A. The transmitted power is

$$\Delta L = \Delta Z_A I^2 = [4\pi/\mu_0]\Omega_F(\Omega_H - \Omega_F)R^2 B_\perp \Delta\Phi/4\pi c. \tag{9.37}$$

This has a maximum as a function of Ω_F at $\Omega_F = \Omega_H/2$ in which case the transmission efficiency $\eta' \equiv \Omega_F/\Omega_H = \frac{1}{2}$ (Section 9.6). In addition, if $\Omega_F = \Omega_H/2$ we have $\Delta Z_A = \Delta Z_H$ and the 'battery' (the hole) and the 'load' (the acceleration region) are impedance matched. It also follows in this case that the power dissipated on the horizon $I^2\Delta Z_H$ equals the power transmitted, which, of course, is consistent with 50% efficiency. From (9.37) we obtain a disc luminosity of order

$$\Delta L \sim \left[\frac{4\pi}{\mu_0}\right]c\left(\frac{\Omega R}{c}\right)^2 \frac{B^2}{16\pi}(2\pi R\Delta\lambda)$$

which, integrated over the area of the hole, gives an output comparable with that obtained from the disc in Section 9.8.

The foregoing argument is not, in fact, peculiar to a black hole magnetosphere, except in as much as its resistance is known. We can apply a similar argument to a circuit of current between the disc and the acceleration region. Equations (9.34) and (9.35) remain unchanged while in (9.33) and (9.36) we replace Ω_H by Ω_D, the angular velocity at the base of the magnetic surfaces in the disc. As before, the energy output from this region of the disc is

$$\Delta L = \Delta Z_A I^2 = I \Delta V_A = ([c]/2\pi c) \Omega I \Delta \Phi.$$

For comparison with our previous results we use the analogue of (9.31) for $\Delta \Phi$ in terms of B_P and (9.22) for I in terms of $B_\perp$, then

$$\Delta L = \left[\frac{4\pi}{\mu_0}\right] \frac{R\Omega B_P B_T}{4\pi} 2\pi R \, dR$$

and the flux per unit area is in agreement with our previous calculation, equation (9.24), so we obtain the same disc luminosity.

10
Thick discs

10.1 Introduction

In previous chapters we have discussed extensively the theory and applications of thin ($H \ll R$) accretion discs. We hope the reader will by now be convinced that this theory is reasonably well understood, and that it rests on a fairly firm observational basis. The case for thick ($H \sim R$) accretion discs however is less compelling as the theory is still under development and the relevant observations are few, difficult and indirect. Furthermore, since the publication of the first edition of this book much work has been done on the dynamical stability of thick discs. The wealth of these investigations is a testimony to the interest generated by these structures. The results obtained so far virtually rule out the reality of thick discs as non-accreting toroidal equilibria but leave open the more exciting possibility of the existence of closely related accreting flows which could be of astrophysical interest.

The current interest in the theory of the structure, evolution and stability of thick accretion discs is due to the possibility that thick discs may be relevant to the understanding of the central power sources in radio galaxies and quasars (see Chapters 7–9). Thick discs may also occur during early stages of star formation, they could be involved in the generation of bipolar outflows and jets, and they might form during the merger of two stars as the system attempts to handle its excess angular momentum. The effects of self-gravity in these cases are likely to be relatively more important than in AGN models. The observational evidence on powerful radio sources suggests that their energy is generated in the central galaxy, beamed to large distances by well-collimated jets and converted into radiation as the beam interacts with the external medium. It is possible that the gas that fuels AGNs (Section 7.6) forms a thick disc as it accretes on to a massive black hole sitting at the centre of the host galaxy. As we shall see, these thick discs may possess a narrow funnel along the rotation axis where large radiative pressure gradients can accelerate matter in collimated jets. Thus the appeal of thick discs resides in the feasibility of a model which derives its power from accretion and intrinsically generates jets without any extra assumptions (e.g. magnetic fields, a confining cloud, etc).

There are other theoretical reasons for pursuing the study of thick discs. As the reader will recall from Chapter 5, in the theory of thin discs the radial pressure gradients are neglected and the vertical pressure balance is solved separately. It was shown that this approximation is adequate as long as the disc is geometrically thin.

This condition may be violated in the innermost regions of accretion discs around stellar black holes and neutron stars (Section 5.6). The study of thick discs provides a better theoretical understanding of the thin disc as a limiting case and enables us to deal with intermediate situations. Furthermore, the techniques developed can be useful in the investigation of supercritical accretion. The study of their stability has furthered our understanding of the stability of differentially rotating fluids and has had a stimulating impact on other fields such as planetary discs and the formation of the solar system. In the past five years or so, thick discs were first discarded as they seemed hopelessly unstable but a couple of stabilizing effects have been discovered in the meantime, the implications of which are not fully understood yet. Despite the remaining theoretical uncertainties about their structure and stability, thick discs are still with us and may – despite early pessimism – be central to the most energetic phenomena observed in the universe.

A survey of the recent literature on thick discs will reveal that there is no universally accepted 'standard' model for a thick *accretion* disc. The work published so far demonstrates that it is, in principle, possible to account for the high luminosities inferred from observations of AGNs, and to generate the well-collimated and persistent jets observed in powerful radio sources. But all the models proposed suffer from serious internal inconsistencies and, in most of the cases where radial motion is absent, they are demonstrably unstable dynamically. We have chosen to present an account of the most frequently used models, which have been the subject of the stability studies, and to discuss some idealized examples to illustrate the assumptions and physics involved in modelling thick discs.

A brief outline of this chapter is as follows. In Section 10.2 we discuss the equilibrium and structure of fluids in *pure* rotation in general with particular emphasis on differentially rotating compressible tori. The implicit assumption is that realistic thick *accretion* discs will resemble such configurations if the radial and vertical velocity components are small compared with the velocity of rotation. It is not clear to date whether this assumption is justified, and further investigation is necessary. In Section 10.3 we show how the critical Eddington luminosity can be exceeded by large factors in toroidal figures in radiative equilibrium. In Section 10.4 we discuss a simple Newtonian equilibrium thick disc (or torus) designed to illustrate some of the most important properties common to the majority of models. In Section 10.5 we introduce viscosity, we allow for radial drift and dissipation, we obtain an expression for global energy conservation and we provide examples and criticisms. In Section 10.6 we present the main results of over half a dozen years of stability studies. Finally, in Section 10.7 we summarize the present status and future prospects and discuss astrophysical applications.

10.2 Equilibrium figures

We begin by addressing the question of the shape and internal stratification of a steady thick disc, with no viscosity, in a state of pure rotation. This problem is a special case of the theory of the shape and structure of rotating masses of fluid. Therefore many of the results derived here apply to other situations as well and

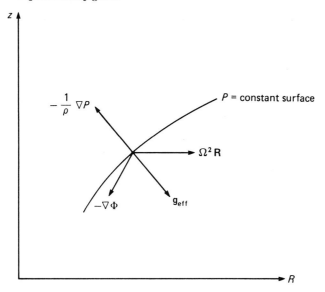

Figure 67. Balance of forces acting on an element of fluid in the z–R plane.

were first obtained in classical works on the shape of the Earth and the structure of rotating stars. A good summary of these and other results can be found in J.-L. Tassoul (1978).

Consider an axially symmetrical mass of fluid in pure rotation about its axis of symmetry. We choose a cylindrical system of coordinates (R, ϕ, z) with the z-axis coincident with the rotation (symmetry) axis. Then the velocity field has the components.

$$v_R = 0, \quad v_\phi = R\Omega, \quad v_z = 0 \tag{10.1}$$

where Ω is the angular velocity, in general a function of R and z.

The equations of motion of an inviscid fluid subject to gravitational forces (including external bodies) reduce in this case to the following

$$\frac{1}{\rho}\frac{\partial P}{\partial R} = -\frac{\partial \Phi}{\partial R} + \Omega^2 R, \tag{10.2a}$$

$$\frac{1}{\rho}\frac{\partial P}{\partial z} = -\frac{\partial \Phi}{\partial z}, \tag{10.2b}$$

where P is the pressure and Φ is the gravitational potential. These equations represent the balance of forces acting on an element of fluid (see Fig. 67) and can be written in the equivalent vector form:

$$\frac{1}{\rho}\nabla P = -\nabla\Phi + \Omega^2 \mathbf{R} = \mathbf{g}_{\text{eff}} \tag{10.2c}$$

where we have introduced the effective gravity $\mathbf{g}_{\text{eff}}$ as the vectorial sum of gravitational and centrifugal acceleration. Thus the effective gravity must be orthogonal to surfaces of constant pressure (*isobaric surfaces*). In particular, if the body under consideration is immersed in an external medium of constant pressure, its (free) surface must be perpendicular to $\mathbf{g}_{\text{eff}}$ at every point.

This property can be exploited to obtain a general relationship between the shape of the surface (any isobaric surface) and the rotation field on it. Suppose we know the equation of the surface. Because of the assumed symmetry all we need to specify is the equation of the meridional cross-section:

$$z_s = z_s(R), \tag{10.3}$$

where the subscript s indicates quantities on the (isobaric) surface. Thus we must have

$$g_{\text{eff},R}\,dR_s + g_{\text{eff},z}\,dz_s = 0. \tag{10.4}$$

Using equations (10.2*a*) and (10.2*b*) and the definition of effective gravity, we get

$$(\Omega^2 R)_s = \left(\frac{\partial \Phi}{\partial z}\right)_s \frac{\partial z_s}{\partial R_s} + \left(\frac{\partial \Phi}{\partial R}\right)_s. \tag{10.5}$$

In general, Φ will depend on the distribution of mass inside the disc. However, if we assume that the mass of the disc is relatively small and that the gravitational forces are dominated by a central point mass (or any other *external* sources), then we know Φ everywhere independently of the disc shape and stratification. Under these conditions, equation (10.5) enables us to find the required equilibrium rotation field on the surface without reference to the interior. Conversely, if the coordinates of some point on the surface are known, equation (10.5) can be integrated to find the cross-section for a given rotation field.

Let us illustrate some of these points with an example: suppose a cloud of negligible mass rotates around a point mass M with an angular velocity Ω such that

$$\Omega^2 = \frac{GMe^2}{r^3} = \frac{GMe^2}{(R^2+z^2)^{3/2}}. \tag{10.6}$$

This rotation law is chosen artificially just for mathematical convenience. With $\Phi = -GM/r$ appropriate for a point mass, direct integration of equation (10.5) yields for the meridional cross-section of our 'disc',

$$(1-e^2)R^2 + z^2 = z_P^2 = \text{constant}. \tag{10.7}$$

As this can be done for any isobaric surface, we conclude that in such a cloud the pressure is constant on oblate spheroids of eccentricity e. No rotation ($e = 0$) produces spherical surfaces as expected. As e is increased from no rotation to 'Keplerian' ($e = 1$), the isobaric surfaces become increasingly flatter. For $e \geqslant 1$ the equilibrium surfaces are no longer closed. Physically, this is due to the fact that the centrifugal force near the equator always exceeds gravity.

Consider a case with $e < 1$. From what we have just said, the pressure must be some function of the constant of integration in (10.7):

$$P = P(z_P), \tag{10.8}$$

where z_P is the semi-minor axis of the ellipse. We may now learn something about the density stratification from our basic set of equations. From (10.2b) we may solve for the density:

$$\frac{1}{\rho} = -\frac{\partial \Phi}{\partial z} \bigg/ \frac{\partial P}{\partial z} \qquad (10.9a)$$

and using (10.8) we write

$$\frac{\partial P}{\partial z} = \frac{\partial P}{\partial z_P}\frac{\partial z_P}{\partial z} = \frac{dP}{dz_P}\frac{z}{z_P}. \qquad (10.9b)$$

Thus we obtain finally

$$\frac{1}{\rho} = -\frac{GM z_P}{r^3 (dP/dz_P)}. \qquad (10.9c)$$

We extract two conclusions from this result. First, physically plausible equilibria require $dP/dz_P < 0$ (pressure must decrease outwards to balance gravity). Second, in this example, surfaces of constant density (*isopycnic surfaces*) do not coincide with isobaric surfaces unless $e = 0$. This last conclusion follows because ρ is not a function of z_P alone except when $r = z_P$ (see 10.7).

The case we have discussed illustrates how the equilibrium equations ((10.2a)–(10.2b) or (10.2c)) determine the geometry of the disc: its shape and stratification. At this point, we are not much further than in the theory of non-rotating stars when we say that they must be spherically symmetrical. For a complete model, more physics has to be invoked and appropriate boundary conditions specified.

From the example, one also notices that, in general, the density and pressure stratifications do not coincide. As we shall see below there is a simple condition on the rotation field that governs the relationship between these two families of surfaces. Let us take the curl of the equilibrium equation (10.2c), bearing in mind the form of the velocity field (10.1):

$$\nabla\left(\frac{1}{\rho}\right) \wedge \nabla P = 2\Omega \nabla \Omega \wedge \mathbf{R} = 2\frac{\partial \Omega}{\partial z}\mathbf{v}. \qquad (10.10)$$

The vectors $\nabla(1/\rho)$ and ∇P are orthogonal to isopycnic and isobaric surfaces respectively. If $\partial\Omega/\partial z = 0$ everywhere, these vectors must be parallel everywhere, which implies that surfaces of constant pressure and constant density coincide. Conversely, if they coincide, equation (10.10) implies $\Omega = \Omega(R)$ for non-vanishing rotation. This is an important result because it simplifies considerably the task of finding the structure of a disc when the angular velocity is constant on cylinders.

Another immediate consequence of the identity of isopycnic and isobaric surfaces is that the pressure and the density must be functionally related,

$$P = P(\rho). \qquad (10.11)$$

The systems for which a relationship of this type holds are known as *barotropes* or *barotropic systems*. It must be emphasized that, in general, (10.11) is not the

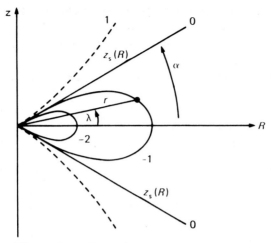

Figure 68. Equipotentials in the z–R plane of a disc with surface $z_s(R)$.

equation of state of the gas: if $\partial\Omega/\partial z = 0$ there exists a *geometric* relationship of the form (10.11) whatever the equation of state. Sometimes these systems are termed *pseudo-barotropic* to distinguish them from true barotropes, for which $P = P(\rho)$, for physical reasons independent of the state of rotation.

When $\partial\Omega/\partial z = 0$ it is possible to introduce a rotational potential ψ_{rot} such that

$$\Omega^2(R)\mathbf{R} = \nabla\psi_{\text{rot}} \tag{10.12}$$

where

$$\psi_{\text{rot}} = \int^R \Omega^2(R')\, R'\, dR'. \tag{10.13}$$

Then equation (10.2c) states that the effective gravity can be derived from a potential $\Phi_{\text{eff}} = \Phi - \psi_{\text{rot}}$,

$$\mathbf{g}_{\text{eff}} = -\nabla\Phi_{\text{eff}} \tag{10.14}$$

and the equation for isobaric/isopycnic surfaces becomes simply

$$\Phi_{\text{eff}} = \text{constant}. \tag{10.15}$$

Thus in a barotropic configuration the surfaces of constant pressure, density and effective potential all coincide. Therefore finding the shape and stratification of a barotropic disc is reduced to obtaining the equipotentials of Φ_{eff}. The latter task is considerably simplified if the gravitational potential is known *a priori*, as in the case where Φ is due to external sources. As all level surfaces satisfy $\Phi_{\text{eff}} = \text{constant}$, we have

$$\Phi_{\text{eff}} = \Phi(R, z) - \psi_{\text{rot}}(R) = \text{constant}. \tag{10.16}$$

We may now solve (10.16) for z as a function of R to get an expression for the cross-section of the equipotentials for every value of Φ_{eff}. Conversely, we may use (10.16) to obtain the equilibrium rotation field for a given equipotential shape.

$$\psi_{\text{rot}}(R) = \Phi(R, z_s(R)) \tag{10.17}$$

Equilibrium figures

where $z = z_s(R)$ is the equation for the cross-section of the surface, and the constant has been absorbed in the potential. Notice that this method would also apply to self-gravitating discs if we knew a general expression for the potential generated by a density stratification of the form $z_s(R)$.

As an example, consider a disc the surface of which is generated by the revolution of two straight lines at angles $\pm\alpha$ to the equatorial plane (see Fig. 68). Let us assume that this disc has negligible mass, it is in equilibrium with the gravitational field of a point mass M at the origin, and it is barotropic.

The equation for the cross-section of this disc of angular half-thickness α is

$$z_s(R) = \pm(\tan\alpha)R. \tag{10.18}$$

As all the conditions required for the validity of (10.17) are met, we substitute (10.18) into the potential of the point mass and obtain

$$\psi_{\rm rot}(R) = GM\cos\alpha/R, \tag{10.19}$$

from which the angular velocity can be derived

$$\Omega^2(R) = \frac{1}{R}\frac{\partial\psi_{\rm rot}}{\partial R} = \frac{GM\cos\alpha}{R^3}. \tag{10.20}$$

Thus the equilibrium rotation field is 'Keplerian' but corresponds to a mass $M\cos\alpha < M$. The velocities required for equilibrium are sub-Keplerian because the pressure distribution necessary to support such a disc vertically has a radial gradient. As we let $\alpha \to 0$, we recover the pure Keplerian field assumed in the standard thin disc theory because pressure forces vanish.

The internal pressure and density stratifications can now be obtained easily from (10.16) as a one-parameter family with $\Phi_{\rm eff}$ as a parameter. Without loss of generality we have chosen $\Phi_{\rm eff} = 0$ for the surface $z_s(R)$. After a little manipulation we obtain the following general expression for the equipotentials:

$$r = [GM/(-\Phi_{\rm eff})](1 - \cos\alpha/\cos\lambda) \tag{10.21}$$

where λ is the latitude (see Fig. 68). A few equipotentials labelled in arbitrary units are shown in Fig. 68. Only negative values of $\Phi_{\rm eff}$ permit discs of finite extent. The singular behaviour at the origin is due to the form of Φ and arises only because we allowed the disc to extend to the origin. This was done solely for mathematical simplicity, but one can, by the same method, easily study discs which do not reach the centre by suitably modifying (10.18).

We conclude this section with some comments on the transition from geometrically thick to thin discs. The example studied above shows, either by inspection of Fig. 68, or by direct manipulation of (10.21), that $|\partial P/\partial z| \gg |\partial P/\partial R|$ for small values of α. This also holds for a variety of similar models which in some sense tend to thin discs. Therefore one is justified in neglecting radial pressure gradients in the standard theory while at the same time keeping vertical pressure gradients to obtain the thickness (cf. Chapter 5).

252 Thick discs

Clearly, the standard model is not in a strict sense the limit of any single sequence of thick discs. More physics and additional assumptions would be necessary to construct such a sequence and there is no unique way of doing so.

10.3 The limiting luminosity

In Chapter 1 we discussed the maximum luminosity that could be steadily emitted by a spherically symmetrical, non-rotating body: the Eddington limiting luminosity $L_{\rm Edd}$ (equation (1.2a)). In this section we shall show that thick-discs can exceed this limit by large factors. This can happen in a large class of thick discs which contain a hollow axial region commonly known as the 'funnel'. In such regions the centrifugal force always exceeds gravity and it is mainly the balance between pressure gradients and rotation that determines the equilibrium. The maximum radiative output in these funnels is related to rotation rather than to the central mass and therefore can exceed $L_{\rm Edd}$.

We shall first obtain an expression for the critical luminosity allowing for differential rotation and then conclude this section by discussing an example for which the luminosity can be obtained explicitly.

The following argument was first presented for thick discs by Abramowicz, Calvani & Nobili, 1980. Consider a stationary and electromagnetically neutral body in mechanical equilibrium; the only forces present being gravitation, rotation and pressure gradients. The maximum flux that can be emitted from the surface of such a configuration is achieved when the effective gravity is balanced by the gradient of radiation pressure alone; in other words, when the body is in equilibrium and the pressure is almost entirely radiative.

In radiative equilibrium the flux of radiation is simply related to the radiative pressure gradient (see the Appendix, or e.g. Schwarzschild, (1965)):

$$\mathbf{F} = -\frac{c}{\kappa\rho}\nabla P_{\rm rad} \tag{10.22}$$

where κ is the opacity per unit mass. Therefore, by referring to equation (10.2c) we obtain

$$\mathbf{F}_{\rm max} = -\frac{c}{\kappa}\mathbf{g}_{\rm eff} = \frac{c}{\kappa}\nabla\Phi - \frac{c}{\kappa}\Omega^2(R,z)\mathbf{R}. \tag{10.23}$$

Note that no assumption of barotropic structure is made. The maximum luminosity in equilibrium is calculated simply by integrating $F_{\rm max}$ over the entire surface of the body; assuming $\kappa = $ constant for simplicity, we get

$$L_{\rm max} = \frac{c}{\kappa}\int_S \nabla\Phi\cdot d\mathbf{S} - \frac{c}{\kappa}\int_S \Omega^2\mathbf{R}\cdot d\mathbf{S}. \tag{10.24}$$

We use Gauss' theorem to transform the surface integrals to volume integrals:

$$L_{\rm max} = \frac{c}{\kappa}\int_V \nabla^2\Phi\, dV - \frac{c}{\kappa}\int_V \nabla\cdot(\Omega^2\mathbf{R})\, dV. \tag{10.25}$$

The limiting luminosity

The first integral is easily performed after substituting Poisson's equation $\nabla^2 \Phi = 4\pi G\rho$ where ρ is the matter density of the *sources* of Φ. Let us examine in more detail the divergence in the second integral. It can be written as follows,

$$\nabla \cdot (\Omega^2 \mathbf{R}) = \mathbf{R} \cdot \nabla \Omega^2 + \Omega^2 (\nabla \cdot \mathbf{R}) = 2R\Omega \frac{\partial \Omega}{\partial R} + 2\Omega^2. \tag{10.26}$$

By completing squares and regrouping, after some algebra we get

$$\nabla \cdot (\Omega^2 \mathbf{R}) = \tfrac{1}{2}\left[\frac{1}{R}\frac{\partial}{\partial R}(R^2 \Omega)\right]^2 - \tfrac{1}{2}\left(R\frac{\partial \Omega}{\partial R}\right)^2. \tag{10.27}$$

The first term in brackets can be shown to be the square of the z-component of the vorticity $\boldsymbol{\omega} = \nabla \wedge \mathbf{v}$, and the second term is related to the shear. Thus, collecting results,

$$L_{\max} = \frac{4\pi GMc}{\kappa} + \frac{c}{2\kappa}\int_V \left(R\frac{\partial \Omega}{\partial R}\right)^2 dV - \frac{c}{2\kappa}\int_V \left[\frac{1}{R}\frac{\partial}{\partial R}(R^2 \Omega)\right]^2 dV \tag{10.28}$$

where M is the total mass enclosed by the surface S. The first term can be recognized as the usual Eddington luminosity for the mass M. Notice that in the case of thick discs with axial funnels M is the mass of the disc and not the central object. The mass of the central object will appear only indirectly in the second term through rotation. It is this second term, containing the shear, which makes the dominant contribution and enables $L_{\max}$ to exceed L_{Edd} of the central body. The third term containing the vorticity always reduces $L_{\max}$ unless the angular momentum per unit mass $l = Rv_\phi = R^2\Omega$ is independent of R. The configurations which have the property that the axial component of the vorticity vanishes everywhere can radiate the highest luminosities. In the next section we shall discuss an example of such a vorticity-free disc.

Let us work out the maximum luminosity that can be emitted between two radii R_1 and R_2, by the disc given as an example at the end of Section 10.2 (see Fig. 68). $L_{\max}$ can be obtained by direct integration of $F_{\max}$ over the surface (both sides of the disc)

$$L_{\max} = 2\int_{R_1}^{R_2} F_{\max}(2\pi R/\cos\alpha)\,dR. \tag{10.29}$$

$F_{\max}$ can be evaluated using (10.23) and the equilibrium rotation field (10.20), but in this case it is easier to obtain $F_{\max}$ by simple geometry (Fig. 69):

$$F_{\max} = \frac{c}{\kappa}g_{\mathrm{eff}} = \frac{c}{\kappa}g\tan\alpha \tag{10.30}$$

where $g = GM/r^2$ is the gravitational acceleration and $R = r\cos\alpha$. Then

$$L_{\max} = L_{\mathrm{Edd}}\sin\alpha\,\ln(R_2/R_1). \tag{10.31}$$

Clearly, $L_{\max}$ can exceed L_{Edd} by a large amount if R_2/R_1 is sufficiently large. This

254   *Thick discs*

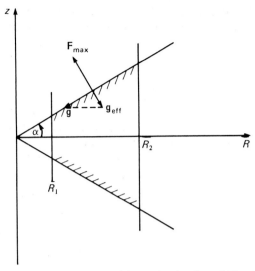

Figure 69. Balance of forces in the disc of Fig. 68.

property is common to a large variety of physically reasonable disc shapes. It arises because of the large radiative pressure gradients necessary to support the walls of the funnel. Of course, energy must be resupplied at the same rate at which it is radiated away if the disc is to remain steady. Otherwise these radiation-pressure supported configurations would cool rapidly and collapse to a thinner, subcritical disc. We shall discuss energy generation by viscosity in Section 10.5.

Finally, we note that gas-pressure supported thick discs are equally possible but they are clearly much less luminous. This however need not always be a disadvantage (Section 10.7).

10.4   Newtonian vorticity-free torus

We shall present here a simple model of a thick disc which, although in many respects unrealistic, illustrates the main features of the subject. It is fair to say at this point that a 'realistic' model with dissipation driving accretion does not exist at present. Furthermore, the main body of the stability studies so far is confined to barotropic toroidal equilibria in a state of pure rotation with simple power laws of the form $\Omega(R) \propto R^{-q}$. The vorticity-free tori described here ($q = 2$) were the first equilibria discussed in the literature and the focus of the earliest stability studies (see Section 10.6).

Motivated in part by the discussion in the preceding section and by simplicity, we shall consider the equilibrium of a vorticity-free configuration in rotation around a compact object (e.g. a black hole) of mass M, much greater than the disc mass. By vorticity-free we mean $\omega = \nabla \wedge \mathbf{v} = 0$. This implies $l = R^2\Omega = $ constant throughout and consequently barotropic structure, since $\partial\Omega/\partial z$ must vanish. We assume Newtonian dynamics for simplicity. Although this assumption is likely to be violated near the inner boundary of the disc where l is sufficiently small, some of the most important features remain unaffected.

We an use the methods developed in Section 10.2 to study this system. By virtue of the assumptions made, $\Omega(R) = lR^{-2}$ and there exists an effective potential (see equations (10.13)–(10.15))

$$\Phi_{\rm eff} = -\frac{GM}{r} + \frac{l^2}{2R^2}. \tag{10.32}$$

$\Phi_{\rm eff}$ turns out to be the energy per unit mass in this case. This is an exclusive property of the $l = $ constant rotation field. The pressure and density stratification is simply obtained by setting $\Phi_{\rm eff} = $ constant in equation (10.32) and solving for z as a function of R with parameter $\Phi_{\rm eff}$. But before we do so it is convenient to introduce some notation.

Given l, one can always define a Keplerian radius $R_{\rm K}$ such that fluid elements rotating with the local Keplerian velocity have specific angular momentum l:

$$l^2 = GMR_{\rm K}. \tag{10.33}$$

We may express all lengths in units of $R_{\rm K}$, and rewrite (10.32) as follows:

$$E = 2R_{\rm K}\Phi_{\rm eff}/GM = (R_{\rm K}/R)^2 - 2R_{\rm K}/(R^2+z^2)^{1/2}, \tag{10.34}$$

where E is the specific energy of the fluid in units of the binding energy of a Keplerian circular orbit of radius $R_{\rm K}$. For every value of E we have an equipotential (isobaric and isopycnic) surface, the cross-section of which is implicitly given by (10.34). It may be easily shown from this equation that E reaches its minimum value $E = -1$ at the circle $R = R_{\rm K}$, $z = 0$. For the fluid to remain bound to the central object we require $-1 \leqslant E < 0$.

From equation (10.34) we may get more explicit forms for the equipotentials. Introducing the polar angle θ with the usual definition $R = r\sin\theta$, we obtain

$$\sin^2\theta = R_{\rm K}^2/(2rR_{\rm K} + Er^2), \tag{10.35}$$

which is a quadratic equation for r as a function of θ with parameter E. Fig. 70 shows a meridional cross-section through some equipotentials labelled by their E values. The resultant disc is toroidal, with the most tightly bound fluid elements moving around in an equatorial circle of radius $R_{\rm K}$. This is also the locus of highest density and pressure, the actual values of which are still arbitrary. As we move away from this limiting equipotential the fluid becomes less tightly bound and the pressure and density drop.

It can be shown from (10.35) that a given equipotential surface extends between the inner and outer equatorial radii: $R_{\rm in}$ and $R_{\rm out}$ respectively. Expressions for these may be obtained by solving (10.35) with $\sin^2\theta = 1$:

$$R_{\rm in} = R_{\rm K}[1+(1+E)^{1/2}]^{-1}, \tag{10.36a}$$
$$R_{\rm out} = R_{\rm K}[1-(1+E)^{1/2}]^{-1}. \tag{10.36b}$$

As expected, when $E = -1$, $R_{\rm in} = R_{\rm out} = R_{\rm K}$, showing that the most tightly bound toroidal surface degenerates into the equatorial circle of radius $R_{\rm K}$. When $E \to 0$ one sees that $R_{\rm in} \to R_{\rm K}/2$ and $R_{\rm out} \to \infty$. Thus the limiting equipotential for bound discs is an open surface. For $E > 0$ all equipotentials are open (an example is shown in Fig. 70 by dotted lines).

256 Thick discs

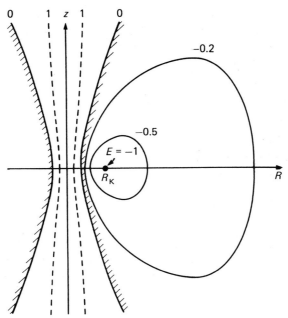

Figure 70. Meridional cross-section through equipotentials of the toroidal disc discussed in the text.

An important feature of this disc and other thick disc models is the narrow axial funnel. Matter in steady rotation supported by pressure can 'fill' all equipotentials up to a small negative value of E. (The precise value of E can be obtained, as in models of stellar structure, by providing appropriate boundary conditions and the mass of the disc.) The resultant configuration is a fat distorted torus with a narrow axial channel along which large radiation-pressure gradients are set up. This can be seen by direct inspection of Fig. 70: near the equator the equipotentials are closely packed implying large pressure gradients and therefore large fluxes of radiation. As one moves along the funnel away from the equatorial plane and towards the 'openings' of the funnel at both ends, the radiative flux emitted by the walls decreases. Therefore the density of radiation along the funnel is maximal near the equator and falls towards both ends. If, as a result of a small viscosity or simply 'overfilling', gas drifts into the funnel, it will either be sucked in by the central object or expelled along the axis by radiation pressure.

We can estimate the aperture of the funnel by differentiating (10.35) with respect to r and finding that the minimum value of θ occurs at $r = -R_K/E$ for a given (negative) value of E. Substituting back into (10.35) we obtain

$$\sin^2 \theta_{\min} = -E. \tag{10.37}$$

If the surface of the disc is the equipotential E then a cone of aperture $\theta_{\min}$ with vertex at the origin is contained in the funnel and it is tangent to the surface of the disc (see Fig. 71). Thus the collimation provided by the funnel is of the order of $\theta_{\min}$. A narrow funnel requires small $|E|$ and consequently such a funnel is defined by rapidly rotating weakly bound walls. Because the binding energy of material near

Thick accretion discs

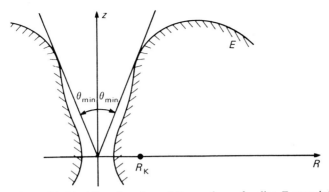

Figure 71. Meridional section of the surface of a disc E containing a conical funnel.

the surface of the funnel is relatively low, it is, in principle, easy for matter entrained from these walls to escape to infinity propelled by radiation pressure. This is the basic mechanism by which a pair of well-collimated jets in opposite directions could be generated by a thick disc.

10.5 Thick accretion discs

So far we have considered the equilibrium of pressure supported rotating fluids in the gravitational field of a central point mass. Such equilibrium models do not allow for radial motions and are not, in principle, 'accretion' discs. However, by analogy with thin discs, we can impose certain conditions on the radial motion to guarantee that the departures from equilibrium are everywhere small. This should tell us in which circumstances equilibrium models may constitute a reasonable approximation to thick accretion discs.

In the disc model we envisage, the field of motions is dominated by a rapid rotation around the axis of symmetry specified by the angular momentum distribution $l = l(R)$. Superimposed on this general rotation we allow for a small poloidal velocity $v_p \ll v_\phi$. This poloidal velocity must have $v_z \sim v_R$ in some regions of the disc as a consequence of the geometry. If viscosity is responsible for a slow radial drift, as in thin discs, we may assume as a first approximation that the relationship (5.27) derived for α-discs still holds:

$$v_p \sim v_z \sim v_R \approx \alpha c_s^2 / v_\phi \approx \alpha c_s (H/R). \tag{10.38}$$

Thus the condition of subsonic poloidal motions $v_p \ll c_s \Rightarrow \alpha \ll 1$ imposes a more severe restriction on α than in the thin case because here $H \sim R$. Even if the viscous stresses can be written following the α-prescription (see Section 4.7, equation (4.30)), one expects α to be, in general, a function of both R and z. However, no theory has been developed to support a particular choice of viscosity distribution. Models in which viscous stresses and poloidal motions are confined either to a thin surface layer or to the equatorial plane have been discussed by Paczyński and collaborators (Paczyński (1982)).

Let us assume for the moment that the restriction $v_p \ll v_\phi$ is sufficient to ensure that the techniques developed in previous sections give us a reasonable approximation to the shape and internal structure of thick discs. There are two main differences from the thin disc case that we have to be aware of when we come to build a model.

First, we are free to choose the specific angular momentum distribution (within certain restrictions based on dynamical equilibrium and stability). As we know from preceding sections, this determines the shape of the disc and the maximum luminosity that can be emitted in radiative equilibrium. By contrast, in the thin disc case the rotation field is uniquely determined by the balance with gravity. Thus an extra condition will be necessary to remove this apparent arbitrariness from thick disc models.

Second, we cannot assume 'local' balance of energy, because the heat generated by dissipation can travel in any direction before emerging to the surface. By contrast, in a thin disc the geometry is such that vertical gradients dominate and radiative transport is essentially in the z-direction. For thick discs it is easier to impose a global energy balance: in steady state the total luminosity L must equal the total rate of energy generation by viscosity L_{gen}.

We shall now discuss this global energy balance in more detail. If the disc atmosphere can be assumed to be in radiative equilibrium, then the radiative flux is (equation (10.22))

$$\mathbf{F} = -\frac{c}{\kappa\rho}\nabla(1-\beta)P, \tag{10.39}$$

where P is the total pressure and $P_{rad} = (1-\beta)P$ following usual notations. If $\beta = 0$ the disc is fully supported by radiation pressure and the discussion of L_{max} in Section 10.3 applies. If β = constant or more generally if $\mathbf{n} \cdot \nabla\beta = 0$ (where $\mathbf{n}$ is the unit vector orthogonal to the surfaces P = constant), we can obtain in analogy with equation (10.23) the following expression for the flux of radiation across the surface:

$$\mathbf{F}_n = -\frac{c}{\kappa}(1-\beta)\mathbf{g}_{eff} = \frac{c}{\kappa}(1-\beta)(-\nabla\Phi + \Omega^2\mathbf{R}). \tag{10.40}$$

If the gravitational potential is known *a priori* (as in the case of negligible self-gravity), all we need to know in order to compute the luminosity is the rotation field on the surface. Given $l(R)$ we can get the equation of the surface by the methods described earlier, and we can, in principle, integrate (10.40) over the entire surface to find L:

$$L = 2\int_{R_{in}}^{R_{out}} 2\pi R|\mathbf{F}_n|[1+(dz_s/dR)^2]^{1/2}\,dR, \tag{10.41}$$

where the factor 2 arises from equatorial symmetry.

On the other hand, we must evaluate L_{gen}. We assume that all the energy is generated by viscous dissipation. This assumption can always be checked afterwards: once the density and temperature of the disc are known we can show that nuclear energy generation is negligible in cases of interest.

Thick accretion discs

By analogy with our discussion of thin discs (see Chapter 5 and Section 4.6 on viscous torques), we introduce the surface density, the accretion rate and the torque on the material inside a certain radius R:

$$\Sigma = \int_{-z_s}^{z_s} \rho \, dz, \tag{10.42a}$$

$$\dot{M} = \int_{-z_s}^{z_s} 2\pi R \rho(-v_R) \, dz, \tag{10.42b}$$

$$G = 2\pi R^3 \frac{\partial \Omega}{\partial R} \int_{-z_s}^{z_s} \eta \, dz, \tag{10.42c}$$

where $z_s(R)$ is the equation of the disc surface and η is the dynamical viscosity.

As shown in Section 4.6 (equation 4.25 and following paragraphs), the total rate of dissipation in a cylindrical shell between radii R and $R+dR$ is given by

$$dL_{\text{gen}} = G(R)(d\Omega/dR) \, dR. \tag{10.43}$$

This is also valid for thick discs provided that $\partial \Omega/\partial z = 0$ or, more generally, if dissipation and torques due to vertical shear are negligible. The torque can be obtained from the integrated form of angular momentum conservation valid in a steady state (see e.g. equation (5.4) and section 5.3):

$$G(R) = G_0 - \dot{M}(l-l_0); \tag{10.44}$$

where G_0 and l_0 are the torque and specific angular momentum at some arbitrary reference radius. (Note that (10.44) reduces to (5.13) with $C = -G_0 - \dot{M}l_0$.) Thus the total luminosity generated by viscous dissipation between two arbitrary cylindrical radii R_1 and R_2 is

$$L_{\text{gen}}(R_1, R_2) = \int_{R_1}^{R_2} (G_0 - \dot{M}(l-l_0))(d\Omega/dR) \, dR. \tag{10.45}$$

To proceed further, we note that after a trivial integration of the constant terms in (10.45) the remainder can be integrated by parts, following the substitution $l = R^2 \Omega$:

$$\int_{R_1}^{R_2} R^2 \Omega \, d\Omega = [\tfrac{1}{2} R^2 \Omega^2]_{R_1}^{R_2} - \int_{R_1}^{R_2} \Omega^2 R \, dR. \tag{10.46}$$

The first term on the right hand side of (10.46) is simply the difference in specific kinetic energies at the two radii, whereas the second term is the corresponding change in ψ_{rot} (see equation (10.13)). Collecting results we obtain an expression for the viscous energy generation rate between R_1 and R_2:

$$L_{\text{gen}}(R_1, R_2) = G_0(\Omega_2 - \Omega_1) + \dot{M}[l_0(\Omega_2 - \Omega_1) - \tfrac{1}{2}(v_2^2 - v_1^2) + (\psi_{\text{rot},2} - \psi_{\text{rot},1})]. \tag{10.47}$$

When we apply this expression to the whole disc, by taking $R_1 = R_{\text{in}}$ and $R_2 = R_{\text{out}}$, since its surface is defined by $\Phi_{\text{eff}} = \text{constant}$ (see Section 10.2), we have

$\psi_{\text{rot},2} - \psi_{\text{rot},1} = \Phi_2 - \Phi_1$. To eliminate Φ we introduce the energy per unit mass $e = \frac{1}{2}v^2 + \Phi$ and rewrite (10.47), using $v^2 = \Omega l$, as

$$L_{\text{gen}} \equiv L_{\text{gen}}(R_{\text{in}}, R_{\text{out}}) = G_{\text{in}}(\Omega_{\text{out}} - \Omega_{\text{in}}) + \dot{M}[(e_{\text{out}} - e_{\text{in}}) - \Omega_{\text{out}}(l_{\text{out}} - l_{\text{in}})] \tag{10.48}$$

where the innermost boundary of the disc has been chosen as reference radius. For most cases, the torque at the inner boundary can be assumed to vanish and also $l_{\text{out}} \Omega_{\text{out}} \to 0$ as $R_{\text{out}} \to \infty$. Thus for large discs with $R_{\text{out}} \gg R_{\text{in}}$ one has approximately

$$L_{\text{gen}} = \dot{M}(-e_{\text{in}}). \tag{10.49}$$

This result also holds for a standard Keplerian disc as the reader may recall from our discussion of thin accretion discs (equation 5.20)). In fact, equations (10.47) and (10.48) hold for the standard disc because all the assumptions concerning torques are valid and $\psi_{\text{rot}}(R) = \Phi(R, 0)$ for a Keplerian distribution of angular momentum.

With the definition of e_{in} and assuming $\Phi = -GM/r$, we may rewrite (10.49) as

$$L_{\text{gen}} = \frac{GM\dot{M}}{2R_{\text{in}}} \left(2 - \frac{l_{\text{in}}^2}{GMR_{\text{in}}} \right). \tag{10.50}$$

So, if the disc is Keplerian near its inner radius, the luminosity per unit mass transferred is equal to the standard value. However, the Newtonian potential does not adequately describe the gravitational field near a black hole. As some models do invoke a thick disc with $R_{\text{in}} \sim R_{\text{Schw}} \equiv 2GM/c^2$, one must modify the above expressions for the luminosity. We shall not derive them here but just quote one of the results (see Paczyński (1982) and references therein). It turns out that $e_{\text{in}} \to 0$ as $R_{\text{in}} \to 2R_{\text{Schw}}$ for 'Keplerian' orbits, where 'Keplerian' has been generalized to mean balance between rotation and gravity alone. Therefore a very large accretion rate may be necessary to support a disc with a given luminosity, the inner boundary of which lies close to the so-called marginally bound radius $2R_{\text{Schw}}$ (the radius at which the binding energy vanishes; see Section 7.7).

We return now to the requirement of global balance of energy $L = L_{\text{gen}}$. Given the mass of the central object M, the distribution of angular momentum $l = l(R)$, the accretion rate $\dot{M}$ and the fractional gas pressure β, both luminosities can be computed using (10.41) and (10.48). Equating them, one obtains a relationship between $\dot{M}$, M and β. Just as in the thin disc case, the viscosity does not enter explicitly in (10.48). However, to compute the density and the temperature everywhere, a knowledge of v_R is necessary which in turn requires a value for the viscosity.

One serious problem related to the arbitrariness of $l(R)$ remains: global energy balance is not sufficient to ensure local energy conservation. As we have seen, given $l(R)$ we can compute both the distribution of energy generation and the distribution of radiative flux emerging from the surface. Clearly, these need not be consistent with local energy conservation throughout the configuration. This is closely related to the von Zeipel paradox for rotating stars in radiative equilibrium. The paradox is resolved by allowing either configurations with $l = l(R, z)$, or meridional

Dynamical stability

circulation. It remains to be shown that it is possible by radiative and perhaps other transport mechanisms to carry the energy from where it is generated to where it is radiated away. In the thin disc all physical quantities exhibit a strong vertical stratification, and it is natural to assume that radiative transport occurs in that direction. By contrast, in thick discs the stratification has comparable radial and vertical gradients.

If one insists on keeping $l(R)$ arbitrary, a contradiction will generally occur in the energy conservation at a local level. Suppose a certain distribution of volume dissipation rates, opacities, gradients, etc., does indeed produce the emergent flux. As the local rate of shearing is essentially fixed by $l(R)$, the local viscosity would have to be adjusted to match the desired distribution of dissipation. This would be in contradiction, in general, with a viscosity which depends on local density and temperature. Therefore this aspect of the theory of thick discs requires further investigation with the aim of removing the present arbitrariness of $l(R)$.

10.6 Dynamical stability

The very existence of thick discs was thrown into doubt by the discovery by Papaloizou & Pringle (1984) that the vorticity-free tori described in Section 10.4 are *dynamically* unstable to global non-axisymmetric modes. It would be impossible for such an object to exist in an equilibrium state in nature since it would be destroyed or drastically distorted by the instability during the time over which equilibrium is established. This discovery stimulated a great number of detailed analytical and numerical studies which considered tori with different rotation laws and entropy distributions. It is impossible to do justice to all the work done in this field in the space available. The interested reader should consult recent review papers by Narayan (1991) and Narayan & Goodman (1989). Here we shall aim to introduce the main concepts, to summarize the current status and to indicate the likely directions of future work.

The term *global* above in the context of stability studies as opposed to *local* refers to the fact that the character of the modes and their stability depends on the global spatial dependence of the eigenmode functions obtained by solving the linearized perturbation equations subject to appropriate boundary conditions. Unlike local stability studies where only local quantities are involved, a global instability involves a high degree of coherence over the entire spatial extent of the equilibrium configuration. A typical *non-axisymmetric* mode has the form $\delta Q = \delta Q(R, z) \exp(im\phi - i\omega t)$, where Q is any of the hydrodynamic quantities (pressure, density, velocities), $m \geq 1$ is the azimuthal wave number, and ω is a complex eigenfrequency. The eigenmode equation is a partial differential equation for $Q(R, z)$ on R and z, with ω as the eigenmode, linking spatial dependence and character of the mode. The results of linear studies are meaningful only when the perturbations remain small. As the mode grows, the linearized equations no longer describe the evolution correctly and one has to employ the full set of hydrodynamic equations. To follow the evolution of the perturbations into the nonlinear regime requires the use of numerical techniques, and some very interesting work has already been done, but the details are beyond the scope of this summary.

Most of the extant stability work is limited to polytropic equilibrium (therefore barotropic) discs with power law rotation distributions of the form $\Omega(R) = \Omega_K (R/R_K)^{-q}$, with Newtonian gravity entirely dominated by the central object (i.e. non self-gravitating discs), and in absence of accretion (pure rotational equilibria). Departures from this set of assumptions have also been considered in a few studies. The physical relevant range of rotation power laws stretches from $q = \frac{3}{2}$, which yields an infinitesimally thin Keplerian accretion disc, to the torus models with $q = 2$ described in Section 10.4. The $q = 2$ models are the thickest tori worth considering since $q > 2$ violates the well-known local stability criterion for axisymmetric perturbations due to Rayleigh. The discs and tori with $\frac{3}{2} < q \leqslant 2$ have a pressure maximum at the Keplerian radius R_K. The Keplerian disc with $q = \frac{3}{2}$ is a degenerate case, and discs with $0 \leqslant q < \frac{3}{2}$ have a pressure minimum at R_K and require pressure confinement, or have a pressure maximum at the central object and are thus akin to rotating stars and Roche lobe filling equilibria.

Thick discs are divided for descriptive purposes into 'slender' and 'wide' (or 'fat', if also $H \gg R_K$) depending on the ratio of the radial width of the disc $(R_{out} - R_{in})$ to the Keplerian radius R_K. The stability studies soon revealed that in slender tori the modes were to a very good approximation independent of z. This simplified the stability analysis considerably, reducing it to a two-dimensional problem with cylindrical symmetry. Such configurations are termed 'annuli' in the literature and their stability has been extensively studied. In fact, most of the nonlinear calculations have been devoted to annuli, and they have shown that the instability results in fragmentation of the annulus into several blobs or 'planets'. The term 'tori' is henceforth used to describe true three-dimensional thick discs.

The main results of the stability studies can be summarized as follows. Slender annuli and tori are very similar in their stability properties. The so called 'principal branch' modes are the fastest growing modes with $\mathrm{Im}(\omega) \sim 0.3\Omega_K$, whose pattern speed is equal to Ω_K – thus with 'corotation radius' $R_c = R_K$. These modes are two-dimensional even in tori, they appear whenever $q > \sqrt{3}$, for any polytropic index, and for lower m they grow faster. There exist also higher order modes with slower growth rates, which persist for rotation power-laws flatter than $q = \sqrt{3}$. Numerical simulations of the nonlinear development show that both slender annuli and slender tori break up into m planets. For wide annuli, the principal branch becomes stable at higher values of q and is stable even for $q = 2$ if the annulus is sufficiently wide. However, the higher order modes do survive and, although their growth rates are somewhat lower, they are still dynamical. In wide tori the principal branch is stable and the surviving higher order modes are clearly three-dimensional, particularly near the outer radius where the eigenfunctions have many nodes. Numerical integrations of the initial value problem for the linear stability equations indicate that very wide tori become stable even for higher order modes as $(R_{out} - R_{in})/R_K \to \infty$. This result is potentially very interesting since the tori expected in AGNs should be extremely fat. However, these results must be confirmed by further simulations, since effects of numerical dissipation might mimic stability.

Other stabilizing effects studied in the literature include self-gravity and accretion flows. Self-gravity reduces the strength of the principal branch but higher order modes survive in slender tori. Also new instabilities – probably related to fission – appear in the self-gravitating case. Concerning the effects of accretion, a very interesting result was obtained by studying an annulus in a pseudo-Newtonian potential which simulates very well several relativistic effects. It is possible to set up an annulus whose inner radius is smaller than the radius of a critical 'cusp' radius such that accretion on to the central object occurs without any viscosity. This is analogous in some respects to Roche lobe overflow in binaries. It was found that a relatively small rate of accretion was sufficient to completely stabilize all modes. The physical explanation for this effect is still a matter of some debate and it remains to be seen if the effect is also present in tori. In fact, according to some physical interpretations of this effect, its expected influence on unstable modes in tori should be weaker. At this point this is still an open question.

Thus, the most astrophysically interesting tori – at least in the context of AGN theory – are likely to be fat accretion tori with $q = 2$. We find ourselves faced with the following dilemma: should we believe that they cannot exist since $q = 2$ is the most violently unstable rotation law, or should we hope that they are probably stable since they are fat and accreting? The present calculations are tantalizing in that they touch closely on several of the expected properties of astrophysically relevant discs, but it is not possibly to predict which effect is going to dominate when everything is put together in a 'realistic' torus model. Future studies of this problem will probably involve large-scale numerical simulations of the nonlinear evolution of fat tori. There are severe limitations of resolution and dynamical range which make progress difficult, but the rewards might include a better understanding of the central engine in AGNs and the mechanism generating viscosity in discs.

10.7 Astrophysical implications

Originally the attraction of thick disc models in the context of AGN theory stemmed from the ability of these discs to generate the energy by accretion and to collimate a substantial fraction of this energy into narrow jets. It was thought initially that these jets could be accelerated by the strong radiation pressure gradients along the funnel. When this idea was examined in detail by modelling numerically the motion of test particles in the funnels of $l = $ const discs, the results were disappointing. The calculations show that a light optically thin plasma composed entirely of electrons and positrons could not be accelerated beyond Lorentz factors $\gamma \sim 3$. When a small fraction of protons was mixed in, then the terminal speeds dropped to $v \sim 0.4c$. This is too slow for models of superluminal expansion where values of γ from 5 to 10 have been invoked (Section 9.3).

Although optically thick jets have not been sufficiently studied, it appears that radiatively accelerated jets cannot achieve the velocities required by observations of superluminal expansion. It has also been pointed out in the meantime that Compton drag by disc radiation would limit the achievable plasma terminal Lorentz factors even if some effective acceleration mechanism existed. Recently, the investigations of jet acceleration have favoured hydromagnetic mechanisms, but a definitive model has not emerged yet. Assuming there is a mechanism generating ultrarelativistic jets near the central black hole, model calculations have shown that Compton drag in the funnels of a thick disc would decelerate the jets to values consistent with superluminal expansion.

In our discussion of the equilibrium equations we neglected the pressure of the radiation emitted by one side of the funnel on the opposite side. By analogy with binaries, this effect has been termed the *reflection effect*. The treatments of this effect in the literature include analytic estimates and iterative numerical procedures. In general it is possible to balance the component of the radiative force normal to the surface of the funnel but leaving tangential components unbalanced. Therefore atmospheric layers of the funnel could be accelerated and mixed into the jet material. Once radiative effects get the outer layers of the funnel in motion more material is likely to be entrained by hydrodynamical instabilities. Thus the assumption of an empty funnel – both for calculations of jet acceleration and for the reflection effect – is probably wrong and deserves further study.

The spectrum emitted by a thick disc can be calculated in analogy with thin discs by adding local black body contributions, allowing for simple opacity effects which shift the spectrum to slightly higher effective temperatures. The reflection effect in the funnels also raises the effective temperature when one looks down the funnel and Comptonization of ambient radiation by relativistic jet electrons will likely produce soft X-rays. It has been proposed recently that the soft X-ray excesses in some optically selected radio quiet QSOs could be due to this effect. In this picture, when the accretion torus is viewed at high inclination (i.e. nearly perpendicular to the jet axis) an optical/UV bump is observed, whereas at low inclinations an excess in the UV/soft X-ray band is detected.

Another interesting application of tori models suggested recently makes use of the angular distribution of ionizing due to the funnel mechanisms already mentioned to explain the alignment of extended emission line regions with radio and optical structures in Seyfert galaxies. Using the spectrum of ionizing radiation emerging along the funnel it has been shown using detailed ionization codes that it is possible to get good fits to observed emission line spectra.

The low efficiency of conversion of mass into energy in most models of accretion tori (Section 10.5) implies that a large amount of 'dead' mass must accumulate in the central object over the lifetime of an AGN. If quasars occur during early phases of galactic evolution then it should be possible to detect massive remnants of an active phase in nearby 'normal' galaxies (Section 7.4). Current observational limits on black hole masses seem to be consistent with efficiencies $\sim 10\%$ but much lower values would pose serious problems.

A number of difficulties can be alleviated if we abandon the requirement that the high luminosities and the collimation rise jointly from the thick disc without extra ingredients. If we consider a thick disc supported by *gas* pressure, then we no longer have large unbalanced radiative forces in the funnel and we can deal with the energetics of the jet and the disc separately. Two temperature plasma tori supported by ion pressure can exist if the electrons and ions are not coupled effectively by collisions (Rees *et al.* (1982)). These ion tori emit a relatively small fraction of their luminosity in directions other than the funnel in agreement with observations of

Astrophysical implications

double radio sources with relatively weak core luminosities. Recent work on plasma instabilities suggest however that electrons and ions might couple effectively even if collisions are inefficient.

The problem with energy balance in radiation pressure dominated tori mentioned in Section 10.5 has not been addressed properly yet. As in radiatively supported rotating stars, it is likely that a large-scale meridional circulation will be driven by the thermal imbalances. Because of the high rotational velocities involved in tori, it is likely that these circulations are dynamically important and have an effect on the stability properties of thick discs. Another possibility which has remained entirely unexplored consists of baroclinic equilibria with $\Omega = \Omega(R, z)$. Neither the equilibrium properties nor the stability of these configurations has been studied. Nature may have ways of setting up thick discs somewhat different or maybe even radically different from what theorists have been able to come up with so far. The study of thick discs is still in its infancy and perhaps the astrophysically more interesting insights still lie ahead.

Despite the theoretical uncertainties involved in the structure and stability of thick discs, which make it unclear whether these models are relevant to astrophysical objects, there are clearly a number of areas worth further study. In addition to those we have already mentioned, possible applications of the theoretical ideas developed in the course of these studies include pressure supported discs formed during early stages of massive star formation, merging of double degenerate binary stars, bipolar outflows in young stellar objects, generation of jets in objects such as SS 433, planet formation, fission of protostellar clouds, and last but not least, the origin of viscosity in accretion discs.

Appendix: radiation processes

In this appendix we collect together the concepts and results that have been used in the text. It is not intended as an *ab initio* introduction to the subject. A thorough treatment can be found in Rybicki & Lightman (1979) and a good reference work is Tucker (1975).

The (unpolarized) radiation field is defined by the *specific intensity* $I_\nu(\mathbf{r}, \mathbf{n}, t)$ as a function of position $\mathbf{r}$, direction $\mathbf{n}$, time t and frequency ν, which gives the flux of energy dE per second per unit solid angle per unit frequency interval in direction $\mathbf{n}$ at $\mathbf{r}, t$ crossing a unit area normal to $\mathbf{n}$. Thus

$$dE = I_\nu \, dA \cos\theta \, d\nu \, d\Omega \, dt.$$

The radiation *flux* is the rate at which energy crosses unit area independent of direction and is obtained by integrating over solid angle ($F_\nu = \int I_\nu \cos\theta \, d\Omega$). The *specific luminosity* of a source is the flux integrated over an area enclosing the source ($L_\nu = \int F_\nu \, dA$). The variation of I_ν in a medium which is emitting and absorbing is given by the *radiative transfer equation* which simply expresses energy conservation for the radiation field; in the time-independent case:

$$\mathbf{n} \cdot \nabla I_\nu = -\mu_\nu I_\nu + j_\nu \tag{A1}$$

where $\mu_\nu = \kappa_\nu \rho / N$ is the absorption coefficient and j_ν the emission coefficient (energy per second per steradian per unit volume). These coefficients depend on the processes occurring in the medium and these processes in turn depend on the radiation field, since this may control the temperature, state of ionization, etc. This complexity is reduced in several simplified limits.

It is customary to characterize the medium by the *source function* $S_\nu = j_\nu / \mu_\nu$ and the *optical depth* τ_ν, along a path $\mathbf{r} = \mathbf{r}(s)$ from source to observer defined by

$$d\tau_\nu = \mu_\nu \, ds. \tag{A2}$$

(There is a minus sign in equation (A2) if the path is in the reverse direction from observer to source.) Then (A1) becomes

$$\frac{dI_\nu}{d\tau_\nu} = -I_\nu + S_\nu. \tag{A3}$$

If the radiation field corresponds to thermal equilibrium at a temperature T, we

Appendix   267

know that $I_\nu = B_\nu(T)$, the Planck (blackbody) function, and I_ν is then independent of position (hence of optical depth) so, from (A3),

$$S_\nu = B_\nu(T) \equiv 2h\nu^3/c^2[\exp(h\nu/kT) - 1] \qquad (A4)$$

whatever the radiation mechanism. If the medium can be characterized by a temperature T (i.e. the emission is thermal) then

$$S_\nu \equiv j_\nu/\mu_\nu = B_\nu(T) \qquad (A5)$$

holds (Kirchhoff's law), independent of whether the radiation field is Planckian since (A5) is a relation between atomic properties. In general, the source function is determined by the state populations in the medium.

Conversely, S_ν will satisfy (A4) if the states in the medium which contribute to emission and absorption, j_ν and μ_ν, are populated according to a Boltzmann distribution ($N(E) \propto \exp(-E/kT)$) at some temperature T, even if other states (e.g. bound electron states) which do not contribute are not so populated. Collisions amongst free electrons are so rapid that the free electron distribution can usually be taken as Boltzmann. Thus free-free emission and absorption have j_ν, μ_ν obeying (A5) even when bound electron states are not Boltzmann distributed. In the case where the matter distribution is assumed to be described by the Boltzmann formula (and its generalizations to include ionization, etc.) but the radiation field is not in equilibrium with the medium the gas is said to be in local thermodynamic equilibrium (LTE). A necessary condition for a thermal radiation spectrum (derived by integrating (A3) with $S_\nu = B_\nu(T)$) is that the optical depth $\tau_\nu \to \infty$. In this case, we say that the medium is *optically thick*. In general, in an optically thick medium $I_\nu = S_\nu$. On the other hand, if $\tau_\nu \to 0$ we can neglect absorption and (A1) gives

$$I_\nu = \int j_\nu \, ds.$$

Such a medium is said to be *optically thin*. The optically thick and thin limits are those required in the text. Finally, if $S_\nu = 0$ we have a purely absorbing medium and from (A3),

$$I_\nu = I_\nu(0) e^{-\tau_\nu}.$$

This is a valid approximation for a scattering medium also if scattering *into* the line of sight can be neglected.

In the case of a star, or any other optically thick medium, the state of the matter can be characterized locally by a temperature T which varies only slowly with the position (i.e. we have approximate thermal equilibrium locally, a special case of LTE). In this approximation, (A3) is equivalent to

$$F_\nu = -\frac{4\pi}{3\kappa_\nu \rho} \frac{dB_\nu(T)}{dr} \qquad (A6)$$

(obtained by multiplication of (A3) by $\cos\theta$ and integration over solid angle with $ds = dr/\cos\theta$). Integrating (A6) over frequency we obtain

$$F = -\frac{ac}{3\kappa_R \rho} \frac{d}{dr}(aT^4) \tag{A7}$$

where κ_R is the *Rosseland mean opacity* defined by

$$\frac{1}{\kappa_R} = \int \frac{1}{\kappa_\nu} \frac{\partial B_\nu}{\partial T} d\nu \bigg/ \int \frac{\partial B_\nu}{\partial T} d\nu,$$

and we have used $\int B_\nu d\nu = (ac/4\pi)T^4 \equiv (\sigma/\pi)T^4$. This form of the radiative transfer equation is required in Chapter 5.

Except in the optically thick case the emerging intensity is specified by a knowledge of j_ν and κ_ν or μ_ν. Fortunately, there is only a relatively small number of emission and absorption mechanisms of importance in most astrophysical applications. We summarized some of the processes and coefficients that occur in the text.

(i) Free–free emission (or thermal bremsstrahlung): this is produced by the scattering of a *thermal* distribution of free electrons off ions at sub-relativistic temperatures. The emission coefficient is

$$4\pi j_\nu \cong 6.8 \times 10^{-38} N_i N_e T^{-1/2} \exp(-h\nu/kT) \, \text{erg s}^{-1} \, \text{cm}^{-3} \, \text{Hz}^{-1}$$

and

$$4\pi j \cong 1.4 \times 10^{-27} N_i N_e T^{1/2} \, \text{erg s}^{-1} \, \text{cm}^{-3}.$$

This is frequently valid even when the system is not in overall thermal equilibrium because the electrons relax quickly to a Maxwell–Boltzmann distribution.

(ii) Cyclotron emission: an electron in a circular orbit at subrelativistic velocity in a magnetic field B gives a spectral line at the Larmor or cyclotron frequency $\nu_{cyc} = eB[c]/m_e c$.

(iii) Synchrotron emission: a relativistic electron of energy E in a magnetic field can be considered in a zeroth order approximation to radiate at a frequency $\nu_s \cong \gamma^2 \nu_{cyc} (\gamma = E/m_e c^2)$.

For an isotropic distribution of pitch angles an electron of energy E radiates an average power

$$-\frac{dE}{dt} \sim \tfrac{4}{3}\sigma_T c \left(\frac{B^2}{8\pi}\right)\left(\frac{E}{m_e c^2}\right)^2 \sim 2 \times 10^{-3} B^2 E^2 \, \text{erg s}^{-1},$$

where σ_T is the Thomson cross-section. The emission coefficient for a number density $n(E) dE$ of electrons between E and $E+dE$ is given by

$$4\pi j_\nu \cong 2.5 \times 10^{-2} E^2 B^2 n(E) \frac{dE}{d\nu}; \quad E \cong 4 \times \nu^{1/2} B^{-1/2} \, \text{erg}. \tag{A8}$$

For non-relativistic electrons $(E/m_e c^2)$ is replaced by $(v/c)E/m_e c^2$.

(iv) Inverse Compton emission: a relativistic electron scatters a lower-energy photon of frequency ν_0 and boosts it to a higher frequency $\nu \cong \gamma^2 \nu_0$. An electron of energy E radiates a power

Appendix

$$\frac{dE}{dt} \cong \tfrac{4}{3}\sigma_T c U_{rad}\left(\frac{E}{m_e c^2}\right)^2$$

in an isotropic radiation field of energy density U_{rad}. The emission coefficient is obtained by summing over the electron distribution as in (A8). For non-relativistic electrons the radiated power is

$$-\frac{dE}{dt} \sim \tfrac{8}{3}\sigma_T c U_{rad}\left(\frac{E}{m_e c^2}\right). \tag{A9}$$

(v) Line emission: a density $N(X_s^{(m+)})$ of excited atoms or ions of species $X^{(m+)}$ in a state s, labelled by a set of quantum numbers s making spontaneous transitions to a state with quantum numbers s' radiates line photons with

$$4\pi j_\nu = N(X_s^{(m+)}) A_{s\to s'} h\nu$$

where $h\nu = (E_s - E_{s'})$ and $A_{s\to s'}$ is the Einstein A-coefficient for the transition. Ions may also be stimulated to radiate by the presence of a radiation field at the line frequency, but this is usually taken into account as negative absorption in the absorption coefficient. Of course, $N(X)$ has to be determined by balancing the rates of transition into and out of each state.

Absorption processes are the time-reverse of the emission processes. In general, they are governed by absorption cross-sections σ_ν, with $\mu_\nu = \sigma_\nu N$ in a medium of absorbers of density N. Pure scattering by electrons is sometimes important and in the non-relativistic limit is governed by the Thomson cross-section

$$\mu = \sigma_T N_e = N_e \tfrac{8}{3}\pi(e^2/m_e c^2)^2 [4\pi\varepsilon_0]^{-2}.$$

For a *thermal* medium μ_ν can be obtained directly from Kirchhoff's law (A4).

An important non-thermal case in the text is synchrotron self-absorption. If $h\nu \ll E$ the absorption coefficient is

$$\mu_\nu d\nu = -\frac{c^2}{8\pi h\nu^2}\sum 4\pi j_\nu(E)g(E)\frac{\partial}{\partial E}\left(\frac{N(E)}{g(E)}\right)\frac{dE}{d\nu}d\nu \tag{A10}$$

(see e.g. Rybicki & Lightman (1979)) where $N(E)$ is the spectral density of electrons of energy E and $g(E)$ is the statistical weight of the state of energy E. For free electrons $g(E) \propto E^2$ and, in general, we must interpret the sum in (A9) as an integral. For a power law electron distribution, $N(E) = N_0 E^{-\alpha}$. In the approximation that only synchrotron radiation of frequency $\nu \propto E^2 B$ is emitted by an electron of energy E, the sum contains only one term and

$$\mu_\nu \propto \nu^{-2}[E^2 B^2 (dE/d\nu) E^{-\alpha-1}]_{E=\nu^{1/2}B^{-1/2}}$$

or

$$\mu_\nu \propto \nu^{-(\alpha+4)/2} B^{(\alpha+2)/2}.$$

The source function is $S_\nu = j_\nu/\mu_\nu$ with j_ν given by (A8), hence $S_\nu \propto \nu^{5/2} B^{-1/2}$. Hence if the medium is optically thick to synchrotron self-absorption we have $I_\nu = S_\nu \neq B_\nu$ and $I_\nu \propto \nu^{5/2} B^{-1/2}$ as in the text (Section 9.4).

Problems

Chapter 3

1. If cooling behind a shock front is very effective, the gas temperature may rapidly return to its pre-shock value. If we are not interested in the details of the cooling region, we can regard it and the shock proper as a single discontinuity, usually called an *isothermal* shock, and can treat it in the manner used for adiabatic shocks in Section 3.8. Thus we retain the mass and momentum jump conditions (3.53), (3.54), but replace the energy condition (3.55) by

$$T_1 = T_2$$

where T_1, T_2 are the gas temperatures each side of the discontinuity, or equivalently the equality of isothermal sound speeds:

$$c_1 = c_2 = c.$$

where $c_1 = (P_1/\rho_1)^{1/2}$ etc.
Show that

$$v_2 = \frac{c}{\mathcal{M}_1}, \quad \rho_2 = \mathcal{M}_1^2 \rho_1$$

and

$$P_2 = \rho_1 v_1^2,$$

where $\mathcal{M}_1$ is the pre-shock Mach number.

Strong isothermal shocks result in very large compressions, and in all cases the post-shock gas pressure is equal to the pre-shock ram pressure.

Chapter 4

1. A degenerate star has a mass–radius relation of the form

$$R_2 = K M_2^{-1/3}$$

where K is a constant and M_2 is measured in solar masses. Show that if the star fills the Roche lobe in a close binary with $q < 1$ we have

$$P \propto M_2^{-1}.$$

If $K = 2 \times 10^9$ cm show that this relation and the main-sequence relation (4.9) intersect at about $P = 0.6$ h, $M_2 = 0.07$. This shows that there is a minimum orbital period for CVs, since the secondary cannot be smaller than its radius when fully degenerate. The *actual* minimum period is somewhat greater because mass transfer makes the secondary deviate from its thermal-equilibrium radius.

2 For the degenerate secondary of problem 1, show that

$$-\frac{\dot{M}_2}{M_2} = -\frac{\dot{J}/J}{2/3-q}.$$

3 The evolution of a close binary is driven by angular momentum loss to gravitational radiation, so that

$$\frac{\dot{J}}{J} = -\frac{32}{5}\frac{(GM_\odot)^3}{c^5}\frac{M_1 M_2(M_1+M_2)}{a^4}.$$

Use problems 1, 2 above and the results of Section 4.4 to show that for main sequence and degenerate secondaries we have

$$\dot{M} \simeq 10^{-10}(P_{\rm hr}/2)^{-2/3}\, M_\odot\, {\rm yr}^{-1}$$

and

$$\dot{M} \simeq 1.6 \times 10^{-12}(P_{\rm hr}/2)^{-14/3}\, M_\odot\, {\rm yr}^{-1}$$

respectively.

4 Some binary pulsars appear to be evaporating their companion stars, which do not fill the Roche lobe, at a rate $\dot{M}_2 < 0$ driven by the stream of high-energy particles from the pulsar. If the lost mass has specific angular momentum βj_2, where j_2 is the companion's specific angular momentum and β is a constant, and the orbit remains circular, show that the binary period evolves as

$$P \propto M_2^{3(\beta-1)}(M_1+M_2)^{(1-3\beta)}$$

where M_1 (= constant) is the pulsar mass and M_2 the companion mass.

5 Consider the Roche potential Φ_R (eq. 4.5) in Cartesian coordinates (x, y, z) with $z = 0$ the orbital plane and the x-axis along the line of centres, $x = 0$ being the centre of the secondary star. Let the inner Lagrange point L_1 have coordinates $(x_1, 0,)$. Show that, to order of magnitude, the potential at a nearby point $(x_1, \Delta y, 0)$ is

$$\Delta\Phi_R \simeq \omega^2 \Delta y^2,$$

where ω is the binary frequency $2\pi/P$. (*Hint*: recall that L_1 is a saddle point, so that $\nabla\Phi_R = 0$, and use Kepler's law to express lengths in terms of ω.) Use an energy argument to show that matter escapes from a lobe-filling star in a patch of radius

$$H = \Delta y \sim c_{\rm s}/\omega$$

around L_1, and hence that the instantaneous mass transfer rate is

$$-\dot{M}_2 \sim \frac{1}{4\pi}\rho_{L_1} c_{\rm s}^3 P^2,$$

where $c_{\rm s}$ and ρ_{L_1} are the stellar sound speed and density near L_1.

6 For the case of a lower main-sequence star filling the Roche lobe (P = a few hours, surface temperature 3000–4000 K) use the result of problem 5 to show that

$$-\dot{M}_2 \sim 10^{-8}(\rho_{L_1}/\rho_{\rm ph})\, M_\odot\, {\rm yr}^{-1}$$

where $\rho_{\rm ph} \sim 10^{-6}$ g cm^{-3} is the photospheric density. (This result shows that except at high transfer rates the L_1 point is outside the stellar photosphere: see problems 7, 8.)

272 Problems

7. Lobe-filling binaries adjust the instantaneous value of $-\dot{M}_2$ to that driven by angular momentum losses (cf. problem 3 above) or nuclear evolution (cf. eq. 4.14) by adjusting the binary separation and thus the ratio $\rho_{L_1}/\rho_{\rm ph}$ in the equation of problem 6 above. Show that near L_1, ρ varies as $\exp(-x^2/H^2)$ along the line of centres, where H is defined in problem 5, while away from this line it varies as $\exp(-x/H_*)$, where $H_* = c_s^2 R_2^2/GM_2$ is the usual stellar scaleheight. Evaluate H, H_* for the case $P_{\rm h} = 2$, $M_2 = 0.22$, $R_2 = 0.22 R_\odot$, $T_{\rm eff} = 3500$ K and show that the required adjustment is always very small compared with the stellar radius.

8. A binary has $P_{\rm h} = 2$, $M_2 = 0.22$ and total mass 1 $M_\odot$. If the mass transfer rate is 10^{-10} $M_\odot$ yr^{-1}, find by how much the separation changes in one year. Show that the timescale for the separation to change by the scaleheight H_* is of order 6×10^4 yr. This is a lower limit to the timescale on which the mass transfer rate can change through evolutionary effects alone.

9. Show that in Roche lobe overflow, matter tends to escape the binary entirely rather than being captured by the primary star once $v_\parallel$ exceeds a certain value $v_{\rm crit}$. Use the formulae of Section 4.5 to express $v_{\rm crit}$ in the alternative forms

$$\left[\frac{2b_1(1+q)}{aq^2}\right]^{1/2} v_1, \quad \left[\frac{2b_1(1+q)}{a}\right]^{1/2} v_2$$

where v_1, v_2 are the orbital velocities of the primary and secondary. $v_{\rm crit}$ marks the boundary between conventional Roche lobe and wind accretion. Show that $v_{\rm crit}$ is always supersonic for the mass-losing star. Find $v_{\rm crit}$ for a binary with $P = 50$ days, $M_1 = 1\ M_\odot$ and $M_2 = 8\ M_\odot$.

Chapter 5

1. In a certain Keplerian disc, matter is neither accreted nor expelled, and the disc simply extracts angular momentum from the central star at a constant rate. Show that the disc's effective temperature varies with radius as

$$T_{\rm eff} \propto R^{-7/8}$$

2. Matter is injected at a constant rate $\dot{M}_0$ into Kepler orbits at radius R_0 around a central mass. Within R_0 it spirals inwards under the action of viscosity v, accreting at radius R_* on to a central object, to which it gives up all its Keplerian angular momentum. Outside R_0 a steady distribution of mass extends to very large radii and removes the angular momentum of the matter accreting within R_0 (cf. problem 1). Show that in a steady state

$$v\Sigma = \frac{\dot{M}_0}{3\pi}\left[1 - \left(\frac{R_*}{R}\right)^{1/2}\right]$$

for $R < R_0$ and

$$v\Sigma = \frac{\dot{M}_0}{3\pi}\left[\left(\frac{R_0}{R}\right)^{1/2} - \left(\frac{R_*}{R}\right)^{1/2}\right]$$

for $R > R_0$.

3. Consider a spherical star of radius R_* and uniform temperature T_* radiating like a blackbody. Assume the star is surrounded by an infinitesimally thin disc of optically thick material and calculate the temperature distribution of this 'passive' disc illuminated by the central star. Compare the T distribution obtained with that

Problems

of the standard accretion disc. How would you distinguish observationally an accretion disc from an illuminated passive disc around a young star?

4 Derive some of the equations governing the structure of *slim discs* around a black hole (M. Abramowicz et al. (1988), *Astrophys. J.* **332**, 646). Assume that the general relativistic effects can be described by the pseudo-Newtonian potential

$$\Psi = -\frac{GM}{r - R_{\text{Schw}}}.$$

The radial component of the gravitational force can be written conveniently as $\Omega_K^2 R$, where R is the cylindrical radius and Ω_K is the regular velocity on circular Keplerian orbits.

Starting from the basic equations for fluid motion in cylindrical geometry, assuming equatorial symmetry and neglecting second order terms in small velocities and disc thickness, write down the equations for mass, r- and z-momentum conservation for a steady, geometrically thin, axially symmetrical disc where the local rotation velocity is allowed to depart from the Keplerian value.

(a) Show that

$$\frac{P}{\rho} = B_1 \Omega_K^2 H^2$$

$$\dot{M} = B_2 4\pi R H \rho v_R$$

and

$$v_R \frac{dv_R}{dR} + \frac{1}{\rho}\frac{dP}{dR} + (\Omega_K^2 - \Omega^2)R = 0$$

where all variables are defined on the equatorial plane, the B_i are constants which depend on vertical structure and the usual notation has been used.

(b) Obtain integral expressions for the constants B_i above.

(c) Combine the last two equations in (a) to show that a critical sonic point exists. Compare with spherical accretion in Section 2.5.

5 Consider an optically thin layer above an optically thick Keplerian accretion disc. Assume that there is dissipation caused by an α-viscosity in the layer. Show that in hydrostatic equilibrium the energy balance between viscous heating, Comptonization, and radiative cooling in the layer can be written

$$\tfrac{3}{2}\alpha\rho\frac{v_K}{R}c_s^2 + \frac{\kappa\rho F}{m_e c^2}(E - 4kT_1) - \rho^2 \Lambda(T_1) = 0,$$

where ρ is the density, v_K the Kepler velocity at radius R, c_s the local sound speed in the layer, κ the electron scattering opacity coefficient, F the total radiative flux generated locally in the optically thick disc, m_e the electron mass, Λ the local cooling function at the local temperature T_1, and E the flux-weighted mean photon energy of photons from the optically thick disc. Find $T_1(R)$ for $\rho \to 0$. Use the formulae for a Shakura–Sunyaev disc with surface temperature $T(R)$, and $E \simeq 3.83T(R)$ to show that there is no solution for T_1 unless the condition

$$\frac{R^{3/2}}{f^4} < \frac{\kappa\mu m_H \dot{M}(GM)^{1/2}}{\pi m_e c^2 \alpha}$$

holds, where $f^4 = 1 - (R_*/R)^{1/2}$, M and R_* are the mass and radius of the central star, μ the mean molecular mass, and m_H the hydrogen atom mass. Show that the left hand side of this equation has a minimum for $R_* = 16R/9$, and hence that the disc cannot have a hydrostatic atmosphere at all for accretion rates $\dot{M}$ exceeding a certain value. (This leads to the possibility of mass outflow from discs – see M. Czerny & A. R. King (1989), *Mon. Not. R. astr. Soc.* **236**, 843.

6 Taking $A = 10^{18} A_{18}$ cm² as an estimate for the cross-section of the accretion stream in an LMXB or a CV (see Ch. 4, problem 5), calculate the mean density of the stream material assuming that part of it reaches the circularization radius R_{circ} in an approximately ballistic orbit. Estimate the thermal pressure this material would reach if shocked at that radius. Show that if this material is illuminated by radiation from a central source with ionizing luminosity $L_{\text{ion}} = 10^{36} L_{36}$ erg s^{-1} the ionization parameter $\Xi = P_{\text{ion}}/P_{\text{gas}}$ is likely to be in the range of two-phase instability (See Fig. 61) for LMXBs. The geometry of LMXBs and CVs is similar but the ionizing luminosities in CVs are a factor $\sim 10^3$ lower. This could be the reason why narrow dips in the X-ray flux are observed in LMXBs but not in CVs. How do these considerations depend on the binary orbital period and on the accretion rate?

Chapter 6

1 Consider a boundary layer of radial thickness b at the inner edge of an accretion disc around a white dwarf of mass M_* and radius R_*. Assume that viscosity ν heats the gas at a volume rate $\Gamma = \nu \rho v_\phi^2/b^2$, where ρ is the density and $v_\phi \sim (GM_*/R_*)^{1/2}$. Assuming an α viscosity (eq. 4.30) with $H = b$ and a volume cooling rate $\rho^2 \Lambda(T)$, express the timescales t_{heat}, t_{exp}, t_{rad} for heating, cooling by adiabatic expansion, and cooling by radiation of a gas element in terms of its temperature T and the combination ρb. The cooling function can be taken as $\Lambda \sim T^{-1}$ for $T \lesssim 10^6$ K, $\Lambda \sim T^{1/2}$ for $T > 10^6$ K. By considering the ratios $t_{\text{heat}}/t_{\text{exp}}$, $t_{\text{rad}}/t_{\text{heat}}$ and $t_{\text{rad}}/t_{\text{exp}}$ show that gas with ρb sufficiently small tends to heat up to the virial temperature

$$T_v \simeq \alpha \mu m_H v_\phi^2 / 3k \sim 10^8 \text{ K}$$

and forms a hard X-ray emitting corona around the white dwarf.

2 The impact behaviour of blobs of plasma in free-fall along magnetic field-lines on to the polar cap of a magnetic degenerate star can be represented approximately as a shock tube problem. Consider the impact of gas of density ρ_b (blob) moving supersonically with velocity v towards gas of density ρ_a (atmosphere) initially at rest. After the two gases have come into contact at $t = 0$, a moving contact discontinuity and two shocks propagating in opposite directions relative to the contact discontinuity are generated. Find the condition for the backward shock propagating through the blob gas to hover motionless as the shocked blob material pushes further into the atmosphere: assume that both the atmosphere and the blob material are cold so that all motions are at infinite Mach number and the shocks are strong. Using mass and momentum conservation and pressure continuity across a contact discontinuity, show that the velocity of the backward shock is given by

$$s_1 = \frac{1 - \frac{1}{3}\sqrt{(\rho_a/\rho_b)}}{1 + \sqrt{(\rho_a/\rho_b)}}.$$

Therefore the net expansion of the shocked blob material is backwards when $\rho_a > 9\rho_b$.

3 The white dwarfs in AM Her systems can be taken as having masses $\sim 0.7 M_\odot$ and magnetic moment $\lesssim 3 \times 10^{34}$ G cm³. For orbital periods $\gtrsim 3$ hours the average accretion rates are $\sim 4 \times 10^{16}(P_{\text{hr}}/3)^{5/3}$. Show that the white dwarfs in any

Problems

progenitor systems at periods $\gtrsim 4$ h would not rotate synchronously with the orbit. (The apparent lack of such non-synchronous strong-field systems is not fully understood. See A. R. King & J. P. Lasota (1991), *Astrophys. J.*, **378**, 674.)

4 The torque between the two component stars in an AM Her system can be represented as the interaction of two dipoles μ_1, μ_2. Show that there is an orbital torque

$$\mathbf{T}_{\text{orb}} = \frac{3}{a^3}[(\mathbf{n}\cdot\boldsymbol{\mu}_2)(\mathbf{n}\wedge\boldsymbol{\mu}_1)+(\mathbf{n}\cdot\boldsymbol{\mu}_1)(\mathbf{n}\wedge\boldsymbol{\mu}_2)]$$

between the two stars (i.e. tending to change their orbital angular momentum rather than their spins), where a is the orbital separation and $\mathbf{n}$ the unit vector pointing from the white dwarf to the centre of the secondary. The accretion torque is normal to the orbital plane; show that in equilibrium with the magnetic torque, both dipoles must lie in the orbital plane. (See A. R. King, J. Frank & R. Whitehurst (1990) *Mon. Not. R. astr. Soc.* **244**, 731.)

Chapter 7

1 If $n(S, z)\,\mathrm{d}S\,\mathrm{d}z$ is the number of quasars per steradian with flux between S and $S+\mathrm{d}S$ and redshift between z and $z+\mathrm{d}z$, show that energy emitted by all quasars per unit volume in a homogeneous and isotropic universe is given by (A. Soltan (1982), *Mon. Not. R. astr. Soc.* **200**, 115)

$$E = \frac{4\pi}{c}\int \mathrm{d}z(1+z)\int \mathrm{d}S\,Sn(S,z).$$

Current estimates give $E \sim 10^{76}$ erg Mpc^{-3}. What is the expected average dead (black hole) mass per host galaxy assuming that quasars are powered by accretion on to a massive black hole at the centre of a bright galaxy? Express your answer as a function of the efficiency η and assume a space density of bright galaxies $n_G \sim 2\times 10^{-3}h^{-3}$ Mpc^{-3} ($h = H_0/100$ km^{-1} s^{-1} Mpc^{-1}).

2 A spherical cloud of hydrogen of radius r_A is in steady free-fall on to a central black hole, mass M, radius R_{Schw}. Assume that the Mach number of the inflow is unity (which is approximately true for adiabatic flow for $r \ll r_A$) and that matter is resupplied to the cloud at r_A to keep this radius fixed. If the medium is optically thin, show that the bremsstrahlung luminosity is

$$L = C\int_{R_{\text{Schw}}}^{r_A} n^2 T^{1/2} r^2\,\mathrm{d}r,$$

where C is a constant and n and T the gas density and temperature. Show that $T = 2GMm_H/kr$ and that n has the form $n = n_A(r_A/r)^{3/2}$, and hence, when $r_A \gg R_{\text{Schw}}$, find an approximate expression for r_A as a function of L, n_A and physical constants. Estimate the Compton y-parameter of the medium (assuming most of the scattering occurs near R_{Schw}) and hence show that if the condition $y < 1$ is satisfied,

$$L \lesssim \frac{CGMm_e}{16m_H^{1/2}k^{1/2}c\sigma_T^2}.$$

(This is a variation on the efficiency problem discussed in Section 7.8. The next problem shows how it can be overcome in spherical accretion if a shock is present.)

3   If gas in free fall towards a black hole of mass $10^8 m_8 M_\odot$ accretes at a rate $\dot{m}\dot{M}_{crit}$ show that the density satisfies $n \sim 10^{11}(r/R_{Schw})^{-3/2}\eta^{-1}m_8^{-1}\dot{m}$. Below a spherical shock at radius r_s electrons are held at a temperature $T_e \sim m_e c^2/k$ by radiation losses, but the proton temperature, T_p, reaches the virial temperature; show that $T_e \ll T_p$.

Under these circumstances the electron–proton energy exchange time for relativistic particles is approximately $t_{e-p} \sim (m_p c^2 \theta_e)/(n_e \sigma_T m_e c^3 \log \Lambda)$ where $\theta_e = kT_e/m_e c^2$. Show that $t_{e-p} \gg t_{ff}$. (So matter diffuses in much more slowly than free-fall and can radiate efficiently.)

If $T_e \propto r^{-1}$ above the shock and constant below it show that the emitted bremsstrahlung spectrum is $\nu^{-1/2}$ below a certain frequency ν_0 and falls off exponentially above ν_0. (See Mészáros & Ostriker (1983).)

4   The flux from the surface of a standard thin disc has the form $F \propto r^{-3} \mathcal{Q}(r)$ where $\mathcal{Q} = (1-(r_*/r)^{1/2})$ (cf. equation 5.39); show that it is a maximum at $r = (49/36)r_*$. The flux from a disc round a Schwarzschild black hole has a similar form, $F \propto r^{-3} \mathscr{C}^{-1/2}(r) \mathcal{Q}(r)$, where $\mathcal{Q} \to 0$ as $r \to r_{ms}$ and $\mathscr{C} \to \frac{1}{2}$; here r is measured in units of the Schwarzschild radius and $r_{ms} = 6$ is the radius of the last stable orbit in these units. The function $\mathscr{C}$ is given by $\mathscr{C} = 1 - 3/r$ but $\mathcal{Q}$ is complicated and we cannot differentiate directly to find where F is a maximum. However, the following allows us to find this radius approximately. Let

$$\mathscr{L} = \frac{1}{(1-3/r)^{1/2}} - \frac{2\sqrt{3}}{r^{1/2}}.$$

If $\delta = r - r_{ms} = r - 6$ show that $\mathscr{L} \sim \delta^2$ as $\delta \to 0$.
Now

$$\mathcal{Q} = \mathscr{L} - \frac{3}{2r^{1/2}} \mathscr{I} \int_{r_{ms}}^{r} \frac{\mathscr{L}}{\mathscr{I}} \frac{1}{(1-3/r)} \frac{dr}{r^{3/2}}$$

where $\mathscr{I}$ tends to a finite limit as $r \to 6$. It is easy to show that $\mathcal{Q} \sim \delta^2$ as $\delta \to 0$.

Now differentiate $\log F$ and evaluate the result approximately near $r = r_{ms}$ to show that if the maximum flux occurs near r_{ms} then it occurs approximately for

$$\frac{1}{2} + \frac{1}{18\sqrt{2}} = \frac{2}{\delta}.$$

Deduce that the maximum occurs at about 10 Schwarzschild radii. (For the Kerr case the algebra is yet more complicated: for the relevant formulae see I. Novikov & K. Thorne in *Black Holes* ed. by C. DeWitt & B. DeWitt, Gordon & Breach (1973).)

Chapter 8

1   Starting from equations (5.45) appropriate for accretion on to stellar mass objects, scale up the variables to values appropriate for AGNs (cf. Chapter 8). What would be the temperature distribution of a passive disc of the standard shape with constant α illuminated by a central point source? At what radius would the illumination by the central source dominate over local viscous dissipation?

2   A gas of two-level atoms is photoionized by a flux of radiation F_ν. Ignoring

collisional processes show that the ratio of the number densities of ions and atoms in the ground state satisfies

$$\frac{N_+}{N_1} = \frac{1}{\alpha N_e} \int_{v_0}^{\infty} \frac{\sigma_v F_v}{hv} dv,$$

where N_e is the free electron density, α the recombination coefficient to the upper level and σ_v the photoionization cross-section above a threshold frequency v_0. (Recall that only ionizations from the *ground* state are significant.)

If $F_v = F_0 (v_0/v)^2$ and $\sigma_v = \sigma_0 (v_0/v)^3$ find the condition on the ionization parameter such that $N_+ \gg N_1$ and show that, under this condition, the luminosity in the spectral line, at frequency v_e, is $\alpha N_e^2 h v_e V$, where V is the volume of the emitting plasma. Hence formulate a method by which the mass and filling factor of the line emitting material may be obtained.

3 Assuming that the broad line clouds are not bound to the central mass and are optically thick to the Lyman continuum, estimate the minimum mass flow rate through the broad line region in terms of the ionizing luminosity L, the gas density in the clouds N, the ionization parameter Γ, the covering factor f and the black hole mass M_{bh} (and physical constants). If this mass loss rate is assumed to be less than the accretion rate and the active nucleus is taken to radiate below the Eddington limit, show that

$$M_{bh} \gtrsim \frac{c^8 m_p \sigma_T^3 \Gamma^3 f^4 \varepsilon^4}{\alpha^4 N h v_0 G},$$

where ε is the accretion efficiency, α the hydrogen recombination coefficient and v_0 the Lyman limit frequency. Does this give a useful limit, or could the result be used to estimate a different parameter?

4 Let the collisional excitation rate coefficient for a model two level atom be q_{12} (so the rate of upward transitions for the atom in its ground state in a medium of electron density N_e is $q_{12} N_e$ per second), and let the radiative recombination rate to the ground state be α. Show that the ratio of line emission from radiative recombination to that from collisional excitation is $\alpha N_+ / q_{12} N_1$ where N_+ and N_1 are the ionized and ground state populations. Use the results of problem 2 and the following data to estimate whether the C IV $\lambda 1549$ line arises from collisional excitation or radiative recombination in a broad line cloud of density 10^{10} cm^{-3}, temperature 10^4 K. ($q_{12} = 8.36 \times 10^{-6} \Omega T^{-1/2} e^{-\chi/kT} (2J'+1)^{-1}$, $\Omega = 8.88$, $\chi = 8$ eV, $J' = 1/2$, $\alpha = 8.45 \times 10^{-12}$ cm^3 s^{-1}, $\sigma_v = 0.68 \times 10^{-18} (v/5.2 \times 10^5 \text{ cm}^{-1})^{-2}$ cm^2.)

5 A disc emits isotropically a flux $l(r)$ in a line of rest frequency v_0. If the emitting material moves with circular velocity $v(r)$ in the plane of the disc, the normal to which makes an angle i to the line of sight, show that an element of angular extent $d\phi$ in the disc emits in a frequency interval $dv = -(v/c) \sin i \cos \phi \, d\phi$, where ϕ is measured from the plane of the normal and the line of sight. Hence show that the emission line has a profile

$$L_v \, dv = \int \frac{l(r) r \, dr}{[(v^2/c^2) \sin^2 i - (v-v_0)^2]^{1/2}} dv,$$

where the integral is taken over radii such that $(v/c) \sin i \geq |v-v_0|$.

By considering the behaviour of the integrand as a function of v explain why, provided $l(r)$ does not fall off too steeply with r, a disc profile will exhibit a double peak at a velocity separation $v \sin i$ corresponding to the outer edge of the disc.

Problems

6 Show that a spherical shell of gas, radius r, thickness dr, moving radially with velocity v, which emits a line at frequency v_0 in its rest frame with emissivity $4\pi j_{v_0}$ gives rise to a 'top hat' line profile,

$$L_v = \frac{4\pi j_{v_0}}{2v/c} 4\pi r^2 dr \quad \text{for} \quad -\frac{v}{c} \leqslant v - v_0 \leqslant \frac{v}{c},$$

$L_v = 0$ otherwise.

Hence show that a spherical distribution of n_c clouds per unit volume each emitting a luminosity L_c in the line gives a profile

$$L_v = \int \frac{L_c n_c}{(2v/c)} 4\pi r^2 \, dr,$$

where the integral is over radii such that $|v - v_0| \leqslant v(r)/c$.

If clouds are thick to the Lyman continuum then recombination lines have $L_c \propto AL/r^2$ where L is the ionizing luminosity of the central source. If ε is the cloud covering factor (so $d\varepsilon = n_c A \, dr$) show that the line profile is

$$L_v \propto \int \frac{d\varepsilon}{v}.$$

According to long held views, broad lines in active nuclei have logarithmic profiles, at least in the wings. Deduce that these are obtained if $\varepsilon \propto v$.

Chapter 9

1 A synchrotron radio source which has the form of a spherical shell of inner radius r and given volume with a uniform equipartition magnetic field, B, emits a spectral luminosity

$$L_v = \frac{L}{2v_{max}} \left[\frac{v_{max}}{v}\right]^{1/2}, \quad \text{for } v_{min} < v < v_{max},$$

and is zero otherwise. Show that the equipartition field is related to the total luminosity, L, by $B \propto L^{2/7}$.

Show that the Compton catastrophe will not occur if $L < r^2 Bc/2$, and hence that

$$L < br^{14/3},$$

where b is a constant (independent of B, L and r, but a function of v_{max} and v_{min}).

2 The Compton temperature is given by

$$T_c = \frac{4h \int v L_v \, dv}{k \int L_v \, dv}.$$

Find the Compton temperature of the radiation from the synchrotron source in problem 1 as a function of v_{max} and v_{min}.

Let the synchrotron source surround a central black hole of mass M and define the outer source radius as that at which a fully ionized hydrogen plasma at temperature T_c has a thermal velocity sufficient to escape the (Newtonian) gravity of the hole. Show that $R \lesssim GMm_H/kT_c$ and hence find an upper limit for L. Compare the result with the Eddington limit.

(In this problem the luminosity of the source is limited by the driving of a thermally

Problems

heated wind in contrast to an Eddington limited source where a wind is generated by the deposition of *momentum*. For a more realistic application of this process to the loss of material from the surface of an accretion disc see M. C. Begelman, C. F. McKee & G. A. Shields (1983), *Astrophys. J.* **271**, 89.)

3 A non-relativistic axisymmetric jet has constant velocity v and radius $r \propto z^\alpha$, α being a given constant and z the distance along the jet axis from the origin. A relativistic electron of energy E at a distance z loses energy at a rate

$$-\frac{dE}{dt} = kE^2 B^2 + \alpha \frac{E}{z}\frac{dz}{dt},$$

where $k = 2\sigma_T/3m_e^3 c^4 \mu_0$ (and the second term on the right represents expansion losses). If the magnetic field B is parallel to the jet axis, show that it falls off as $B \propto r^{-2}$ and that the energy of an electron that starts from z_0 with energy E_0 satisfies

$$\frac{1}{Ez^\alpha} - \frac{1}{E_0 z_0^\alpha} = \frac{kB_0^2 z_0^{4\alpha}}{v(1-5\alpha)}(z^{1-5\alpha} - z_0^{1-5\alpha}).$$

Hence find an expression for the maximum frequency v_{max} of synchrotron emission as a function of z (which corresponds to $E_0 \to \infty$) for $z \gg z_0$.

By considering v_{max} as $z \to \infty$ deduce that a perturbation in the input electron density will produce a perturbation in the observed spectrum propagating from high to low frequencies for all values of α.

4 An axisymmetric relativistic jet has cross-section area $A \propto z^{2\varepsilon}$, where z is the distance along the jet axis and ε is a constant. The Lorentz factor of the bulk motion is $\gamma_b \propto z^\varepsilon$, whilst the Lorentz factor of internal random relativistic motion of an electron of given energy satisfies $\gamma_e \propto z^{-\varepsilon}$ as a result of adiabatic expansion of the jet. The jet contains a magnetic field B parallel to its axis and can be assumed to emit synchrotron radiation only at the self-absorbed frequency v_a at each z. If the jet is viewed from a direction perpendicular to its axis, show that (i) $B \propto z^{-2\varepsilon}$, (ii) the flux received is given by

$$F_v dv \propto v^2 T_B r \frac{dz}{dv} dv \quad \text{at} \quad v = v_a,$$

where T_B is the brightness temperature.
Hence show that the observed spectrum rises with v if $\varepsilon > 1/4$.

(The restriction to a completely self-absorbed jet viewed at right-angles is, of course, a highly artificial simplification. For a detailed survey see D. L. Band (1987), *Astrophys. J.* **321**, 80.)

5 In the jet of problem 4 let t_c be the timescale for inverse Compton losses from the *bulk* motion of the electrons through the local synchrotron generated radiation field. By considering the ratio of t_c to the flow timescale (defined as z/c for a relativistic jet), show that a jet with a rising spectrum cannot be stopped by inverse Compton losses in its own local radiation field.

6 The maximum cross-section for the production of $e^\pm$ pairs from two photons occurs for photons of energy $x = hv/m_e c^2$ scattering on photons of energy $1/x$. The optical depth for photons of energy x is therefore approximately $\tau_{\gamma\gamma} = r\sigma_T u(1/x)/m_e c^2$ where $u(x)$ is the radiation energy density at energy x. Show that $\tau_{\gamma\gamma} \sim 1$ if the compactness parameter $l \sim 1$.

Show that the rate of production of $e^\pm$ pairs, $\dot{n}_{\gamma\gamma}$, is $\propto u(1/x)u(x) \propto L^2$ (where L is the source luminosity). If the maximum annihilation rate of pairs (which occurs for equal densities of positons and electrons) is $\sim \sigma_T c n_e^2$ deduce a value of L above which pair equilibrium is not possible.

7 The average energy exchange in a single Compton scattering of a photon of energy ε in an ionized medium having (non-relativistic) electron temperature T is of the form

$$\Delta \varepsilon = \frac{\varepsilon^2}{m_e c^2} - \frac{\alpha k T}{m_e c^2}\varepsilon$$

where α is a constant to be determined. If photons in equilibrium with the medium (for which $\langle \Delta \varepsilon \rangle = 0$) acquire a Wien spectrum ($u(\varepsilon) \propto \varepsilon^2 e^{-\varepsilon/kT}$) show that $\alpha = 4$. Show that Comptonization has a significant effect on the spectrum of radiation if the Compton y-parameter, $y \equiv (4kT/m_e c^2) \max(\tau, \tau^2)$, satisfies $y \gtrsim 1$.

If a photon of energy ε is scattered by a relativistic electron of Lorentz factor γ the energy exchange in the case $\varepsilon \ll \gamma m_e c^2$ is approximately

$$\Delta \varepsilon \sim \tfrac{4}{3}\gamma^2 \varepsilon.$$

If the electrons have a relativistic thermal distribution with relativistic Maxwell–Boltzmann distribution $f(E) \propto E^2 e^{-E/kT}$, find the average energy exchange per scattering for a photon of energy ε and hence show that the appropriate y-parameter in this case is $y = 16(kT/m_e c^2)^2 \max(\tau, \tau^2)$.

Chapter 10

1 Most of the physically relevant features of a relativistic torus can be reproduced by adopting the following pseudo-Newtonian potential (B. Paczyński & P. Wiita (1982), *Astr. Astrophys.* **88**, 23).

$$\Phi = -\frac{GM}{r - R_{\text{Schw}}},$$

where r is the spherical radius.

(a) Find an expression for the effective potential Φ_{eff} assuming that the angular velocity of the fluid in the torus follows a simple power-law of the form

$$\Omega(R) = \Omega_K (R/R_K)^{-q},$$

where R is the cylindrical radius and the allowed range for q is $\tfrac{3}{2} < q \leqslant 2$.

(b) Show that for the potential given above, the 'Keplerian' angular momentum for circular orbits in the equatorial plane is given by

$$l_K^2 = GMR^3/(R - R_{\text{Schw}})^2.$$

Show that for a given angular momentum value ($l > \sqrt{27}/2$), two circular orbits are possible with radii R_1 and R_K. Are they both stable? Show that for $q = 2$ (constant angular momentum) Φ_{eff} has extrema at these radii. How do these properties generalize for any q in the allowed range?

2 Using the results of problem 1 for Φ_{eff}, find an expression for the shapes of tori and their internal equipotentials for any $\tfrac{3}{2} < q \leqslant 2$. Show that they can be written as follows

$$z = \left\{ \left[\frac{B(1+\phi)(R_K - R_{\text{Schw}})^2 - R_{\text{Schw}} R_K^{1-B} R^B}{B\phi(R_K - R_{\text{Schw}})^2 - R_K^{1-B} R^B} \right]^2 R_{\text{Schw}}^2 - R^2 \right\}^{1/2}$$

where $\phi = -2\Phi_{\text{eff}}/c^2$ is a constant for a given surface and $B = 2 - 2q$.

Problems

3 For a torus where electron scattering is the dominant source of opacity and radiation pressure dominates ($\beta \ll 1$), the radiative flux is given approximately by

$$\mathbf{F} = -\frac{c}{\kappa_T}(1-\beta)\mathbf{g}_{\text{eff}}.$$

Show that the magnitude of the flux can be calculated everywhere from Φ_{eff}. For a pseudo-Newtonian torus with power-law rotation, it is given by (P. Madau (1988), *Astrophys. J.* **327**, 116)

$$F = \frac{c}{\kappa_T}(1-\beta)\frac{GM}{R_{\text{Schw}}^2}\left[\frac{1}{(r-1)^4} + \frac{1}{(R_K-1)^4}\left(\frac{R}{R_K}\right)^{2-4q}\right.$$

$$\left. - \frac{2R_K}{(R_K-1)^2(r-1)^2 r}\left(\frac{R}{R_K}\right)^{2-2q}\right]^{1/2}$$

where r, R and R_K are in units of R_{Schw}. Note that to obtain the effective temperature on the surface it will suffice to set $F = \sigma T_{\text{eff}}^4$ with r on the surface.

4 Find an approximate expression for the conical opening angle of the funnel of a torus of the type described in problem 2. How does the opening angle depend on q and ϕ?

5 Consider a situation in which waves propagating in a differentially rotating medium possess energies of different signs on different sides of an evanescent (forbidden) region or 'potential barrier'. Physically this could arise if the energy of the perturbed flow is greater/smaller than in the unperturbed case. A wave of amplitude A_i propagating towards the barrier is transmitted with amplitude τA_i where τ depends on the height and width of the barrier as in analogous quantum mechanical problems.

(a) Use energy conservation to show that the amplitude of the reflected wave is given by $A_r = (1 + \frac{1}{2}\tau^2)A_i$ for small τ.

(b) Assume that a perfectly reflecting boundary exists at one side of the forbidden region, and show that a wave of arbitrary amplitude generated on this side will grow exponentially with a growth rate

$$\omega = \frac{a}{4d}\tau^2,$$

where d is the distance between the barrier and the reflecting boundary and a is the speed of propagation of the wave (P. Goldreich & R. Narayan (1985), *Mon. Not. R. astr. Soc.* **213**, 7P).

Bibliography

Chapter 2
J. Dyson & D. Williams (1980), *Physics of the Interstellar Medium*, Manchester University Press. An excellent introduction to astrophysical gas dynamics, with applications to supernovae, stellar winds, etc.
L. D. Landau & E. M. Lifschitz (1959), *Fluid Mechanics*, Pergamon. A classic reference for all branches of fluid mechanics.

Chapter 3
T. J. M. Boyd & J. J. Sanderson (1969), *Plasma Dynamics*, Nelson. A comprehensive introductory text in plasma physics. Chapter 10 on plasma kinetic theory is particularly useful.
T. E. Holzer & W. I. Axford (1970), 'Theory of stellar winds and related flows', *Ann. Rev. Astron. Astrophys.* **8**, 31–60, gives a thorough discussion of spherical winds and accretion.
J. D. Jackson (1975), *Classical Electrodynamics*, (2nd edn) Wiley.
J. I. Katz (1987), *High Energy Astrophysics*, Addison-Wesley. Discusses much of the physics of Chapters 2 and 3 in more detail, and some aspects of accretion.
M. S. Longair (1981), *High Energy Astrophysics*, Cambridge University Press. Gives a very readable account of much of the physics of this chapter.

Chapter 4
P. P. Eggleton (1983), *Astrophys. J.* **268**, 368.
R. Kippenhahn & A. Wiegert (1990), *Stellar Structure and Evolution*, Springer-Verlag.
S. L. Shapiro & S. A. Teukolsky (1983), *Black Holes, White Dwarfs, and Neutron Stars*, Wiley-Interscience. Deals mainly with the internal structure of stellar-mass compact objects, but touches on some areas of accretion.
W. H. G. Lewin & E. P. J. van den Heuvel (eds.) (1983), *Accretion-driven Stellar X-ray Sources*, Cambridge University Press. An extremely useful collection of research-level chapters covering theory and observations. Chapters 8 (E. P. J. van den Heuvel) and 9 (G. J. Savonije) discuss some of the evolutionary questions raised in Section 4.4 and elsewhere. An updated version is in preparation.
A recent review of close binary evolution is
A. R. King (1988), 'The evolution of compact binaries', *Q. J. R. astr. Soc.* **29**, 1–25.

The complex dynamics of accretion from stellar winds is studied by
J. M. Blondin, T. R. Kallman, B. A. Fryxell & R. E. Taam, (1990) 'Hydrodynamic simulation of stellar wind distribution by a compact X-ray source', *Astrophys. J.* **356**, 591.

Chapter 5

A standard review of disc theory is
J. E. Pringle (1981), 'Accretion discs in astrophysics', *Ann. Rev. Astron. Astrophys.* **19**, 137–62.
Much useful material is contained in three collections of reviews:
K. O. Mason, M. G. Watson & N. E. White (eds.) (1986), *The Physics of Accretion onto Compact Objects*, Springer-Verlag;
F. Meyer, W. J. Duschl, J. Frank & E. Meyer-Hofmeister (eds.) (1989), *Theory of Accretion Disks*, NATO ASI, Kluwer;
C. Bertout, S. Collin-Souffrin, J. P. Lasota & J. Trân Than Vân (eds.) (1991), Structure and Emission Properties of Accretion Discs, IAU Coll. No. 129, Editions Frontières.
Eclipse mapping of discs in both lines and continuum is reviewed by
K. Horne (1991) 'Variability and structure of accretion discs in cataclysmic variables', in Bertout *et al.*, op cit.
For the discussion of disc structure in low-mass X-ray binaries see
N. E. White (1989) 'Accretion disks in low-mass X-ray binaries', in Meyer *et al.* op. cit., pp. 269–82;
M. G. Watson & A. R. King (1991) 'Accretion discs in low-mass X-ray binaries', in Bertout *et al.*, op. cit.
Disc instabilities and dwarf nova outbursts are reviewed in
F. Verbunt (1986) 'Theory and observations of time-dependent accretion disks', in Mason *et al.*, op. cit., pp. 59–75.
For a discussion of the superhump phenomenon see
R. Whitehurst & A. R. King (1991), 'Superhumps and resonances in accretion discs', *Mon. Not. R. astr. Soc.* **249**, 25.
Evidence for discs around young stars is discussed in
S. V. W. Beckwith, A. I. Sargent, R. S. Chini & R. Güsten (1990), 'A survey for circumstellar disks around young stellar objects', *Astron. J.* **99**, 924–945,
and their structure is discussed in
L. Hartman, S. J. Kenyon & P. Hartigan (1991) in *Protostars and Planets III*, eds. E. H. Levy, J. Lunine & M. S. Matthews, University of Arizona Press.
Star formation is reviewed in
F. H. Shu, F. C. Adams & S. Lizano (1987) 'Star formation in molecular clouds: observation and theory', *Ann. Rev. Astron. Astrophys.* **25**, 23–81.
For a discussion of spiral shocks and other angular momentum transport mechanisms, see
G. Morfill, H. C. Spruit & E. H. Levy (1991) in Levy *et al.*, op. cit.

Chapter 6

A review of boundary-layer theory is
O. Regev (1991) 'Modelling accretion disk boundary layers', in Bertout *et al.*, op. cit.
Both high- and low-mass X-ray binaries are discussed in
N. E. White, 'X-ray binaries' (1990) *Astr. Astrophys Rev.* **1**, 85.
For a review of QPOs see

M. van der Klis (1989) 'Quasi-periodic oscillations and noise in low-mass X-ray binaries', *Ann. Rev. Astron. Astrophys.* **27**, 517.

All aspects of AM Herculis systems (polars) are reviewed in

M. Cropper (1990) 'The polars', *Space Science Reviews*, **54**, 195,

while the possible relation to intermediate polars and evolutionary questions are discussed by

A. R. King & J. P. Lasota (1991) 'Magnetic cataclysmic variables', in the review volume *Interacting Close Double Stars*, eds. Y. Kondo, G. E. McCluskey, Jr. & J. Sahade, Kluwer.

For X-ray bursters see

W. H. G. Lewin & P. C. Joss (1983), Chapter 2 of Lewin and van den Heuvel, op. cit.

Binary black hole candidates are discussed by

J. E. McClintock (1986) 'Black holes in X-ray binaries', in Mason *et al.*, op. cit., pp. 211–28.

Chapter 7

Quasar surveys are reviewed in

D. W. Weedman (1986), *Quasar Astronomy*, Cambridge University Press.

For discussions of continuum emission rather more wide ranging than suggested by the title see

J. R. P. Angel & H. S. Stockman (1980), 'Optical and infrared polarisation of active extragalactic objects', *Ann. Rev. Astron. Astrophys.* **18**, 165–218, and, more recently,

D. B. Sanders, E. S. Phinney, G. Neugebauer, B. T. Soifer & K. Mathews (1989), 'Continuum energy distribution of quasars: shapes and origins', *Astrophys. J.* **347**, 29–51.

For the X-ray properties see e.g.,

T. J. Turner & K. A. Pounds (1989), 'The EXOSAT spectral survey of AGN', *Mon. Not. R. astr. Soc.* **240**, 833.

and for a general review,

D. E. Osterbrock (1991), 'Active galactic nuclei', *Rep. Prog. Phys.* **54**, 579–631.

The arguments leading to the black hole model are summarized in

M. J. Rees (1978), 'Accretion and quasar phenomenon', *Physica Scripta*, **17**, 183.

For gravitational lenses consult

Y. Mellier, B. Fort & G. Soucail (eds.) (1990), *Gravitational Lensing*, Lecture Notes in Physics, 360, Springer.

A review of the properties of the galactic centre can be found in

R. Genzel & C. H. Townes (1987), 'Physical conditions, dynamics and mass distribution in the centre of the Galaxy', *Ann. Rev. Astron. Astrophys.* **25**, 377.

For the problem of the gas supply see

I. Shlosman, M. C. Begelman & J. Frank (1990), 'The fuelling of active galactic nuclei', *Nature*, **345**, 679, and

M. J. Rees (1988), 'Tidal disruption of stars by black holes of 10^6–10^8 solar masses in nearby galaxies', *Nature*, **333**, 523.

A recent conference discussing many observational and theoretical aspects of the environmental and morphological characteristics of active galaxies is

J. W. Sulentic, W. C. Keel & C. M. Telesco (eds), *Paired and Interacting Galaxies*, IAU Coll 124, Tuscaloosa, Alabama, *NASA Conf. Pub. 3098*.

Bibliography

J. M. Bardeen (1973) 'Timelike and null geodesics in the Kerr metric', in *Black Holes. Ecole d'été de Physique Théorique, Les Houches, 1972*, eds. C. DeWitt & B. S. DeWitt, Gordon and Breach, is a useful reference for the effective potential method.

Accretion efficiency in spherical accretion is considered in

R. McCray (1979), 'Spherical accretion onto supermassive black holes' in *Active Galactic Nuclei*, eds. C. Hazard & S. Mitton, Cambridge University Press;

P. Meszaros & J. P. Ostriker (1983), 'Shocks and spherically accreting black holes: a model for classical quasars', *Astrophys. J.* **273**, L59–63.

and the efficiency constraint in

A. Fabian (1979), 'Theories of the nuclei of active galaxies', *Proc. Roy. Soc.* **A366**, 449.

Compton scattering is discussed in

G. B. Rybicki & A. P. Lightman (1979), *Radiative Processes in astrophysics*, Wiley.

Chapter 8

The discussion of iron line emission comes from

L. M. George & A. C. Fabian (1991), 'X-ray reflection from cold matter in active galactic nuclei and X-ray binaries', *Mon. Not. R. astr. Soc.* **249**, 352–67.

Cloud motion in the narrow line region is considered by

M. M. De Robertis & R. A. Shaw (1990), 'Line profiles and the kinematics of the narrow-line region in Seyfert galaxies', *Astrophys. J.* **348**, 421–33.

A very good and concise summary of optical spectroscopy of active nuclei can be found in

D. E. Osterbrock (1984), 'Active galactic nuclei', *Q. J. R. astr. Soc.* **25**, 1–17.

and at greater length in

M. C. Smith (1980), 'Quasars: observed properties of optically selected objects at large redshifts', *Vistas in Astronomy*, **22**, 321–62.

An excellent recent review of emission lines can be found in

H. Netzer in *Active Galactic Nuclei*, Saas-Fee Advanced Course 20, ed. R. D. Blandford, H. Netzer & L. Woltjer (1991), Springer,

and more briefly in, for example,

S. Collin-Souffrin, Theoretical studies of the line emission spectrum, in *Quasars*, IAU Symp. 119, ed. G. Swarup & V. K. Kapahi (1986), 295–303, Reidel,

while the 'classic' reference is

K. Davidson & H. Netzer (1979), 'The emission lines of quasars and similar objects', *Rev. Mod. Phys.* **51**, 715–66.

The 'standard model' goes back to

J. Krolik, C. F. McKee & C. B. Tarter (1981), 'Two phase models of quasar emission line regions', *Astrophys. J.* **249**, 422.

For radio sources see

G. Miley (1980), 'The structure of extended extragalactic radio sources', *Ann. Rev. Astron. Astrophys.* **18**, 165–218;

M. Begelman, R. D. Blandford & M. J. Rees 'Theory of extragalactic radio sources', *Rev. Mod. Phys.* (1984), **56**, 255–351.

References in the text are to

G. A. Shields (1978) in *Proceedings of the Pittsburgh Conference on BL Lacertae Objects*, pp. 257–76, ed. A. Wolfe, University of Pittsburgh Press;

D. J. Raine (1988), Broad emission lines in active galaxies in *Vistas in Astronomy*, **32**, 321–39;

M. D. Smith & D. J. Raine (1988), 'The duelling winds model for the emission line region of active galaxies', *Mon. Not. R. astr. Soc.* **234**, 297.

Chapter 9
For the unified model see
P. D. Barthel (1989), 'Is every quasar beamed?' *Astrophys. J.* **336**, 606.
Comptonization is treated in
G. B. Rybicki & A. P. Lightman (1979), *Radiative Processes in Astrophysics*, Wiley, and
R. A. Sunyaev & T. Titarchuk (1980), 'Comptonisation of X-rays in plasma clouds. Typical radiation spectra', *Astr. Astrophys.* **86**, 121.
The two temperature disc was introduced in
S. I. Shapiro, A. P. Lightman & D. Eardley (1976), 'A two-temperature accretion disk model for Cygnus X-1: structure and spectrum', *Astrophys. J.* **204**, 187.
See also
M. C. Begelman (1989) in *Accretion discs and Magnetic Fields in Astrophysics*, ed. G. Belvedere, Kluwer, pp. 255–66.
Slim discs are described in
M. A. Abramowicz, B. Czerny, J. P. Lasota & E. Szuszkiewicz (1988), 'Slim accretion discs', *Astrophys. J.* **332**, 646.
The elaboration of the cauldron model is based on work of Svensson & Zdziarski, for example
A. P. Lightman & A. A. Zdziarski (1987), 'Pair production and Compton scattering in compact sources and comparison to observations of active galactic nuclei', *Astrophys. J.* **319**, 643.
Magnetized disc winds are treated in
R. V. Lovelace, J. C. Wang & M. E. Sulkanen (1987), 'Self-collimated electromagnetic jets from magnetised accretion disks', *Astrophys. J.* **315**, 504;
A. Konigl (1989), 'Self-similar models of magnetised accretion discs', *Astrophys. J.* **342**, 208.
The Blandford–Znajek model is discussed further in
R. D. Blandford & R. L. Znajek (1977), 'Electromagnetic extraction of energy from Kerr black holes', *Mon. Not. R. astr. Soc.* **179**, 433;
D. Macdonald & K. S. Thorne (1982), 'Black-hole electrodynamics: an absolute space/universal time formalism', *Mon. Not. R. astr. Soc.* **198**, 345.
It may be useful to consult
K. S. Thorne, R. H. Price & D. Macdonald (1988), *Black Holes: The Membrane Paradigm*, Yale, for a technical approach to the analysis of black holes based on standard physical ideas.
Other references in the text are
R. D. Blandford (1976), 'Accretion disc electrodynamics – a model for double radio sources', *Mon. Not. R. astr. Soc.* **176**, 465;
R. D. Blandford & D. G. Payne (1982), 'Hydromagnetic flows from accretion discs and the production of radio jets', *Mon. Not. R. astr. Soc.* **199**, 883;
R. D. McCray (1979), 'Spherical accretion onto supermassive black holes' in *Active Galactic Nuclei*, eds. C. Hazard & S. Mitton, Cambridge University Press.

Bibliography

Chapter 10

The best review of the material of this chapter is

R. Narayan & J. Goodman (1989), Non-axisymmetric shear instabilities in thick accretion disks, in *Theory of Accretion Disks*, eds. Meyer *et al.*, NATO ASI, Kluwer, pp. 231–47, and this paper is updated by

R. Narayan (1991), in *Structure and Emission Properties of Accretion Discs*, IAU Coll. No. 129, eds. C. Bertout, S. Collin-Souffrin, J. P. Lasota & Tran T. Van, Editions Frontières.

Of particular importance is

J. C. Papaloizou & J. E. Pringle (1984), 'The dynamical stability of differentially rotating discs with constant specific angular momentum', *Mon. Not. R. astr. Soc.* **208**, 721.

Rotating stars are treated in

J.-L. Tassoul (1978), *Theory of Rotating Stars*, Princeton University Press.

Other useful references are

M. A. Abramowicz, M. Calvani & L. Nobili (1980), 'Thick accretion disks with super-Eddington luminosities', *Astrophys. J.* **242**, 772;

B. Paczyński (1982), *Astron. Gesells. Mitteilungen*, **57**, 27;

M. J. Rees, M. C. Begelman, R. D. Blandford & E. S. Phinney (1982), 'Ion-supported tori and the origin of radio jets', *Nature*, **295**, 17;

M. Schwarzschild (1965), *Structure and Evolution of Stars*, Dover.

Appendix

G. B. Rybicki & A. P. Lightman (1979), *Radiative Processes in Astrophysics*, Wiley.

W. H. Tucker (1975), *Radiation Processes in Astrophysics*, MIT Press.

List of symbols

(*See index for references to definitions of important quantities*)

a_x	X-ray albedo	P	gas pressure
b	boundary layer thickness	P_{ph}	photospheric gas pressure
b	impact parameter	P_{mag}	magnetic pressure
B	magnetic field	P_{rad}	radiation pressure
B_ν	Planck function	P_{ram}	ram pressure
c	speed of light	q	conductive heat flux
c_s	sound speed	q	mass ratio
e	electron charge	r	radial coordinate in spherical polars
$\mathbf{f}$	force density		
f	filling factor	R	radial coordinate in cylindrical polars
F, F_{rad}	flux of radiation		
F_ν	specific flux	r_{acc}	accretion radius
$\mathbf{g}, g_{eff}$	acceleration due to gravity	r_{coll}	collision radius
G	gravitational constant	r_L	Larmor radius
$h = 2\pi\hbar$	Planck constant	r_M, R_M	Alfvén radius
H	disc height	r_s	sonic point
H_0	Hubble constant	r_t	trapping radius
I_ν	specific intensity (surface brightness)	R_{circ}	circularization radius
		Re	Reynolds number
j_{br}	bremsstrahlung emissivity	R_L	Roche radius
j_ν	emissivity	Rm	magnetic Reynolds number
k	Boltzmann constant	R_m	rotation measure
L	luminosity	R_{Schw}	Schwarzschild radius
L_1	inner Lagrange point	R_Ω	corotation radius
$M_\odot$	solar mass	$R_\odot$	solar radius
m_e	electron mass	R_*	stellar radius
m_p	proton mass	S	flux of radiation
m_i	ion mass	S_ν	source function
m_H	mass of hydrogen atom	T	gas temperature
$\dot{M}$	accretion rate	T_b	blackbody temperature
$\dot{M}_{crit}$	critical accretion rate	T_B	brightness temperature
N	particle number density	T_C	central temperature
N_e, N_i	electron, ion density	t_{Compt}	inverse Compton cooling timescale
P	binary period		
P_{pulse}	spin period of compact magnetic star	t_d	deflection time
		t_E	energy exchange timescale

List of symbols

t_{ff}	free-fall timescale	λ_{Deb}	Debye length
t_s	slowing down time	$\ln \Lambda$	Coulomb logarithm
t_ϕ	dynamical timescale	$\Lambda(T)$	cooling function
t_{th}	thermal timescale	λ_s	stopping length
t_{rad}	radiative cooling timescale	μ	mean molecular weight
T_{rad}	radiation temperature	μ	magnetic moment of compact star
t_{sync}	synchrotron timescale		
t_{visc}	viscous timescale	μ	vertically integrated dynamical viscosity
U	ionization parameter		
U_{rad}	radiation energy density	ν	frequency
v	fluid velocity	ν	kinematic viscosity
z_{em}, z_{abs}	emission, absorption redshift	ν_{cyc}	cyclotron frequency
		ν_p	plasma frequency (Hz)
α	binary separation	ξ, Ξ	ionization parameter
β	ratio of radiation to total pressure	ρ	density
		σ	Stefan–Boltzmann constant
γ	adiabatic index	σ_T	Thomson cross section
γ	Lorentz factor	Σ	surface density
Γ_{Coul}	Coulomb heating rate	τ	optical depth
ΔE_{acc}	accretion yield	ω_p	plasma frequency (rad s^{-1})
ΔE_{nuc}	nuclear burning yield	ω_*	fastness parameter
ε	internal energy per unit mass	Ω	solid angle
η	efficiency	Ω	angular velocity
κ_ν	opacity	Ω_K	Kepler angular velocity
κ_R	Rosseland mean		
λ	mean free path	$\mathcal{M}$	Mach number

Object index

3C 273, 172, 175, 176, 221, 225, 231
3C 120, 216
3U0352+30, 126
4U0115+63, 126
4U0900-40, 126
4U1624-49, 93
4U1626-67, 126
A05353+26, 126
A0620-00, 168–70, 185
AM Her, 145, 146, 150, 151
AO 0235+164, 173
BL Lac, 173, 180, 182
Cen X-3, 124–6, 156
Crab Nebula, 221
Cyg X-1, 168–70, 185, 233
Cyg A, 180, 181, 221
DG Tau, 114
EX Ori, 115
FU Ori, 115
GK Per, 129
GX 1+4, 126
GX 301-2, 126
HL Tau, 114
Her X-1, 126, 132, 133, 156
LMC X-3, 168–70, 185

M 32, 183
M 31, 182
M 87, 182, 221
MXB 1906+00, 162
MXB 1743-29, 162
MXB 1735-44, 162
MXB 1728-34, 162
MXB 1636-53, 162
Mkn 668, 208
NGC 4151, 179, 180, 209
NGC 1068, 188, 208
OH 471, 210
OJ 169, 173, 180
OQ 172, 210
PHL, 957, 173
Rapid Burster (MXB 1730-335), 161, 163, 165, 166
SMC X-1, 126
SS Cyg, 121
SS 433, 265
Sco X-1, 156
Sgr A*, 183
Taurus star-forming region, 114
X1822-371, 92
Z Cha, 91, 109

Index

absorption lines, 173, 177
accretion columns, 117, 133–61, 170
accretion disc, 46, 57, 67–116, 169, 199
accretion from interstellar medium, 19
accretion luminosity, 3, 4, 57, 64, 193, 241, 244, 253
accretion radius, 20, 21, 64, 65, 193
accretion tori, 245–65, 280–1
accretion torque, 125, 275
accretion yield, 1
　see also efficiency
active galactic nuclei (AGN), 4, 171–246
adiabatic flow, 11, 12, 15, 194
Alfvén radius, 121, 122
Alfvén velocity, 196, 235
Algol systems, 56
alpha viscosity prescription, 62, 232, 235
AM Herculis systems, 85, 128–33, 140, 142, 145–51, 169, 274, 275
amplification of waves, 281
angular momentum, 181, 182, 186–8, 189, 191
　losses, 53, 271
anomalous states in AM Herculis systems, 150–51
apsidal precession, 111, 112

BAL quasars, 173
Balmer emission, 200, 214
baroclinic equilibria, 265
barotropes, 249, 262
Bernoulli integral, 18, 19
big bump, 200, 213, 230–2
binary evolution, 50–4, 271
binary pulsars, 271
binary separation, 47
BL Lac objects, 173, 174, 176, 179, 180, 182, 201, 229
black hole uniqueness theorem, 192
black holes, 166–70, 171, 183, 185, 186, 189–92, 194, 242–4, 276
blackbody, 195, 202, 267
blackbody, temperature, 5
Blandford–Znajek model, 239–44
blob accretion, 148–51, 274
Boltzmann, population, 267
Bondi accretion rate, 193
　see also spherical accretion

boundary layer, 72, 73, 117–21, 274
bremsstrahlung (free–free emission), 10, 138, 193, 194, 197, 268, 275, 276
bright spot, 87, 88
brightness temperature, 223, 228
broad line region, 209–17
broad lines, 179, 278
bulk viscosity, 34

cataclysmic variables, 2, 6, 85–92, 168, 169
centre-of-mass frame, 25
Chandrasekhar limit, 168
charge neutrality, 22
circularization radius, 56, 129
clusters
　galaxies, 176, 180
　stars, 183, 185
collision radius, 182
collisional de-excitation, 206
collisional excitation, 209, 215, 277
collisional ionization, 207, 209
collisionless shock, 152, 115–57
collisions, 8, 25–9, 152, 155
compact sources, 223
compactness parameter, 231, 279
Compton catastrophe, 227–8, 278
Compton cooling, 140, 141, 153–4
Compton scattering, 194, 197, 213, 228–31, 280
Compton temperature, 234, 278
Compton y-parameter, 275, 280
conductive flux, 10
conservative mass transfer, 52
contact binary, 50
contact discontinuity, 274
continuity equation, 8, 9
cooling curve, 99
cooling function, 196
Coriolis forces, 48, 49
corotation, 188
cosmic censorship, 192
cosmic rays, 219
Coulomb forces, 22, 25
Coulomb logarithm, 28–9
covering factor, 209, 212, 278
critical density, 204
cyclotron emission, 128, 130–3, 141, 142, 156, 268

cyclotron frequency, 225, 268

Debye length, 22, 24, 25, 134, 153
Debye sphere, 27
deflection length, 28
deflection time, 28
detached binary, 50
differential rotation, 58
diffusion approximation, 267
diffusion equation for time-dependent disc, 69
dipoles, 275
dips, 94–6, 274
disc
    atmospheres, 88–90
    corona, 93, 200, 273
    formation, 54
    instabilities, 103–6, 232, 261–3
    spectrum, 77–9, 202
Doppler favouritism, 229
DQ Her systems, 85
    see also intermediate polars
duelling winds model, 216
dwarf novae, 85, 103–6, 169, 170
dynamical stability, 261
dynamical timescale, 95, 98

eccentric binaries, 65, 66, 125
eclipse mapping of discs, 90–2
Eddington limit, 2–5, 84, 123, 181, 182, 184, 194, 233, 252, 277, 278
EELR, 172, 208
effective potential, 189–90
efficiency, 4, 180, 181, 183, 184, 189, 191–8, 241, 260, 264, 275
efficiency constraint, 198
Einstein coefficient, 269
elastic collision, 25
electron–positron pairs, 231, 233, 242, 279
ellipsoidal variation, 50
elliptical galaxies, 174, 176, 183, 188
emission measure, 207
emission-line eclipses, 86, 87
emissivity, 10, 266
energy balance in thick discs, 256
energy conservation, 9, 10
energy exchange timescale, 30–2, 233
ENLR, see EELR
epicyclic frequency, 111
equipartition, 196, 221, 222, 224, 278
    rate, 141, 142
    timescale, 31
equipotential surfaces, 250, 255, 280
equivalent width, 209
Euler equation, 9, 236

Fanaroff–Riley (FR) class, 218, 219
Faraday rotation, 225, 235
fastness parameter, 123
Ferraro's law, 237
filling factor, 207, 277
fluorescence, 203
forbidden lines, 204

force-free condition, 236, 237, 239
free–free emission, see bremsstrahlung
freezing-in of magnetic fields, 38, 237, 239
funnels, 256, 264, 281
Fuors, 115

galactic bulge sources, 85, 169, 170
galaxy, 183, 187
gamma rays, 172, 183, 231
general relativity, 175, 178, 184, 185, 189
global modes, 261
gravitational lenses, 178
gravitational radiation, 53, 271
Green function for disc, 69–71

high ionization lines, 207, 214, 216
horizon, 190
hot spots, 219, 220, 221
HPQs, 174
hydrostatic equilibrium, 12, 247

impact parameter, 25
inner Lagrange point, 49
instabilities, 187–8, 232, 234
intermediate polars, 85, 129, 140, 145, 275
inverse Compton emission, 268
ion torus, 233
    see also accretion tori,
ionization parameter, 95–7, 177, 211, 213, 215, 277
ionization potential, 208
IRAS, 175
iron lines, 202
isentropic flow, 11
isobaric surfaces, 247
isopycnic surfaces, 247
isothermal flow, 11, 12
isothermal shock, 270
Israel–Carter–Robinson theorem, 192

Jansky, 224
Jeans instability, 234
jets, 187, 221–2, 234, 245, 263–5, 279

Kepler orbits, 47
Kepler's law, 47
Keplerian angular momentum, 74, 75, 255, 273, 260, 280
Keplerian disc, 74, 75, 272
Kerr black hole, 191
Kirchhoff's law, 267
Klein–Nishima cross-section, 230
Kramers' opacity, 80, 82, 135

L_1 point, 49, 271
lacertids, see BL Lac objects
Lagrange points, 49
Landau levels, 133
Larmor frequency, 39, 268
Larmor radius, 39, 134, 153
last stable orbit, 191–2, 276
lifetime of active nucleus, 180, 181

Index

limit-cycle behaviour, 104–6
line emission, 172, 268
line profiles, 215, 216, 277, 278
local thermodynamic equilibrium (LTE), 267
low ionization lines, 214, 215
low-mass X-ray binaries (LMXB), 85, 92–7, 168–70, 274
luminosity distance, 175
Lyman α forest, 178, 200

Mach number, 43, 75
magnetic dipole, 121, 129, 275
magnetic field, 36–40, 128–61, 187
magnetic Reynolds number, 37
magnetic stellar wind braking, 53
magnetoid, 185
magnetosphere, 128
mass function, 170
mass ratio, 47
mean free path, 8, 26
millimetre emission, 201
minimum period, 270
monatomic gas, 9

narrow lines, 205–8
NELGs, 172, 176, 183
non-axisymmetric modes, 261
nova, 2, 170
nova-like variables, 85, 170
nuclear yield, 1

[O III]-problem, 215
old novae, 85
optical depth, 266
optically thick medium, 267
optically thin medium, 267
orbital angular momentum, 52
OVV, 173, 174

P Cygni profiles, 173
passive disc, 272
perfect gas law, 9
perfect magnetohydrodynamic condition, 38, 236, 238
period–mass relation,
   degenerate stars, 270
   main sequence stars, 51
permitted lines, 204
photoelectric absorption, 94, 95
photoionization, 109, 205
Planck function, 267
   see also blackbody
plasma, 22
   frequency, 23, 24
   oscillations, 22–5
polarization, 128, 130, 173, 201, 219–21, 225
polecap fraction, 124
   in AM Her systems, 149–51
poloidal fields, 237
poloidal motions, 257
Price's theorem, 192
primary star, 50

protostars, 114, 115
pseudo-Newtonian potential, 273, 280
$P_{pulse} - P_{orb}$ relation in intermediate polars, 129
pulsing X-ray sources, 62, 168, 169

QSS, *see* quasi-stellar sources
quasar, 4, 171–6, 227, 275
quasi-periodic oscillations (QPOs), 127, 128
quasi-stellar objects, 171, 176
quasi-stellar sources, 219

radial drift velocity, 67
radiation pressure, 76, 82–4
   in neutron star accretion, 156–61
radiative acceleration, 263
radiative equilibrium, 76, 258, 281
radiative flux, 10, 266
radiative transfer equation, 266
radio emission, 172, 179, 181, 183, 219, 223–9
radio lobes, 180, 219, 222, 227
ram pressure, 39, 135
Rankine–Hugoniot conditions, 43
redshift, 175, 177, 179
reflection effect, 264
relativistic beaming, 182
relaxation times, 29–31
resonances, 110–14
Reynolds number, 61
rigid (solid-body) rotation, 59, 60
Roche lobe, 49–55
   overflow, 46–50, 170, 271
Roche, potential, 48–55
Rosseland approximation, 10
Rosseland mean opacity, 268
rotational potential, 250

scaleheight, stellar, 272
scattering, 267
Schwarzschild radius, 33, 189, 193, 260, 276, 280, 281
secondary star, 50
self-gravity, 81
semi-detached binary, 50
Seyfert galaxies, 172, 173, 179, 183, 187, 222
Seyfert types I and II, 174, 175, 208
Shakura–Sunyaev disc, 80–5
shear viscosity, 33, 58
shock jump conditions, 43, 270
shock temperature, 44, 136
shock waves, 14, 40–5
shocks, 196, 276
singularity, 192
slim disc, 234, 273
slingshot, model, 184
slowing-down timescale, 31, 32
soft X-ray excess
   active galaxies, 200, 202, 208
   AM Her systems, 145–51
sonic point, 17
sound waves, 12, 13
source function, 266, 269
specific intensity, 266

specific luminosity, 266
spherical accretion, 14–21, 197
spinar model, 185
spinup rate in X-ray binaries, 125–7
spiral galaxies, 187
spiral shocks, 115, 116
standard model, 211–15
starburst, 175, 184, 187, 188
stellar breeze, 17
stellar collisions, 186
stellar wind, 17
   accretion, 46, 62–4, 272
stopping length, 32, 153
strong shock, 43
SU UMa systems, 106, 108–10, 112
subsonic flow, 13
superhumps, 109–14
superluminal expansion, 180, 226, 227
supermassive stars, 184
supernovae, 184, 185, 186, 219
superoutbursts, 106, 109
supersonic flow, 13
suprathermal particles, 29–33
surface density, 59
symbiotic stars, 2
synchronism in AM Her systems, 274, 275
synchrotron emission, 195, 219, 220–1, 233, 268, 278, 279
synchrotron lifetime, 225
synchrotron self absorption, 223–4, 228, 279
synchrotron self-Compton (SSC) model, 229

T Tauri stars, 114, 115
thermal conductivity, 99, 100
thermal plasmas, 29–31
thermal timescale, 98
thick discs, *see* accretion tori
thin disc approximation, 67, 74, 75
Thomson cross-section, 269
tidal disc radius, 108

tidal disruption, 186
tides, 106–8
time-dependent disc equations, 101
toroidal field, 237
trans-sonic solutions, 17, 18
transport processes, 33–6
trapping radius, 194
Trojan asteroids, 49
turbulence, 61
two-phase instability, 95, 274
two-temperature disc, 233

unified model, 202
unstable mass transfer, 53

variability in active galactic nuclei, 179, 182, 211, 228
vertical disc structure, 75–7
virial temperature, 5, 276
viscosity, 9, 33, 61, 62, 116, 186
viscous dissipation, 60
viscous dissipation rate in steady disc, 73
viscous instability, 102
viscous timescale, 70, 96
viscous torques, 58–60
VLBI, 179, 221, 225, 228
von Zeipel's paradox, 260
vorticity, 254

W UMa binaries, 50
white dwarf, 203

X-ray absorption, 213
X-ray binaries, 6, 168–70
X-ray bursts, 161–6, 170
X-ray production in cataclysmic variables, 120, 121, 274
X-rays, 174, 176, 179, 195, 202, 214, 215, 218, 229, 232